MOBILITY MODELS FOR NEXT GENERATION WIRELESS NETWORKS

WILEY SERIES IN COMMUNICATIONS NETWORKING & DISTRIBUTED SYSTEMS

The 'Wiley Series in Communications Networking & Distributed Systems' is a series of expert-level, technically detailed books covering cutting-edge research, and brand new developments as well as tutorial-style treatments in networking, middleware and software technologies for communications and distributed systems. The books will provide timely and reliable information about the state-of-the-art to researchers, advanced students and development engineers in the Telecommunications and the Computing sectors.

Other titles in the series:

Wright: *Voice over Packet Networks* 0-471-49516-6 (February 2001)

Jepsen: *Java for Telecommunications* 0-471-49826-2 (July 2001)

Sutton: *Secure Communications* 0-471-49904-8 (December 2001)

Stajano: *Security for Ubiquitous Computing* 0-470-84493-0 (February 2002)

Martin-Flatin: *Web-Based Management of IP Networks and Systems* 0-471-48702-3 (September 2002)

Berman, Fox, Hey: *Grid Computing. Making the Global Infrastructure a Reality* 0-470-85319-0 (March 2003)

Turner: *Service Provision. Technologies for Next Generation Communications* 0-470-85066-3 (April 2004)

Welzl: *Network Congestion Control: Managing Internet Traffic* 0-470-02528-X (July 2005)

Raz: *Fast and Efficient Context-Aware Services* 0-470-01668-X (April 2006)

Heckmann: *The Competitive Internet Service Provider* 0-470-01293-5 (April 2006)

Dressler: *Self-Organization in Sensor and Actor Networks* 0-470-02820-3 (November 2007)

Berndt: *Towards 4G Technologies: Services with Initiative* 0-470-01031-2 (March 2008)

Jacquenet: *Service Automation and Dynamic Provisioning Techniques in IP/MPLS Environments* 0-470-01829-1 (March 2008)

Gurtov: *Host Identity Protocol (HIP): Towards the Secure Mobile Internet* 0-470-99790-7 (June 2008)

Boucadair: *Inter-Asterisk Exchange (IAX): Deployment Scenarios in SIP-enabled Networks* 0-470-77072-4 (January 2009)

Fitzek: *Mobile Peer to Peer (P2P): A Tutorial Guide* 0-470-69992-2 (June 2009)

Shelby: *6LoWPAN: The Wireless Embedded Internet* 0-470-74799-4 (November 2009)

Stavdas: *Core and Metro Networks* 0-470-51274-1 (February 2010)

Gómez Herrero, *Network Mergers and Migrations: Junos® Design and Implementation* 0-470-74237-2 (March 2010)

Jacobsson, *Personal Networks: Wireless Networking for Personal Devices* 0-470-68173-X (June 2010)

Barrieros, *QOS-Enabled Networks*, 90-470-68697-3, (December 2010)

Minei: *MPLS-Enabled Applications: Emerging Developments and New Technologies, Third Edition*, 0-470-66545-9 (January 2011)

MOBILITY MODELS FOR NEXT GENERATION WIRELESS NETWORKS

AD HOC, VEHICULAR AND MESH NETWORKS

Paolo Santi

Istituto di Informatica e Telematica del CNR, Italy

A John Wiley & Sons, Ltd., Publication

This edition first published 2012

Registered office
John Wiley & Sons Ltd, The Atrium, Southern Gate, Chichester, West Sussex, PO19 8SQ, United Kingdom

For details of our global editorial offices, for customer services and for information about how to apply for permission to reuse the copyright material in this book please see our website at www.wiley.com.

Library of Congress Cataloging-in-Publication Data

Santi, Paolo.
Mobility models for next generation wireless networks : ad hoc, vehicular and mesh networks / Paolo Santi.
p. cm.
Includes bibliographical references and index.
ISBN 978-1-119-99201-1 (cloth)
1. Wireless communication systems. I. Title.
TK5103.2.S2577 2012
004.6--dc23

2012002026

A catalogue record for this book is available from the British Library.

ISBN: 9781119992011

Typeset in 10.5/13pt Times by Laserwords Private Limited, Chennai, India

Printed and bound in Malaysia by Vivar Printing Sdn Bhd

1 2012

To my wife Elena,
my daughters Bianca and Marta,
and our baby who is on the way

To my families

Contents

List of Figures

List of Tables

About the Author

Paolo Santi has worked at the Istituto di Informatica e Telematica del CNR in Pisa, Italy, since 2001, first as a Researcher, and now as a Senior Researcher. He received his Laura Degree and PhD in computer science from the University of Pisa in 1994 and 2000, respectively. During his career, he visited the School of Electrical and Computer Engineering, Georgia Institute of Technology, in 2001, and the Department of Computer Science, Carnegie Mellon University, in 2003.

His research interests include fault-tolerant computing in multiprocessor systems (during his PhD studies) and, more recently, the investigation of the fundamental properties of wireless multi-hop networks such as connectivity, topology control, lifetime, capacity, mobility modeling, and cooperation issues. He has presented more than 70 papers at highly reputed conferences or published in journals in the field of short-range wireless networking and mobile computing. He also authored the book *Topology Control in Wireless Ad Hoc and Sensor Networks* published by John Wiley & Sons. Dr. Santi was the recipient of the 2004 ITG Outstanding Paper Award for the paper (co-authored with C. Bettestetter and G. Resta) "The Node Distribution of the Random Waypoint Mobility Model for Wireless Ad Hoc Networks," which appeared in *IEEE Transactions on Mobile Computing* in 2003.

Dr. Santi served as the General Co-Chair of ACM VANET 2007 and 2008, the Technical Program Co-Chair of IEEE WiMesh 2009, and on the organizational and technical program committees of several conferences in the field. He was Guest Editor of the *Proceedings of the IEEE*, and Associate Editor for *IEEE Transactions on Mobile Computing* and *IEEE Transactions on Parallel and Distributed Systems*. He is a Member of the IEEE Computer Society, and a senior member of ACM and SIGMOBILE.

Preface

The idea for this book was suggested to me by my colleague and friend Sergio Palazzo, who invited me to deliver a short course on "Mobility Models and Social Networks" at the 2010 Lipari Summer School on Mobile Computing and Communications. At the end of the course, when delivering his closing remarks, Sergio said that in his opinion the topic of the course was very interesting, and concluded with what sounded like "Why don't you write a book on this?" I have to admit that I did not take his suggestion too seriously, since I was sure that, given its importance within the field of mobile computing and networking, several books on the topic had already been published. However, I took his suggestion seriously enough to devote half a working day when I was back in the office to look up the published books on mobility models for short-range wireless networks. To my great surprise, the result of my Web-based research showed that, except for a few book chapters and survey papers, there was no book entirely focused on mobility modeling. After that morning, the idea of writing a book on this topic started to develop further, and I prepared a book proposal for submission and approval by John Wiley & Sons. Thus, a year and a half after the Lipari Summer School (it will be two years at the time of publication), Sergio's suggestion has become a reality.

As mentioned above, the general idea for the book and its organization is adapted from the short course I delivered at Lipari Summer School in 2010. In turn, the short course was an extension of an invited talk on the "Mathematics of Mobility" that I delivered at the Workshop on Mathematical Aspects of Large Networks in Barcelona, in 2009. Of course, the material contained in this book has been considerably extended and revised with respect to these presentations.

The aim of the book is to provide an *exhaustive* coverage of mobility models for *next generation wireless networks*. Some clarification about terminology is in order before proceeding further. By *exhaustive*, I mean that all topics related to mobility modeling are touched upon in this book, ranging from the theoretical characterization of stationary properties of mobility models to the design of mobility models aimed at resembling mobility features observed in the real world. "Exhaustive" does not mean that *each* mobility model presented in the wireless networking and computing literature is reported in this book. Instead, for each topic considered, a few, representative specific models are presented, while other mobility models are simply mentioned for reference at the ends of the relevant chapters. By *next generation wireless networks*, I mean short-range, mostly infrastructure-less wireless networks based on existing or forthcoming technology (e.g., WiFi, Bluetooth, ZigBee, etc.), which will complement traditional, infrastructured wireless networks (e.g., cellular) and make possible the realization of the ubiquitous computing paradigm. Application scenarios of next generation wireless networks include wireless local area network (WLAN)/mesh networks, vehicular networks, wireless sensor networks, and opportunistic networks. These innovative networks will allow us to gain ubiquitous Internet access, to improve safety and traffic conditions on the roads, to realize smart environmental monitoring, messaging with friends through multi-hop, phone-to-phone communication, etc. Since user mobility is a salient feature of next generation wireless networks, how to model user movement becomes a fundamental part of the wireless network performance evaluation process, which explains the considerable efforts devoted in recent years by the research community to understand, characterize, and model mobility. The goal of this book is to give the reader an organic view of this body of the literature, and to present mobility modeling as a *scientific discipline*, encompassing both theoretical and practical aspects related to this challenging task as part of the network performance evaluation process.

Audience

This book is intended for graduate students, researchers, engineers, and practitioners who are interested in acquiring a global view of the discipline of mobility modeling, which is at the basis of wireless network performance evaluation through analysis, simulation, or testbed-based data collection. In general, the book can serve as a reference resource for researchers, engineers, professionals, and developers working in the field of short-range wireless networking.

Although I have tried to make the book as self-contained as possible, including a brief introduction to short-range wireless networking and state of the art presentations of the application scenarios (WLAN/mesh networks, vehicular networks, etc.) considered, the reader is assumed to be familiar with basic concepts in wireless networking, as well as those in graph and probability theory.

Book Overview

The material contained in this book is organized as follows.

The first part of the book (Introduction) presents introductory material that is preparatory for what is described in the rest of the book.

Chapter 1 introduces next generation wireless networks, briefly describing the possible applications of these technologies in the near future, as well as the challenges that the designers of such networks will face.

Chapter 2 presents simple, yet widely accepted, models for next generation wireless networks. In particular, principles underlying radio signal propagation in the air are described in this chapter, as well as basic models for describing the occurrence of wireless links between network nodes, and energy consumption of a wireless node.

Chapter 3 motivates the need for mobility modeling in next generation wireless networks, clearly describing the importance of mobility modeling in the performance evaluation process. The chapter also describes the fundamental difference between mobility models for cellular vs. next generation wireless networks. Finally, the chapter introduces a taxonomy of existing mobility models for next generation wireless networks, and presents the CRAWDAD community resource for archiving wireless data.

The second part of the book ("General-Purpose" Mobility Models) presents the most representative mobility models that are not tailored to specific application scenarios.

Chapter 4 presents the well-known class of random walk models, which have been applied in several scientific disciplines. After describing some of the most representative variants of this model, the chapter discusses features of random walk mobility in the context of next generation wireless networks. Finally, this chapter presents some important theoretical properties of random walk models.

Chapter 5 is devoted to the most widely used mobility model for short-range wireless networks, namely, the random waypoint model. After describing the model and some of its variants, the chapter thoroughly discusses important properties of random waypoint mobility, such as node spatial distribution and average node speed.

Chapter 6 presents representative models of group mobility, in which the mobility pattern of a node is correlated with that of other nodes in the network (typically, with that of surrounding nodes). The chapter also presents other general-purpose mobility models, such as the smooth random mobility model and the Gauss–Markov mobility model, in which directional and speed changes of a node are smooth instead of abrupt, as in the mobility models presented in the previous chapters.

Chapter 7 introduces the important class of random trip models and presents results concerning the existence and characterization of a stationary regime for any mobility model belonging to this class. The class of random trip models includes, among others, some of the mobility models presented in the previous chapters, such as the random waypoint and random walk models.

The third part of the book (Mobility Models for WLAN and Mesh Networks) presents the most representative mobility models for WLAN and mesh networks.

Chapter 8 describes state of the art user scenarios and the prospects for WLAN and mesh networks.

Chapter 9 presents and discusses the extensive research work that has been performed in analyzing real-world WLAN traces and in deriving prominent features of WLAN mobility. The material presented in this chapter is fundamental for providing guidelines on the design of mobility models aimed at resembling movement patterns of WLAN users, which are reported in Chapter 10.

Chapter 10 presents representative WLAN mobility models, belonging to two different categories. The first model, called LH, aims at reproducing user/access point registration patterns observed in real-world WLAN traces. In this model, the interest is in reproducing WLAN usage patterns rather than the physical movement of users. The second model presented in the chapter, called KKK, is instead aimed at modeling the physical mobility of a WLAN user.

The fourth part of the book (Mobility Models for Vehicular Networks) presents the most representative mobility models for vehicular networks.

Chapter 11 describes state of the art user scenarios and the prospects for vehicular networks.

Chapter 12 describes the differences between the macroscopic and microscopic approaches to vehicular mobility modeling, and explains why microscopic mobility models are more relevant for vehicular network performance evaluation.

Chapter 13 presents and discusses in detail representative microscopic mobility models for vehicular movement, and in particular the simulation of urban mobility (SUMO) model. The chapter ends with a discussion of the challenges related to integrating vehicular mobility and wireless network simulation, and presents a tool designed for this purpose.

The fifth part of the book (Mobility Models for Wireless Sensor Networks) presents the most representative mobility models for wireless sensor networks.

Chapter 14 describes state of the art user scenarios and the prospects for wireless sensor networks.

Chapter 15 categorizes mobility models for wireless sensor networks into models for passive and active mobility of sensor nodes. The chapter then presents representative mobility models for passive movement of wireless sensor nodes caused by external forces like ocean flows, animal movements, etc.

Chapter 16 introduces models for active movement of wireless sensor nodes, that is, movement of sensor nodes and/or data collection entities endowed with autonomous motion capabilities, able to control and optimize their movement pattern.

The sixth part of the book (Mobility Models for Opportunistic Networks) presents the most representative mobility models for opportunistic networks.

Chapter 17 describes state of the art user scenarios and the prospects for opportunistic networks.

Chapter 18 describes the basic mechanisms of message routing in opportunistic networks, as well as basic concepts in opportunistic network mobility models.

Chapter 19 introduces basic notions of social network analysis, and presents fundamental properties that have been observed when analyzing human individual mobility data.

Chapter 20 describes mobility models aimed at reproducing human individual mobility patterns, which are known to be heavily influenced by social relationships between individuals.

The seventh part of the book (Case Studies) presents two case studies showcasing the importance of mobility modeling in the wireless network performance evaluation process.

Chapter 21 starts with a discussion of the implications of stationary properties of the random waypoint mobile networks on wireless network simulation accuracy. Then, the chapter introduces techniques aimed at improving simulation accuracy – the so-called "perfect simulation" methodology.

Chapter 22 discusses the effects of different assumptions regarding node mobility on conclusions that can be drawn about the asymptotic performance of opportunist networks.

Finally, this book contains two appendices, collecting basic definitions and notions of the theories underlying the field of mobility modeling. In particular, Appendix A collects basic notions of probability and stochastic process theory, while Appendix B introduces the asymptotic notation and then collects basic definitions from graph theory and miscellaneous notions that have been used in the book.

How To Use This Book

Informally speaking, the first part of the book provides basic concepts and definitions related to next generation wireless networks and mobility modeling that will be used in the rest of the book. While a reader who is familiar with the field of wireless networking can probably skip Chapter 1, she/he should probably not miss Chapters 2 and 3, which introduce the network models and basic mobility concepts used in the remainder of the book.

The second part of the book can be considered as its fundamental part, which the reader should not skip. In fact, in this part important classes of mobility models such as random walk and random waypoint models are introduced and studied, and important notions such as stationarity of a mobility model are presented. The mobility models and notions introduced in this part will be repeatedly called upon in the rest of the book.

The next four parts of the book can be treated as nearly independent, stand-alone parts, each one focusing on a specific application scenario: WLAN/mesh networks (Part Three), vehicular networks (Part Four), wireless sensor networks (Part Five), and opportunistic networks (Part Six). To a large extent, these parts are independent of each other with minimal cross-referencing, and the reader interested in a specific application scenario could focus on the corresponding part only.

Finally, the last part of the book presents two case studies. The first one refers to the well-known random waypoint mobility model, and requires a careful reading of Chapters 5 and 7 at least. The second case study focuses on opportunistic networks, and a careful reading of the corresponding part of this book is a prerequisite for fully understanding the material presented there.

The two appendices at the end are intended to provide a unique reference point for the concepts and notions from probability theory, stochastic process theory, graph theory, etc., used in the book: if the reader is not sure about a certain notion mentioned in the text, she/he can refer to the appropriate appendix to find a formal definition. In the same vein, I have included a list of the abbreviations used in the book.

Acknowledgments

There are several people whose support was fundamental in making the idea for this book become a reality.

My first thought is for my colleague and friend Sergio Palazzo, who suggested the idea of writing a book on mobility modeling, and encouraged me to undertake it. Thanks Sergio – without you this book would simply not exist.

I would also like to thank those colleagues who shared with me the exciting task of studying the realm of mobility modeling over the years: Giovanni Resta, Dieter Mitsche, Alessandro Mei, Giacomo Morabito, Julinda Stefa, Luca Becchetti, Andrea Clementi, Francesco Pasquale, Riccardo Silvestri, and Christian Bettstetter. I am grateful also to Konstantinos Mammasis for carefully reading the text of Chapter 2 on radio channel models, and for providing Figure 2.2. Finally, I wish to thank Massimo Zedda and Aleph Srl for providing Figure 12.2, and Francesca Martelli for providing Figure 12.3.

My special thanks go to all the staff at John Wiley & Sons (Anna Smart, Tiina Rounamaa, and Susan Barclay) for their assistance, encouragement, and support during the proposal, writing, and production phase of the book.

Last but not least, I would like to thank my wife Elena for her constant support and encouragement during these months. Without her support, this book would have not been possible.

Paolo Santi
Pisa

Acknowledgements

List of Abbreviations

ABS	Antilock Braking System
ADSL	Automatic Digital Subscriber Line
AODV	Ad hoc On-demand Distance Vector
AP	Access Point
BER	Bit Error Rate
BSS	Basic Service Set
CALM	Communications Access for Land Mobiles
CDF	Cumulative Density Function
CRAWDAD	Community Resource for Archiving Data at Dartmouth
CSMA/CA	Carrier Sense Multiple Access – Collision Avoidance
CTR	Critical Transmission Range
DS	Distribution System
DSR	Dynamic Source Routing
DSRC	Dedicated Short-Range Communications
DSSS	Directed-Sequence Spread Spectrum
DTN	Delay-Tolerant Network
DUA	Dynamic User Assignment
EEBL	Emergency Electronic Brake Lights
EMT	Expected Meeting Time
ESS	Extended Service Set
FFD	Full-Function Device
GPS	Global Positioning Service
GW	Gateway
HSPA	High-Speed Packet Access
ICW	Intersection Collision Warning
IP	Internet Protocol
ISO/OSI	International Organization for Standardization's Open System Interconnection
ISP	Internet Service Provider
ITS	Intelligent Transportation System
LAN	Local Area Network

LATP	Least Action Trip Planning
LOS	Line Of Sight
LTE	Long-Term Evolution
MAC	Medium Access Control
MANET	Mobile Ad hoc NETwork
MEG	Markovian Evolving Graph (model)
MULE	Mobile Ubiquitous LAN Extensions
OFDM	Orthogonal Frequency-Division Multiplexing
PAN	Personal Area Network
PCD	Personal Communication Device
PDA	Personal Digital Assistant
pdf	probability density function
PER	Packet Error Rate
PSN	Pocket-Switched Network
QoS	Quality of Service
RFD	Reduced-Function Device
RFID	Radio Frequency IDentification
RPGM	Reference Point Group Mobility (model)
RSU	Road Side Unit
RWP	Random WayPoint (model)
RWPB	Random WayPoint Border (model)
SLAW	Self-similar Least Action Walk (model)
SMS	Short Message Service
SNR	Signal to Noise Ratio
SUMO	Simulation Of Urban MObility (model)
SWIM	Small World In Motion (model)
TCP	Transmission Control Protocol
TraCI	Traffic Controller Interface
TTL	Time-To-Live
TVC	Time-Variant Community (mobility model)
UDG	Unit Disk Graph
V2I	Vehicle-to-Infrastructure
V2V	Vehicle-to-Vehicle
VANET	Vehicular Ad hoc NETwork
VFA	Virtual Force Algorithm
VoIP	Voice over Internet Protocol
WAVE	Wireless Access in Vehicular Environment
WiFi	Wireless Fidelity
WLAN	Wireless Local Area Network
WSN	Wireless Sensor Network
ZDO	ZigBee Device Object

Part One
Introduction

1

Next Generation Wireless Networks

Wireless networks nowadays are part of our everyday life, as witnessed by several features: the number of cell (mobile) phone subscriptions reached 4.6 billion in February 2010 (News 2010) and is expected to reach 6 billion (corresponding to about 72% of current world population) by the middle of 2012; short-range radio technologies such as WiFi (Alliance 2011a) and Bluetooth (Bluetooth-SIG 2011) are widespread; innovative technologies based on short-range wireless communication and miniaturized sensor devices have recently been standardized (Alliance 2011b), or are in the final steps of standardization; radio frequency IDs (RFIDs) are becoming a prominent technology in logistics and object tracking; short-range radio technology is being developed and will shortly be deployed onboard vehicles to improve safety conditions on the road and to enable innovative intelligent transportation systems; and so on.

A major outcome of intensive research in both industry and academia in recent years has been the consolidation of a wealth of short-range, inexpensive radio technologies, which promoted, and will promote even further in forthcoming years, the definition of novel network architectures such as ad hoc networks, mesh networks, wireless sensor networks, vehicular networks, and opportunistic networks. It is widely believed that these classes of networks, which we call *next generation wireless networks*, will enable the vision of "ubiquitous computing" to become a reality in the next few years.

Clarification about the term next generation wireless network as used throughout this book is in order. In a search on Google for "next generation wireless network," top hits mostly refer to evolutions of the cellular technology collectively named 4G, including technologies such as LTE (Long-Term

Mobility Models for Next Generation Wireless Networks: Ad Hoc, Vehicular and Mesh Networks, First Edition. Paolo Santi.

Evolution), LTE Advanced, etc. These technologies are essentially aimed at improving capacity and quality of service (QoS) in cellular networks, especially that which concerns data services, which are expected to play an increasing role in coming years. For instance, LTE Advanced is expected to provide maximal data rates of 1 Gbps in downlink and of 500 Mbps in uplink, compared to data rates of 84 Mbps and 22 Mbps of current high-speed packet access (HSPA+) technology. However, the above improvements are mostly obtained through advancements in the physical and medium access layer of the network protocol stack, for example, through extensive use of multi-antenna systems, while the overall architecture of the cellular network is mostly unchanged. Thus, forthcoming 4G networks can be considered as an evolution of the well-established cellular network concept.

On the other hand, the class of networks (ad hoc networks, mesh networks, wireless sensor networks, vehicular networks, and opportunistic networks) that we collectively call next generation wireless networks are characterized by very different features compared to cellular networks:

- Cellular networks rely heavily on a wired, very expensive communication infrastructure – the network of cellular base stations – to perform communication; on the contrary, next generation wireless networks are characterized by a lightweight – or even absent – infrastructure.
- Cellular networks are based on the use of relatively long wireless links, with typical communication ranges in the order of up to a few kilometers; on the contrary, next generation wireless networks use short-range – in the order of a few tenths of meters, or hundreds of meters at most – wireless links.
- Cellular networks rely heavily on the presence of several centralization points, where servers are used for optimizing resource allocation and coordinate radio channel access between the different users in a cell and in adjacent cells; on the contrary, in next generation wireless networks, centralization points are typically lacking, and most network functionalities must be realized in a fully distributed, self-coordinated environment.
- Cellular networks are single-hop wireless networks, since a mobile terminal directly communicates with a base station in the vicinity; on the contrary, next generation wireless networks typically make extensive use of multi-hop communications to compensate for the short radio range and increase coverage.

It is then evident that the class of emerging short-range wireless networks mentioned above displays striking differences with respect to traditional cellular networks, and that the design and realization of a network of this class entails defining brand-new network architectures and networking solutions spanning the entire network protocol stack. This explains our choice of

Table 1.1 Differences between cellular and next generation wireless networks

	Cellular network	Next generation wireless network
Infrastructure	Yes	No, lightweight
Radio range	$\leq$2 km	$\leq$300 m
Centralization points	Yes	Mostly no
Type of communication	Single-hop	Multi-hop
Radio channel access	Coordinated	Mostly uncoordinated

naming the emergent class of short-range, multi-hop, decentralized wireless networks "next generation wireless networks." Unless stated otherwise, in the remainder of this book we will use the term next generation wireless network exclusively to refer to a member of the class of short-range, multi-hop, decentralized wireless networks. The main differences between cellular and next generation wireless networks are summarized in Table 1.1.

In the remainder of this chapter, we will briefly describe different types of next generation wireless networks and relevant application scenarios, as well as the several challenges that the designers of such networks will face.

1.1 WLAN and Mesh Networks

A wireless local area network (WLAN) is a network connecting two or more devices (called *stations* in WLAN terminology) in a relatively small area (e.g., a building, a university campus, etc.), and usually providing a connection to the Internet through one or more access points (APs). A distinguishing feature of WLANs with respect to traditional, wired LANs is that station-to-station and station-to-AP connections exploit wireless technology.

A typical WLAN architecture is reported in Figure 1.1: one or more APs, which are directly connected to the Internet through a wired connection (typically, an Ethernet link), set up and operate wireless links with one or more stations within their coverage area. Wireless connections are typically realized through the establishment of IEEE 802.11 links. The family of IEEE 802.11 protocols (IEEE 2011a) is at the basis of WiFi technology, and allows point-to-point links to be established between different stations, and between stations and APs, at a data rate as high as 600 Mbps (forthcoming IEEE 802.11n).

Two factors have contributed to the success of WLANs: the increasing popularity of laptop computers and powerful handheld devices such as PDAs and smart phones on the one hand, and the maturing of WiFi into a very affordable, short-range wireless technology on the other hand. Due to these

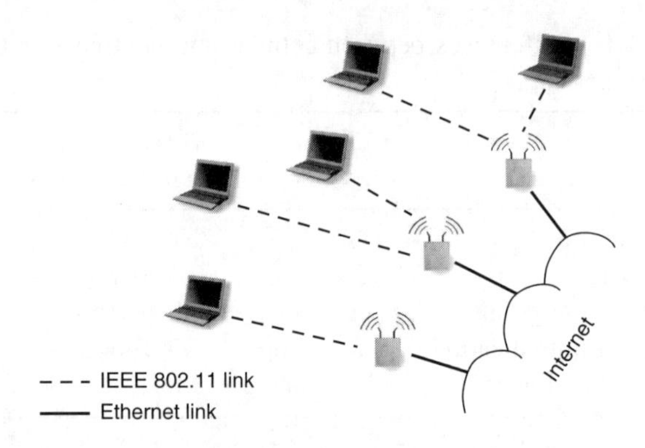

Figure 1.1 A typical WLAN architecture.

factors, WLANs are now widely used to provide Internet connectivity in corporate buildings, university campuses, coffee shops, bookstores, airports, shopping malls, etc. With their high data rates and affordable prices, small-scale WLANs have de facto become the standard way of connecting to the Internet, even in the home.

A major problem facing WLAN designers is radio coverage. In fact, the typical communication range of an IEEE 802.11 radio is up to 100 m (usually much less in indoor environments), implying that many APs must be deployed to cover a relatively large area. For instance, several hundred APs are typically needed to provide indoor and outdoor Internet connectivity to a medium-sized university campus. Considering that in the standard WLAN architecture wireless APs need a wired connection to an Internet gateway, the logistics and wiring cost of setting up a relatively large and dispersed WLAN might be problematic. To overcome this, the IEEE 802.11 working group has recently drafted an amendment to the standard, named IEEE 802.11s, that allows the creation of a WLAN *mesh* network.

The main building block of a WLAN mesh network, as reported in Figure 1.2, is a device called a *mesh station*, which has the ability to establish point-to-point *wireless* links with other mesh stations, thus realizing a multi-hop, mesh network architecture between the mesh stations composing the WLAN. A mesh station, also known as a *mesh router* in the mesh networking research literature, can incorporate additional functionality, namely: (i) AP functionality to provide wireless access to WLAN stations; and (ii) Internet gateway functionality, if the mesh station has a wired connection to the Internet. It is important to observe that, in the WLAN mesh network concept, AP and gateway functionality are optional, and are in general provided only by a subset of the mesh stations. For instance, a mesh station might be deployed with the sole purpose of acting as a bridge

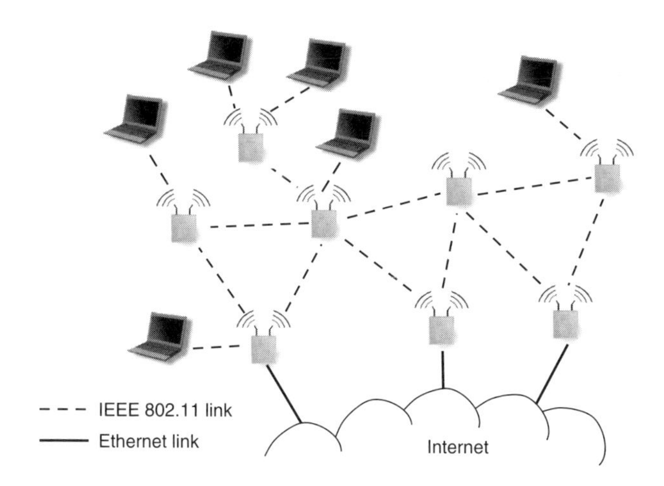

Figure 1.2 A typical WLAN mesh network architecture.

between different parts of the WLAN, in which case neither functionality (i) nor (ii) are provided.

WLAN mesh networks have several potential advantages with respect to traditional WLANs:

(a) *Reduced wiring cost*: since in general only a subset of mesh stations requires wired connection to the Internet, wiring costs can be significantly reduced. This reduction is very notable in application scenarios where the WLAN has to be deployed in a geographically dispersed environment without a pre-existing wired connection to the Internet, for example, in rural scenarios. In these situations, the costs of bringing wired Internet connection to remote locations can be prohibitive, and a mesh network architecture is often the only cost-effective solution.

(b) *Improved coverage*: additional mesh stations acting as a bridge to Internet gateways can be easily deployed, since they do not require a wired connection to the Internet. Hence, the coverage area of a WLAN mesh network can be extended at a limited cost. It should be noted, though, that since mesh stations communicate with each other through wireless IEEE 802.11 links, the radio range between mesh stations can be an issue. However, the radio range between mesh stations can be significantly extended with respect to typical values of WiFi technology through, for example, optimized mesh station deployment (for instance, on the roof of buildings, on top of high poles, etc.), and usage of high-gain directional antennas. Similar solutions have been successfully tested, for instance, in rural WLAN mesh network scenarios (Raman and Chebrolu 2007), where wireless links in the order of a few kilometers have been used.

(c) *Robustness*: due to the presence of several multi-hop paths between an AP and a mesh station providing gateway functionality (recall Figure 1.2), WLAN mesh networks are robust against failures of individual mesh stations.

From the point of view of a network designer, the price to pay for the above advantages is the need to design additional networking solutions with respect to those required in a typical WLAN situation. In particular, since multi-hop wireless communications are heavily exploited in WLAN mesh networks, suitable protocols for routing packets from stations to an Internet gateway through the mesh stations should be designed. Furthermore, resource and bandwidth allocation strategies should be implemented to allow effective and fair sharing of the available bandwidth between the mesh stations and, ultimately, between the stations registered with the APs. Finally, from a technological point of view a major challenge is to increase the capacity of the set of wireless links between mesh stations as much as possible, as this part of the network is likely to become a bottleneck if several stations are registered with the APs.

The notion of mesh network is not exclusive to the WLAN concept. In fact, mesh networking solutions are currently being considered and standardized also for other types of networks, such as within the realm of wireless metropolitan area networks – see WiMax Mesh Mode of operation (WiMax-Forum 2011). For a more detailed description of WLAN and mesh networks, the interested reader is referred to Chapter 8 of this book and references therein.

1.2 Ad Hoc Networks

An ad hoc network is an infrastructure-less, wireless multi-hop network composed of a set of wireless point-to-point links between a set of nodes. The architecture of a typical ad hoc network is shown in Figure 1.3. The term MANET (Mobile Ad hoc NETwork) is commonly used to refer to an ad hoc network whose nodes are mobile.

Differently from WLANs and mesh networks, no infrastructure or centralization point is available to coordinate network functionality and radio resource allocation, so nodes in an ad hoc network must be able to self-organize, set up, and maintain a suitable set of wireless links, as well as to implement all necessary networking protocols in a fully distributed way. Also, the typical traffic pattern in ad hoc network is quite different with respect to those in WLAN and mesh networks: while in the latter types of networks traffic is typically directed to or coming from Internet gateways, ad hoc networks are mostly used to exchange data within the network itself, so the typical traffic pattern is peer-to-peer.

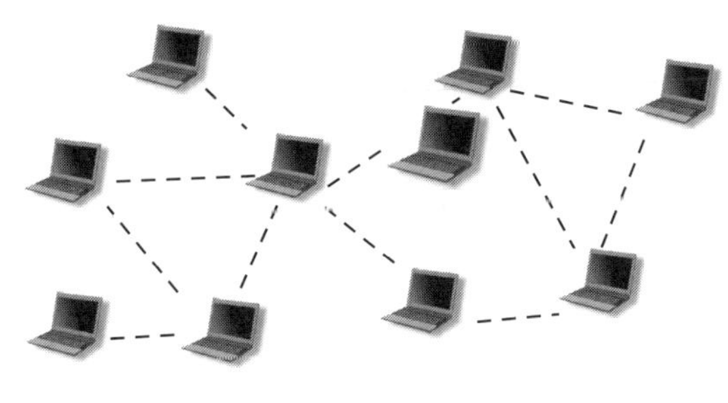

Figure 1.3 A typical ad hoc network architecture.

The prominent wireless technologies used to set up the wireless links in an ad hoc network are IEEE 802.11 and Bluetooth. Both technologies, in fact, allow the creation and operation of wireless links in a point-to-point fashion: in IEEE 802.11, the *ad hoc mode* can be used to set up point-to-point links without the assistance of an AP; in Bluetooth, piconets of up to seven nodes can be formed, with one of the nodes acting as the coordinator; piconets can then be joined through nodes acting as bridges to form larger so-called scatternets. The data rates achievable on the wireless links are those typical of the respective technology, that is, up to 600 Mbps with IEEE 802.11, and up to 24 Mbps with Bluetooth (version 3.0).

Ad hoc networks were originally proposed in military applications – to provide communication between members of a platoon on the battlefield. Another prominent application scenario of ad hoc networks is disaster relief, where this flexible, quick-to-deploy network can be used to provide communication between members of a rescue squad, firefighters, police officers, etc., in case the pre-existing communication infrastructure is no longer working.

Despite the concept of ad hoc network being introduced more than 30 years ago, applications based on the ad hoc networking paradigm outside the military scope are almost completely lacking to date. This is due to the severe challenges that must be faced by an effective ad hoc network design, which can be summarized as follows:

(a) *Energy conservation*: IEEE 802.11 and Bluetooth are relatively power-hungry technologies, so efforts should be made to optimize communications from the viewpoint of energy consumption.
(b) *Unstructured and time-varying network topology*: in general, links in an ad hoc network are relatively unstable, even if nodes are not mobile. This is mainly due to changes in the environment and interference from co-located wireless networks (consider that IEEE 802.11 and Bluetooth are nowadays extremely popular technologies). Given that link instability

can hardly be tamed, ad hoc networking solutions must be designed to properly deal with an unstructured and time-varying network topology.

(c) *Unreliable communication*: another consequence of link instability is that, in general, communication between peer nodes in an ad hoc network, especially if exploiting multi-hop paths, is unreliable: link quality and/or data rates can unexpectedly drop, or experience a sudden increase, in a very unpredictable way. Unreliable link behavior should then be carefully accounted for when designing networking and transport-level protocols.

(d) *Scalability*: given (a), (b), and (c) above, the design of a large-scale ad hoc network is an extremely challenging task. For this reason, ad hoc networks are currently considered an adequate architecture for networks composed of up to a few tens of nodes.

The above-mentioned severe challenges have limited the widespread application of the general ad hoc network concept. However, this networking concept still plays a major role in next generation wireless networks, as it is the basis of more specialized classes of short-range wireless networks, such as vehicular networks, wireless sensor networks, and opportunistic networks, which will be described in the remainder of this chapter.

1.3 Vehicular Networks

Vehicular networks are short-range wireless networks where links are established between nearby vehicles, or between a vehicle and a so-called road side unit (RSU). RSUs are typically constituted of radio devices installed on top of road side poles, which are typically part of a pre-existing road infrastructure (traffic lights, street lamps, etc.).

The architecture of a typical vehicular network is shown in Figure 1.4. As seen from the figure, vehicle-to-vehicle (V2V) and vehicle-to-infrastructure (V2I) links (i.e., links between a vehicle and a RSU) coexist in a vehicular network, with the purpose of forming a highly dynamic, mobile network allowing exchange of messages between vehicles, and possibly between vehicles and external networks (e.g., the Internet), which can be accessed through the RSU. From an architectural point of view, vehicular networks share features with ad hoc and opportunistic networks. In fact, the part of the network formed of V2V links between vehicles can be considered as a special type of MANET, which is named a vehicular ad hoc network, or VANET, in the corresponding literature. On the other hand, if a message has to be delivered to an external network (e.g., the report of a measured travel time), a vehicle often needs to store and carry the message for a relatively long time, till a communication opportunity arises when a RSU comes into reach. Thus, the store, carry, and forward communication paradigm typical of opportunistic networks (see Section 1.5) is exploited.

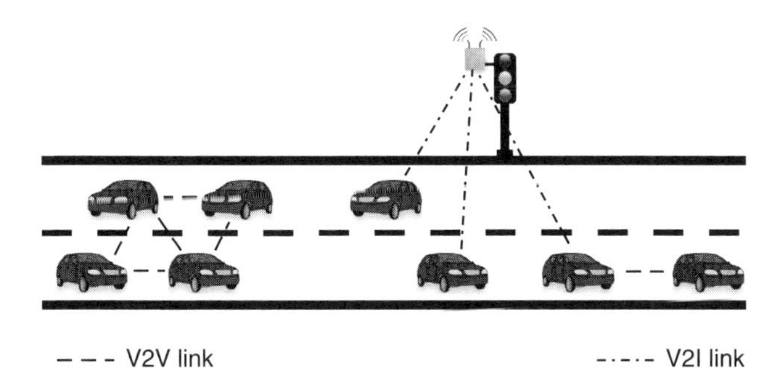

Figure 1.4 A typical vehicular network architecture.

Vehicular networks are attracting increasing interest and investment in research as well as in the automotive industry, and receiving massive funding from governments and various agencies. In fact, they are considered as a prominent means of improving road safety conditions and traffic efficiency in forthcoming years:

- **Vehicular networks for improving road safety conditions**. Direct message exchange between vehicles can be used to let active safety applications running onboard vehicles acquire an up-to-date view of the traffic situation surrounding the vehicle, which is typically achieved through exchange of status messages (i.e., messages reporting information such as vehicle speed, position, direction, and so on) between vehicles. In turn, up-to-date situation awareness enables a suitably designed active safety application to identify early possible hazards and dangerous conditions, which will be immediately communicated to the driver so that she/he can take adequate countermeasures. The primary design goal of active safety applications based on vehicular networks is extending situation awareness *beyond human eyes*, so that a driver can be warned of a potentially dangerous situation well before her/his eyes can see the situation. Considering that reducing a driver's reaction time of a fraction of a second would result in avoiding the majority of accidents, a large fraction of road accidents can potentially be avoided through vehicular networking technologies.
- **Vehicular networks for improving traffic efficiency**. Message exchange between vehicles, and between vehicles and RSUs, can be exploited to quickly inform vehicles about changes of traffic status (e.g., traffic build-up due to an accident). In principle, real-time traffic monitoring and smart navigation systems (i.e., systems that dynamically change the suggested travel path based on real-time traffic information) can lead to considerable reductions in travel time and to increased traffic efficiency in general. Other types of vehicular network applications aimed at improving traffic

efficiency are being designed, such as lane merging assistance, intersection assistance, and so on.

Besides road safety and traffic efficiency applications, other classes of applications are being investigated for vehicular networks, such as peer-to-peer file sharing between neighboring vehicles, chat, localized advertisements, VoIP, etc. However, while for safety/traffic-related applications there is a strong push from both governmental agencies and the automotive industry, "entertainment"-related applications are currently seen as a less important class of applications for vehicular networks. We envision, though, that once the vehicular networking technology has been settled, several entertainment-related applications will be developed and will rapidly become widespread.

Although in principle different radio technologies could be used for V2V and V2I communications, in order to reduce cost and design complexity a single radio technology (and, typically, also a single radio device onboard vehicles) is used for both V2V and V2I communications. The radio technology which is being promoted for vehicular communications is a specialization of the well-known WiFi standard for vehicular environments. More specifically, an amendment to the IEEE 802.11 standard, called IEEE 802.11p (IEEE 2011b), is currently in an advanced status of definition. The main purpose of IEEE 802.11p is to combat severe radio propagation issues arising in vehicular environments, mostly related to a very strong Doppler effect due to high vehicular velocities. Another goal of IEEE 802.11p is to reduce the signaling overhead needed to set up a V2V or V2I link, given the highly dynamic nature of vehicular networks. The current IEEE 802.11p draft allows data rates as high as 54 Mbps on the radio link, although much lower data rates (3 or 6 Mbps) are mandated for active safety application. For a more detailed description of IEEE 802.11p, the reader is referred to Chapter 11.

Several challenges are faced by the vehicular network designer in order to turn the above vision of improved road safety conditions and traffic efficiency into reality. These challenges encompasses both communication and networking aspects. From the communication point of view, a major challenge is to design a radio technology that allows low-delay, reliable communications in a highly dynamic, mobile, and possibly congested radio environment. Low delays and highly reliable communications are prerequisites for an effective active safety application design. From a networking perspective, challenges are related to the design of innovative solutions at the different layers of the network architecture, which are needed to support the above-mentioned classes of vehicular network applications. For instance, concerning the network layer, several routing primitives should co-exist in a vehicular network, such as traditional, IP-based routing (e.g., to access the Internet through a RSU), geo-routing (e.g., to reach vehicles in a specific area to warn them of a change in traffic conditions), geographically scoped broadcast, etc. The above should be achieved in a highly dynamic, mobile environment, where frequent disconnections typically occur, making the design challenge even harder. For a more detailed description of the communication and networking challenges in vehicular networks, the interested reader is referred to Chapter 11.

1.4 Wireless Sensor Networks

Wireless sensor networks (WSNs) are short-range radio networks composed of radio-equipped sensor nodes and a few gathering points, named *base stations* or *sink nodes* in wireless sensor network terminology. A typical wireless sensor network architecture is displayed in Figure 1.5. Similar to vehicular networks, WSNs can be considered as a specific class of ad hoc network, where network nodes are mostly fixed. Another distinguishing feature of WSNs is the number of network nodes which, depending on the application scenario, can grow to several thousands.

In a typical wireless sensor networking application, a certain environmental parameter (e.g., temperature) is monitored at each sensor node and periodically reported to a base station. The base station is in charge of collecting the data of interest and reporting it to the user(s). If the base station has an Internet connection, remote monitoring of the area of interest is possible. Given the vast geographical area typically covered by a WSN, and the short range of wireless communications, multi-hop communication paths are extensively used to report data to the base stations. Depending on the application and on the nature of the monitored information, data can be aggregated on its route to a base station, achieving significant savings in the total number of messages transmitted within the network.

Wireless sensor networks are increasingly being designed and used to achieve fine-grained monitoring of natural phenomena, to control industrial

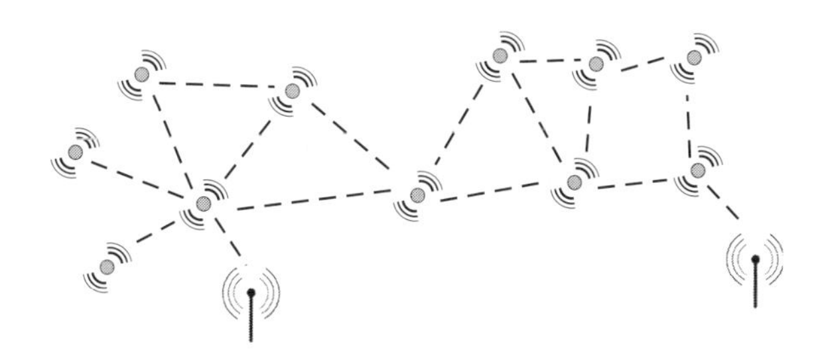

Figure 1.5 A typical wireless sensor network architecture.

processes, to improve energy efficiency of buildings, and so on. The major advantage of WSNs with respect to more traditional sensing technologies lies in the fact that sensor nodes are able to communicate with each other and with the base station through wireless links. On the one hand, usage of wireless technology allows considerable reductions in wiring costs, which in some scenarios (e.g., monitoring of industrial processes) represent a large fraction of the total cost of a monitoring system. Also, brand-new monitoring applications can be enabled by usage of wireless technology, such as remote monitoring of wild animals, herds, and so on. On the other hand, usage of wireless links challenges the WSN designer, especially in those application scenarios (e.g., monitoring of industrial processes) where reliability is paramount. Other challenges facing the wireless sensor network designer are related to scalability of the designed solutions, given the large number of nodes comprising a WSN in most application scenarios. Furthermore, energy conservation is also paramount in WSNs, given that wireless sensor nodes are typically battery operated, and replacing batteries is often very difficult. Although energy harvesting technologies (based, for example, on micro solar panels) are increasingly being used in WSNs, efficient use of the available energy is still a major goal – and it will most likely remain so in the future – in the design of wireless sensor network applications.

In terms of technological development, WSNs are in a phase where the communication technology – mostly based on the IEEE 802.15.4 standard (IEEE 2011c) and ZigBee protocol suite (Alliance 2011b) – is relatively mature, and applications of wireless sensor networks are increasingly being developed by both major IT industries and a plethora of small and medium-size enterprises (often spin-offs of major technical universities). For more a detailed description of the state of the art and prospects of wireless sensor networks the interested reader is referred to Chapter 14.

1.5 Opportunistic Networks

Opportunistic networks are a class of MANETs characterized by a very sparse density of nodes, which, coupled with the relatively short range of radio communication, results in a network topology that is disconnected most of the time. A typical opportunistic network architecture is shown in Figure 1.6. A node which is part of an opportunistic network spends most of its time in isolation, without any wireless connection to other nodes. From time to time, a few other nodes of the network might come into reach, giving rise to a *communication opportunity*, or *contact*. During a contact, nodes exchange the list of messages in the respective buffers, and might decide to forward some of them to another node. This communication mechanism – called *store, carry*, and *forward* – is the basis of opportunistic networking, and enables eventual message reception if and when the destination of a message *m*

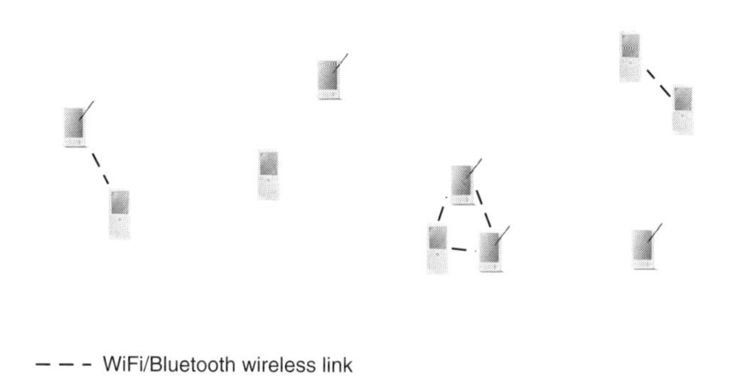

Figure 1.6 A typical opportunistic network architecture.

comes in contact with a node carrying a copy of m. Clearly, opportunistic network applications must be able to tolerate very large delays in message delivery. For this reason, opportunistic networks are also called *delay-tolerant networks* in the literature.

It is important to observe that, since opportunistic networks are disconnected most of the time due to sparse node density, communication opportunities arise typically because of node mobility. In other words, while in other types of next generation wireless networks node mobility is either mostly absent (wireless sensor networks) or perceived as a factor worsening networking performance (in WLANs, mesh networks, vehicular networks, etc.), in opportunistic networks mobility is indeed the only possible communication means: without mobility, network-wide communication would not be possible at all. This is a very distinguishing feature of opportunistic networks with respect to other types of next generation wireless networks, which also explains why so many efforts in the research community have been devoted to understanding and modeling mobility in opportunistic networks (see Part Six of this book).

Although the majority of nodes in an opportunistic network is typically mobile, in some applications a number of fixed nodes are also present in the network. Fixed nodes are typically used either to enable more message-forwarding opportunities within the network, or to provide gateway functionality to external networks. Examples of opportunistic networks where a majority of mobile nodes coexist with a few fixed nodes are instances of vehicular networks, such as DieselNet (see below).

The delay-tolerant network (DTN) concept dates back to the beginning of this century, when Kevin Fall adapted some of the ideas developed around the notion of Interplanetary Internet under investigation at NASA and other agencies to terrestrial networks (Fall 2003). Since then, several applications of the DTN concept have been proposed. Some early examples are ZebraNet (Martonosi et al. 2004), an opportunistic network composed of a small number of radio-equipped zebras living in Sweetwaters Reserve in

Kenya, and DieselNet (Group 2007), an opportunistic network composed of 40 public buses in the city of Amherst, Massachusetts.

Recently, a particular class of opportunistic networks called *pocket-switched networks* is attracting increasing interest in the research community. The notion of a pocket-switched network was first introduced in Hui et al. (2005), where the authors describe an opportunistic network composed of moving individuals carrying a smart phone, PDA, or similar device. The interest in pocket-switched networks stems from the observation that an increasing number of personal mobile communication devices are endowed with several wireless interfaces, allowing not only communication with a cellular network infrastructure, but also direct, peer-to-peer wireless communications. Prominent examples of these wireless interfaces are WiFi and Bluetooth. Hence, very large opportunistic networks composed of individuals moving, say, in a large city can be formed if such wireless interfaces are activated and used to directly exchange messages between mobile devices.

While the communication technology at the base of opportunistic networks in general, and pocket switched networks in particular (mostly WiFi and Bluetooth), is relatively mature, applications for this type of networks are still to be developed, except for a few prototypal examples. The most promising directions seem to be the development of opportunistic networks to provide delay-tolerant Internet connectivity in rural environments, and the extension of social networking applications to mobile environments. In terms of challenges, the major challenge facing the opportunistic network designer is how to reduce message delivery delay as much as possible. In turn, this requires careful optimization of the number of copies of a message circulating in the network, as well as the strategy used to select messages to be forwarded to another node during a contact. With respect to the number of copies of a message circulating in the network, we observe that the higher this number is, the higher the chances for the destination to get in contact with a node carrying a copy of the message; however, a higher number of message copies also entails a larger number of messages circulating in the network, leading to an increased buffer occupation and more chances for a (copy of a) message to be dropped. For more details on the challenges and opportunities of opportunistic networks, the interested reader is referred to Chapter 17.

References

Alliance W 2011a *http://www.wi-fi.org*.

Alliance Z 2011b *http://www.zigbee.org/*.

Bluetooth-SIG 2011 *https://www.bluetooth.org/apps/content/*.

Fall K 2003 A delay-tolerant network architecture for challenged internets. *Proceedings of ACM SIGCOMM*, pp. 27–34.

Group PR 2007 *http://prisms.cs.umass.edu/dome/umassdieselnet*.

Hui P, Chaintreau A, Scott J, Gass R, Crowcroft J and Diot C 2005 Pocket-switched networks and human mobility in conference environments. *Proceedings of the ACM Workshop on Delay-Tolerant Networks*.

IEEE 2011a *http://www.ieee802.org/11/*.

IEEE 2011b *http://grouper.ieee.org/groups/802/11/Reports/tgp_update.htm*.

IEEE 2011c *http://www.ieee802.org/15/pub/TG4.html*.

Martonosi M, Lyon S, Peh LS, Poor V, Rubenstein D, Sadler C, Juang P, Liu T, Wang Y and Zhang P 2004 *http://www.princeton.edu/mrm/zebranet.html*.

News C 2010 *http://www.cbsnews.com/stories/2010/02/15/business/main6209772.shtml*.

Raman B and Chebrolu K 2007 Experiences in using WiFi for rural internet in India. *IEEE Communications Magazine* **45**, 104–110.

WiMax-Forum 2011 *http://www.wimaxforum.org/*.

2

Modeling Next Generation Wireless Networks

Modeling is a substantial part of the network performance evaluation process. Modeling is important not only to devise an analysis through which performance parameters can be optimized, but also to develop accurate network simulation tools. Simulation is the primary performance evaluation tool for next generation wireless networks, for several reasons.

The primary reason for the extensive use of simulation in wireless network performance evaluation is that cost and logistic considerations limit the realization of large-scale testbeds. Large testbeds composed of, say, hundreds or thousands of nodes can be realized only through substantial, well-focused funding programs and R&D efforts, which only very recently are being implemented worldwide. Current state of the art next generation wireless network testbeds are typically composed of less than 100 nodes, most of the times being composed of only a few tens of nodes. Hence, if the scalability of a networking solution has to be assessed, then analysis and simulation are typically the only available tools.

Another reason behind the extensive use of simulation in wireless network performance evaluation is related to the repeatability of experiments, which is the basis of the Galilean scientific method. While exactly reproducing an experiment is difficult for networks in general, it becomes extremely challenging in wireless networks, which are very sensitive to changes in environmental conditions. Not only do properties of the environment (e.g., closing a door during an indoor experiment) influence the outcome of a wireless network experiment, but interference from co-located networks also has a strong influence. This is especially true since most radio technologies exploited in next generation wireless networks operate in unlicensed frequency bands, which are often densely populated (think about the number of

Mobility Models for Next Generation Wireless Networks: Ad Hoc, Vehicular and Mesh Networks, First Edition. Paolo Santi.

different WLANs co-located in a city, a university campus, or a corporate building). While changes in the environment can be somehow minimized, interference generated by external networks operating in the same frequency band is virtually impossible to control, thus compromising the repeatability of experiments.

Modeling next generation wireless networks requires the development of models for at least the following aspects: (i) radio channel; (ii) network topology; (iii) node mobility; and (iv) energy consumption. While the rest of this book is devoted to mobility models, in this chapter we will briefly revise radio channel, network topology, and energy consumption models for next generation wireless networks. Most of the material presented in this chapter is based on Rappaport (n.d.), Molisch (2005), and Chapter 2 of Santi (2005).

2.1 Radio Channel Models

A radio channel is established between a transmitter node u and a receiver node v if and only if the power $P_r(v)$ of the radio signal received at v's antenna satisfies the following conditions:

1. It is above a certain threshold, called the *sensitivity threshold*; formally, $P_r(v) \geq \beta$, where β is the sensitivity threshold.
2. The signal to noise ratio (SNR) is high enough to ensure a sufficient transmission quality; formally

$$\frac{P_r(v)}{N} \geq \gamma,$$

where N is the noise power at the receiver and γ is the SNR threshold.

The sensitivity threshold is a characteristic of the radio: the more sophisticated the radio, the lower the sensitivity threshold enabling successful reception of a transmitted packet. The SNR threshold is instead mandated by the coding scheme and the rate used to transmit data, with higher data rates typically requiring better communication quality, that is, higher SNR thresholds. This also explains why higher data rates result in shorter transmission ranges. Indeed, condition 2 above is a slight simplification of reality, where a certain SNR value corresponds to a bit error rate (BER) which, coupled with coding scheme and data rate, results in a certain probability for a packet to be corrupted (packet error rate – PER). Condition 2 above is equivalent to a situation in which a desired PER describing satisfactory link quality is set at the design stage, and the minimum SNR value ensuring the prescribed PER value is fulfilled is computed and set as the SNR threshold.

From the above discussion, it is evident that whether a radio link between a transmitter node u and a potential receiver node v is established depends mostly on the power $P_r(v)$ of the transmitted signal received by node u. The purpose of radio channel modeling is exactly that of predicting the value of $P_r(v)$ for specific locations of u and v.

Radio channel modeling has been and still is a major research field in wireless communication engineering. This is because modeling a radio channel is a very challenging task. In fact, while in wired communications the transmitted energy can reach the receiver only through a single path (the wire), in wireless communications the energy emitted by the transmitter antenna reaches the receiver antenna through different *radio propagation paths*. This phenomenon, known as *multi-path propagation*, is the very reason why accurate radio channel modeling is a very challenging task.

Multi-path propagation is caused by physical phenomena known as *reflection, diffraction*, and *scattering*, which occur when electromagnetic waves emitted by the transmitter antenna hit surrounding objects. Reflection occurs when the electromagnetic wave hits the surface of an object that has very large dimensions compared to the wavelength of the radio signal. Typically, reflection is caused by the surface of the Earth, large buildings, walls, etc. Diffraction occurs when the path between the transmitter and receiver is obstructed by an object whose surface has sharp edges. For instance, a moving car typically causes diffraction if hit by a radio signal. Finally, scattering occurs when a relatively large number of relatively small (compared to the radio signal wavelength) objects obstruct the radio path between transmitter and receiver. This is the case, for instance, for foliage, street signs in a urban environment, etc.

Among the propagation paths between transmitter and receiver antennas, there might be a *dominant* path, typically corresponding to the direct path between the two antennas when in line of sight (LOS) conditions. In case a dominant propagation path is present, the propagation of the radio signal along this path mostly dictates the amount of power that is received at the v antenna. If no dominant propagation path exists, typically the transmitter and receiver are not in LOS conditions, and the power $P_r(v)$ received at v is the result of the transmitted signal components arriving via different propagation paths.

If the locations of nodes u and v and the geometry of the environment (location, size, and physical nature of objects) are known, accurate predictions of $P_r(v)$ can be performed through so-called ray tracing models, where single paths in the multi-path environment are considered and used to compute the phase and amplitude of the radio signal received at v. However, ray tracing models are computationally intensive and give accurate predictions of $P_r(v)$ only for the specific geometry considered: if the position of u or v, or of one of the objects in the environment, changes even slightly, the value of $P_r(v)$ can change substantially, due to small-scale fading effects (see below). Thus, ray tracing models cannot be used to model situations where mobility (of nodes or of objects in the environment) comes into play, as is the case in most real-life situations.

Given the infeasibility of ray tracing models in most practical situations, statistical radio channel models have been defined and are used extensively in wireless communication system design. In statistical radio channel models, the received power $P_r(v)$ is considered as a random variable, and the goal is to characterize the first and second order moment (i.e., mean and standard deviation) of $P_r(v)$, and its distribution.

The received power $P_r(v)$ depends on the power P_t emitted by the u antenna, and on the so-called *path loss*, which models radio signal degradation with distance. Denoting $PL(u, v)$ as the experienced path loss between u and v, we can write

$$P_r(v) = \frac{P_t}{PL(u, v)}.$$

Note that the term P_t in the above formula incorporates also the effects of transmitter and receiver antenna gains, which are not explicitly reported in the formula to keep the presentation simple. Rewriting the above in *dB* (logarithmic) scale, we have

$$P_r(v)\ (\mathrm{dB}) = 10\log(P_t) - 10\log(PL(u, v)) = P_t\ (\mathrm{dB}) - PL(u, v)\ (\mathrm{dB}).$$

The outcome of several years of intensive research is that $PL(u, v)$ should be modeled as a random variable resulting from the superposition of two different components:

1. A *deterministic* path loss component which depends on the *distance* d_{uv} between nodes u and v.
2. A *random* component accounting for *large-scale fading* – also known as *shadowing* – effects.
   Formally, we can write

$$PL(u, v)\ (dB) = PL(d_{uv})\ (dB) + LS\ (dB),$$

where $PL(d_{uv})$ is the deterministic, distance-related component, and LS is a random variable accounting for shadowing effects. Shadowing is caused by the obstruction of large objects (e.g., buildings), generating radio signal "shadows" in certain regions. Thus, if the receiver is in a shadowed area, a relatively higher path loss and lower signal quality are experienced. On the other hand, components of the radio signal propagating through different paths might add coherently at a certain location, resulting in a relatively higher amplitude of the received signal, that is, in better signal quality. Thus, in general the deterministic, distance-related component $PL(d_{uv})$ of path loss estimates the *average* value of path loss at a certain distance, where averaging must be intended in the (large-scale) spatial domain, while the random shadowing component models variations of the path loss at a specific location due to large-scale fading.

Several path loss models have been developed and validated through measurement campaigns in the wireless communication literature. For specific path loss models, the interested reader is referred to Molisch (2005) and Rappaport (n.d.). We now present the most widely used path loss model, which is the *log-distance path loss* model with *log-normal shadowing*.

The log-distance path loss model dictates that the average long-distance path loss is proportional to the distance d_{uv} raised to a certain exponent α, which is called the *path loss exponent*, or *distance–power gradient*. Formally,

$$PL(d_{uv}) \propto \left(\frac{d_{uv}}{d_0}\right)^{\alpha}, \tag{2.1}$$

where d_0 is the close-in reference distance determined from measurements close to the transmitter. When expressed in dB, Equation 2.1 becomes

$$PL(d_{uv})\ (dB) \propto 10\alpha \log \frac{d_{uv}}{d_0}. \tag{2.2}$$

The value of α depends on the propagation environment, that is, on the density and nature of objects causing multi-path propagation. Some of these values for relevant scenarios are summarized in Table 2.1, as reported in Santi (2005).

The log-normal shadowing model dictates that the fluctuation of the radio signal in the (large-scale) spatial domain due to shadowing can be modeled as a normal random variable (in dB) with zero mean and standard deviation σ, where σ is a parameter depending on the propagation environment. Typical

Table 2.1 Values of the distance–power gradient in different propagation environments

Environment	α
Free space	2
Urban area	2.7 to 3.5
Indoor LOS	1.6 to 1.8
Indoor no LOS	4 to 6

values of σ are in the range 2–6 dB. Formally,

$$LS\ (dB) = N(0, \sigma)$$

Putting everything together, we can write

$$PL(u, v)\ (dB) = PL(d_0)\ (dB) + 10\alpha \log \frac{d_{uv}}{d_0} + N(0, \sigma)\ (dB),$$

where $PL(d_0)$ is a constant representing the path loss experienced at the reference distance (typically obtained through measurements) and includes parameters such as transmitter and receiver antenna gain, system loss factor, etc. Thus, we can conclude that

$$P_r(v)\ (dB) = P_t\ (dB) - \left(PL(d_0)\ (dB) + 10\alpha \log \frac{d_{uv}}{d_0} + N(0, \sigma)\ (dB) \right).$$

As explained above, path loss models have the purpose of estimating the intensity of the radio signal received at a certain distance from the transmitter *when averaged overall a relatively small area*. The geometric interpretation of path loss and shadowing is reported in Figure 2.1. Assume we have a wireless transmitter u, and let us focus our attention on what happens when the potential receiver is located at distance d from u, that is, somewhere on the large circle. Let us fix a certain position at distance d from u, say point A_1. Experience teaches us that the intensity of the radio signal received at A_1 might vary considerably if the position of A_1 is slightly perturbed, due to so-called *small-scale fading* effects which we will describe shortly. To smooth such variations over a relatively small spatial scale, the intensity of the radio signal received at A_1 is computed as the average of the intensity of the received signal when the position of the potential receiver is slightly changed around A_1; say, in a disk of radius r, with $r \ll d$, centered at A_1

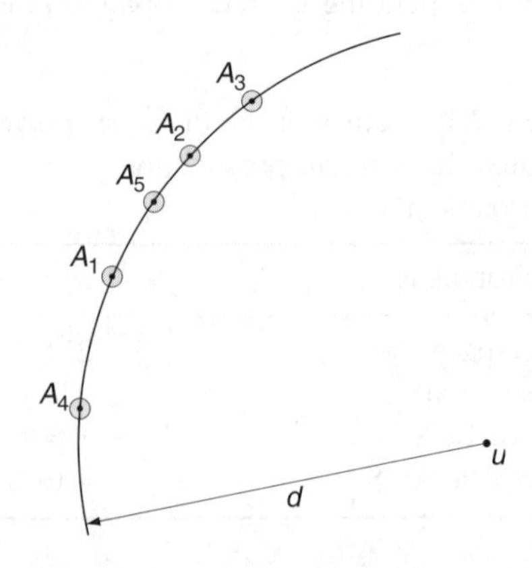

Figure 2.1 Geometrical interpretation of path loss and large-scale fading.

(shaded disk in Figure 2.1). Let $\bar{A}_1$ denote this average value of the radio signal received at A_1, and assume several such values $\bar{A}_1, \ldots, \bar{A}_n$ are sampled by randomly choosing positions on the circle of radius d centered at u. The log-distance path loss model with log-normal shadowing dictates that:

1. As the number n of samples increases, the average of the $\bar{A}_i$ values (which are themselves averages over a small spatial domain) converges to $PL(d)$ as reported in Equation 2.2. Formally,

$$\lim_{n \to \infty} \frac{\sum_{i=1}^{n} \bar{A}_i}{n} \ (dB) = PL(d_0)\ (dB) + 10\alpha \log \frac{d_{uv}}{d_0}.$$

2. If we consider a single sampled value $\bar{A}_j$ for some $1 \leq j \leq n$, its deviation (in dB) from the average path loss value as reported in Equation 2.2 – due to shadowing – is well approximated by a Gaussian distribution with zero mean and standard deviation σ (in dB).

As commented above, experience has taught wireless communication engineers that the intensity of the radio signal often varies significantly even on a small spatial scale, for example, within the shadowed disk in Figure 2.1. This phenomenon, known as *small-scale fading*, is caused by the interference between two or more versions of the transmitted signal arriving at the receiver through different propagation paths, hence at slightly different times. Due to differences in amplitude and, most importantly, phase of the various versions of the signal received, the resulting combined radio signal in general displays considerable variations in both amplitude and phase depending on the distribution of the intensity and relative propagation time of the electromagnetic waves along the different paths. Thus, even slight changes in the propagation environment (e.g., because of movement of the receiver, or of an object in the environment) might cause very significant variations in the received radio signal power.

Small scale fading models have the goal of modeling the fluctuation of the received radio signal power around the mean value (as predicted by a path loss model) in either the (small scale) spatial or temporal domain. Well-known small scale fading models are the Ricean model, which models scenarios in which there is a dominant propagation path (LOS conditions), and the Rayleigh model, which is instead used when no dominant propagation path is present (no LOS conditions). The interested reader is referred to Molisch (2005) and Rappaport (n.d.) for descriptions of small scale fading models. In the remainder of this chapter, we will mostly ignore small scale fading effects, and restrict our attention to path loss models.

The compound effect of path loss, shadowing, and small-scale fading as the distance between transmitter and receiver increases is shown in Figure 2.2:

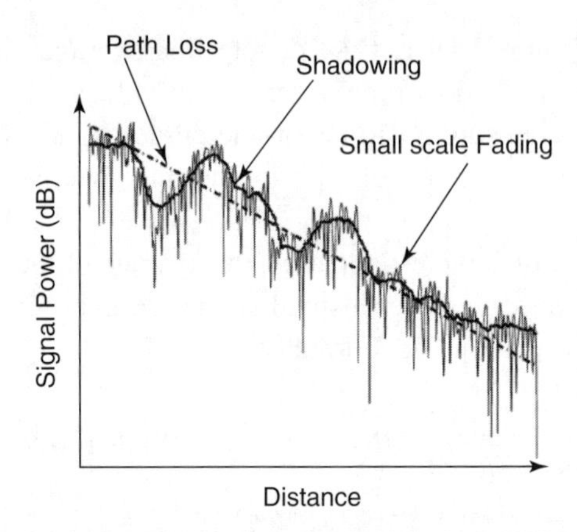

Figure 2.2 Compound effect of path loss, shadowing, and small-scale fading as a function of the separation distance between transmitter and receiver (courtesy of Konstantinos Mammasis).

path loss (dash–dotted plot) gives the decreasing trend of the signal intensity with distance; shadowing (bold curve) describes the variation of the actual received signal intensity around this trend, when averaged over short distances; finally, small-scale fading (light curve) gives the actual intensity of the received signal, considered as a variation over the superimposed effect of path loss and shadowing.

2.2 The Communication Graph

The *communication graph* defines the topology of a wireless network, that is, the set of wireless links that nodes can use to communicate with each other. Given the discussion in the previous section, it is clear that the presence of a wireless link between a pair of units u and v depends on: (i) the location of u and v, and in particular their relative distance; (ii) the power used to transmit data; (iii) the data rate/coding scheme used to send data; (iv) the radio technology; and (v) the radio propagation environment.

To simplify the definition of the communication graph, in what follows we assume that the transmission power used to transmit data is fixed for each node u in the network. To further simplify presentation, and without loss of generality of the presented model, we further assume that the transmit power is the same for each node in the network, and we denote this transmit power value by P_t. Note that this is a simplification of reality, where network nodes can use *transmit power control* to dynamically change transmission power in order to adapt to actual radio channel conditions – transmit power control is a standard technique in cellular networks. Furthermore, in a network composed of heterogeneous devices it is very likely that the transmission power used by different devices, even if not dynamically changed, is set to different values. However, including transmit power control and heterogeneous transmission power values in the model of the communication graph introduced below is a straightforward exercise, which we leave to the interested reader – see also Chapter 2 of Santi (2005). In what follows we further assume that the data rate/coding scheme for each link is fixed and the same for all links; also, the features of the radio technology are incorporated in the thresholds (sensitivity and SNR threshold) used to determine existence/non-existence of a link. With all these simplifications, whether a link between units u and v exists becomes dependent only on (i) and (v) as described above.

Even if we consider a situation where only (i) and (v) play a role in determining the existence of a wireless link, the property defined as "a link between u and v exists" is time varying, that is, the link can exist at a certain time t, and it can no longer exist at a later time $(t + 1)$. It is important to observe that, due to small-scale fading effects, the existence of a wireless link between u and v is a time-varying property even if the positions of u and v are fixed. However, small-scale fading effects are disregarded in the following to simplify the presentation. If such effects are disregarded, the existence of a wireless link becomes a time-varying property only in the presence of node mobility.

Concerning the propagation environment, we start by assuming deterministic path loss with a log-distance power model, for some value $\alpha > 1$ of the distance–power gradient. In the last part of this section, we will generalize the model of the communication graph to random path loss models, bringing log-normal shadowing into the picture.

Let N be a set of wireless nodes, with $|N| = n$. Assume the nodes are located in a certain bounded region R, which for simplicity we assume is two dimensional (extension of the presented model to one- and three-dimensional domains is straightforward). Given any node $u \in N$, the *location* of u in R, denoted $L(u)$, is the position of u in R, expressed in two-dimensional coordinates. Thus, we can define a function $L : N \mapsto R$ that maps a node u to its two-dimensional coordinates in R. In the case of mobile nodes – the scenario of interest in this book – the location function is redefined by adding a time parameter, that us, the location function becomes $L : N \times T \mapsto R$. Summarizing, a mobile wireless network is represented by the pair $M = (N, L)$, where N and L are defined as above.

Given a network $M = (N, L)$, having fixed parameters of the radio link (data rate and coding scheme) and radio technology, disregarding small-scale effects, and given a value $\alpha > 1$ for the distance–power gradient, the topology of the network at a certain time t is represented by the communication graph computed at time t, where the communication graph is defined as follows.

The communication graph at time t is the undirected graph $G(t) = (N, E(t))$, where undirected edge $(u, v) \in E(t)$ if and only if both conditions below are satisfied:

1. $P_t\ (dB) - \left(PL(d_0)\ (dB) + 10\alpha \log(d_{uv}/d_0)\right) \geq \beta\ (dB)$;
2. $P_t\ (dB) - \left(PL(d_0)\ (dB) + 10\alpha \log(d_{uv}/d_0)\right) - N\ (dB) \geq \gamma\ (dB)$.

Note that, under our working assumption of homogeneous transmission power P_t, homogeneous radio technology, and same data rate/coding scheme on each link, the above conditions are satisfied for the wireless link between transmitter node u and receiver node v if and only if the same conditions are satisfied in the reverse link. In other works, the communication graph contains only *symmetric* (or bidirectional) wireless links.

Once specific values for α, β, and γ are set, conditions 1 and 2 above are equivalent to defining a transmission range $r(\alpha, \beta, \gamma)$ – the same for all the nodes– with the property that link $(u, v) \in E(t)$ if and only if $d_{uv}(t) \leq r$ (α, β, γ), where $d_{uv}(t)$ represents the distance between nodes u and v at time t. When the specific values of parameters α, β, and γ are not relevant, we will denote the transmission range simply by r. With all these simplifications, the communication graph as defined here becomes equivalent – once transmission range r is normalized to 1 – to the notion of unit disk graph, which is well known in geometric graph theory (Clark et al. 1990). In a unit disk graph (UDG), a disk of unitary radius is centered at each node, and an edge between nodes u and v is added to the graph if and only if the disks centered at u and v intersect. An example of how the communication graph is computed given node positions and a transmission range r is shown in Figure 2.3.

A communication graph is said to be *connected* at time t if and only if, for any two nodes $u, v \in N$, there exists a path in $G(t)$ connecting u and v

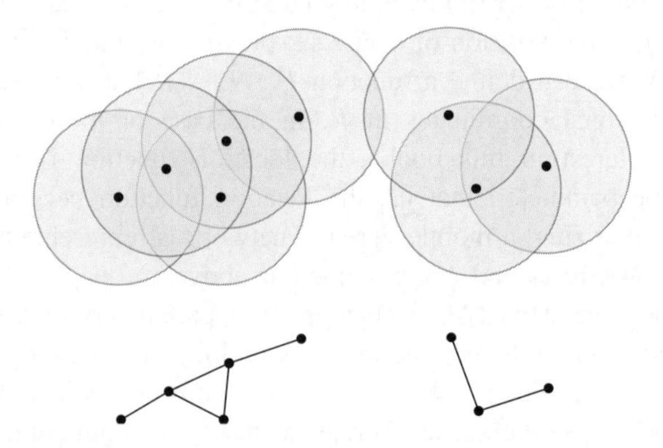

Figure 2.3 A wireless network (top) and the corresponding communication graph (bottom). Transmission range of nodes is represented by a shaded disk.

(and vice versa, given link symmetry). Note that the communication graph in Figure 2.3 is not connected, since nodes on the left hand side of the network cannot reach those on the right hand side.

The major shortcoming of the UDG model is the assumption of perfectly circular radio coverage. As discussed in the previous section, this assumption is hardly met in practice, especially in indoor and urban scenarios where shadowing and small-scale fading play a major role. While including accurate shadowing and small-scale fading models in the network model would make it extremely complex and dependent on scenario, generalizations of the UDG model aimed at accounting for irregular radio coverage areas have been recently proposed (Kuhn et al. 2008; Scheideler et al. 2008). In what follows we present the model introduced in Scheideler et al. (2008), which the authors prove to closely resemble log-normal shadowing.

The main idea of Scheideler et al. (2008) is to introduce a notion of cost related to an arbitrary node pair, and to introduce a link between two nodes if and only if the corresponding cost is below a certain threshold. More formally, consider any cost function $c : N \times N \mapsto \mathbb{R}$ with the property that there is a fixed constant $\theta \geq 0$ so that for all $u, v \in N$,

$$c(u, v) \in \left[\frac{1}{(1+\theta)} \cdot d_{uv},\ (1+\theta) \cdot d_{uv} \right]. \tag{2.3}$$

The cost function c determines the transmission behavior of nodes, and parameter θ bounds the non-uniformity of the environment. In particular, a link between node u and v is considered to be present if and only if $c(u, v) \leq r$, where r is a constant representing the transmission range. Notice that the model does not impose any other constraint on function c, such as being monotonic in the distance, to satisfy the triangle inequality, or symmetric.

In Scheideler et al. (2008), the authors show that, by suitably defining parameter *eta*, the cost function can be defined to closely resemble a specific propagation environment with log-normal shadowing. First, it should be noticed that, since the support of random variable $LS = N(0, \sigma)$ modeling variation in path loss due to shadowing is infinite, the only way to make the cost model defined above resemble a log-normal shadowing environment is to let parameter θ grow to infinity, resulting in a model where nodes even arbitrarily close to each other might not be able to communicate, or where arbitrarily distant nodes might be able to communicate. In other words, the resulting model would correspond to a completely randomized propagation environment, which is not realistic as well. Thus, Scheideler et al. (2008) propose using a path loss model where the random variable modeling large-scale fading has *bounded* support. More specifically, it is assumed that large-scale fading is modeled by a random variable LS' which takes values in $[-h\sigma, h\sigma]$ when expressed in dB, where h is a constant and σ

is the standard deviation of $LS = N(0, \sigma)$. The probability density function (pdf) of the newly defined random variable LS' is obtained from the pdf LS by uniformly distributing the probability density of $N(0, \sigma)$ falling outside $[-h\sigma, h\sigma]$ in the $[-h\sigma, h\sigma]$ interval. For instance, by setting $h = 3$ we get that only 0.0027 of the probability mass of variable $LS = N(0, \sigma)$ falls outside the interval $[-3\sigma, 3\sigma]$, and the pdf of LS' is virtually indistinguishable from the pdf of LS.

The above-described bounded version of log-normal shadowing can be represented by setting parameter θ in Equation 2.3 equal to $10^{h\sigma/(10\alpha)} - 1$, where σ and α are the parameters of the path loss model. For instance, by setting $\alpha = 3$, $\sigma = 6\,dB$, and $h = 3$, we obtain $\theta \approx 1.5$, implying that a transmission between nodes u and v is always successful when $d(u, v) < 0.399r$, and that a successful transmission can only occur at a distance less than or equal to $2.5r$.

An example of a communication graph obtained with the cost model defined above is shown in Figure 2.4. Nodes within the dark shadowed disk centered at u (e.g., node v) always have a wireless link with u; nodes within the light shadowed annulus (e.g., node w) might have a link with u, as determined by the underlying bounded log-normal shadowing model; nodes outside the larger disk centered at u (e.g., node z) cannot have a link with u.

In summary, the cost model proposed in Scheideler et al. (2008) can be used to model a bounded variant of the log-normal shadowing path loss model, under the assumption that the cost function c when defined for node pair (u, v) is defined as a random variable resulting from the product of a deterministic term equal to d_{uv} and a random term which has bounded log-normal distribution.

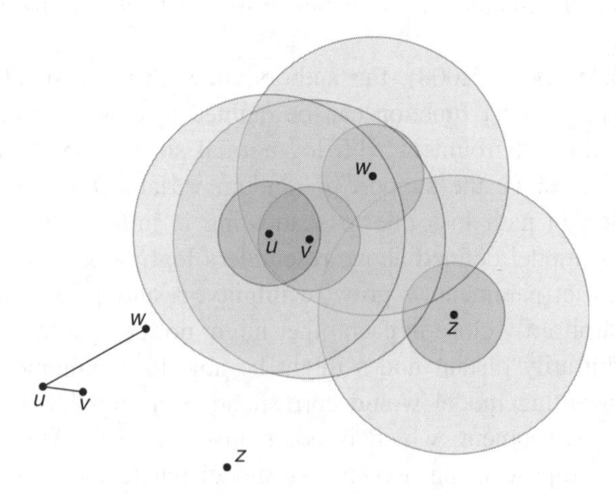

Figure 2.4 A wireless network (top) and the corresponding communication graph (bottom) obtained with a cost-based model.

2.3 The Energy Model

One of the primary design concerns in a vast class of next generation wireless networks (e.g., wireless sensor and opportunistic networks) is the efficient use of energy. Thus, it is fundamental to model the node energy consumption accurately.

Depending on the scenario, next generation wireless networks can be composed of nodes of the most diverse type: laptops, cell phones, PDAs, smart appliances, tiny sensor nodes, and so on. Furthermore, for many application scenarios (e.g., opportunistic networks) the network can be composed of heterogeneous devices. Given this node diversity, a typical approach in the literature is to focus attention on the energy consumption of the wireless transceiver only.

Depending on the type of device, the amount of energy consumed by the transceiver varies from about 15% to about 35% of the total energy dissipated by the node. The former value refers to a laptop equipped with an IEEE 802.11 wireless card, while the latter is typical of a PDA device. An even higher portion of the total energy is consumed by the transceiver in a wireless sensor node.

Energy models for next generation wireless networks typically amount to estimating energy dissipation in the different operational modes of a wireless transceiver, which are:

1. *Idle*: the radio is turned on, but it is not used.
2. *Transmit*: the radio is transmitting a data packet.
3. *Receive*: the radio is receiving a data packet.
4. *Sleep*: the radio is powered down.

Node energy consumption in the various operational modes is typically expressed using the $sleep : idle : rx : tx$ power ratios, where the energy consumption in *idle* mode is conventionally assumed to be 1. Thus, an energy

Table 2.2 Current draw of typical WiFi and ZigBee products

Model	Technology	Current idle (mA)	Current Tx (mA)
Cisco Aironet 350	WiFi	203	450
Texas Instrum. CC2520	ZigBee	2.6	37.2
Model	**Technology**	**Current Rx (mA)**	**Current sleep (mA)**
Cisco Aironet 350	WiFi	270	15
Texas Instrum. CC2520	ZigBee	26.3	0.25

model is defined by assuming power x is consumed when the radio is receiving a message, power y is consumed when the radio is transmitting a message at full power P_t, and power z is consumed when the radio is in sleep mode (the actual values of x, y, and z depend on the specific wireless transceiver).

Typical values of current draw for different radio technologies are given in Table 2.2. Note that, if the wireless transceiver allows transmit power control, different transmit power modes (one for each specific value of the transmission power) might be defined, each resulting in a different power ratio.

References

Clark BN, Colbourn JC and Johnson DS 1990 Unit disk graphs. *Discrete Mathematics* **86**, 165–177.

Kuhn F, Wattenhofer R and Zollinger A 2008 Ad hoc networks beyond unit disk graphs. *Wireless Networks* **14**, 715–729.

Molisch A 2005 *Wireless Communications*. John Wiley & Sons, Ltd, and IEEE Press, Chichester.

Rappaport T n.d. *Wireless Communications: Principles and Practice Second Edition*. Prentice Hall, Upper Saddle River, NJ.

Santi P 2005 *Topology Control in Wireless Ad Hoc and Sensor Networks*. John Wiley & Sons, Ltd, Chichester.

Scheideler C, Richa A and Santi P 2008 An O(log n) dominating set protocol for wireless ad hoc networks under the physical interference model. *Proceedings of ACM MobiHoc*, pp. 91–100.

3

Mobility Models for Next Generation Wireless Networks

In the previous chapter, we motivated the importance of simulation in next generation wireless network analysis and introduced models for the radio channel, network topology, and energy consumption in a wireless node. In this chapter, we will focus on models of wireless node mobility, starting with motivating their importance within the network performance evaluation process. We will then extensively discuss the reasons why the existing know-how on mobility models for cellular networks cannot be applied (at least, directly) to model mobility in next generation wireless networks. Next, a taxonomy of existing mobility models for next generation wireless networks based on different criteria will be introduced. Finally, we will present the CRAWDAD web community resource for archiving mobility traces derived from real-world experiments and testbeds.

3.1 Motivation

We saw in the previous chapter that, given the difficulties related with realization, operation, and maintenance of testbeds, analysis and simulation play a fundamental role in next generation wireless network performance evaluation, especially when the network designer is interested in understanding network behavior when the number of nodes becomes large. Hence, suitable models must be developed to deal with the various aspects that might influence network performance; these models can then be combined and used to estimate expected network performance through either analysis or simulation.

Mobility Models for Next Generation Wireless Networks: Ad Hoc, Vehicular and Mesh Networks, First Edition. Paolo Santi.

Node mobility is a prominent feature of most of the next generation wireless network scenarios and applications briefly described in Chapter 1:

1. In WLAN and mesh networks, wireless stations are mostly mobile devices, such as laptops, PDAs, smart phones, etc., which are typically carried by people. Note that, although WLAN and mesh networks are typically composed of a combination of fixed (WLAN APs and mesh stations) and mobile (wireless stations) devices, the number of mobile devices in a WLAN can be very high (think of a campus network serving several thousand university students). Thus, *human mobility* plays a major role in WLAN and mesh network performance evaluation.
2. In ad hoc networks, network nodes are typically mobile. For instance, nodes can be soldiers of a platoon moving on the battlefield, or members of a rescue squad moving in a disaster area. Thus, human mobility plays a major role also in ad hoc networks.
3. In vehicular networks, *vehicle mobility* is of paramount importance: not only are the vast majority of network nodes moving vehicles, but the degree of mobility of vehicles is typically very high, with absolute speeds in the order of 100 km/h or above, and relative speeds higher than 200 km/h – think of two vehicles traveling in opposite directions on a highway.
4. Mobility plays a role also in wireless sensor networks, at least in some scenarios. For instance, if wireless sensors are mounted on animals for tracking purposes, *animal mobility* is an important aspect to model. Another scenario involving mobility in wireless sensor networks is when floating sensors are deployed to monitor ocean currents. In this scenario, *mobility of ocean flows* plays a role. Finally, even in a scenario where fixed wireless sensor nodes are used to gather environmental data, the data collection might be performed by a mobile data sink – a person, or a vehicle, or a robot. Thus, mobility plays a role also in this case.
5. Mobility is the prominent communication means in opportunistic networks: if node mobility were part of the picture, no network-wide communication could take place, given the very sparse node density resulting in a mostly disconnected network architecture. Depending on the application scenario, different types of mobility – or a combination of them – play a role, the most typical being *human* and *vehicle mobility*.

Summarizing, in the vast majority of next generation wireless network application scenarios, mobility not only is present, but plays a major role in governing network behavior and performance. This explains the extensive research efforts devoted to next generation wireless network mobility modeling in recent years, which is the focus of this book.

3.2 Cellular vs. Next Generation Wireless Network Mobility Models

Since node mobility is an important aspect also of more traditional types of wireless networks, such as cellular networks, the reader might be wondering why the extensive literature on cellular network mobility modeling cannot be applied – at least, directly – to next generation wireless networks. In this section, we will describe the main features of cellular network mobility models and explain why they are not suitable for modeling mobility in next generation wireless networks.

In a cellular network, space is divided into a well-organized cell structure, typically according to an hexagonal pattern as represented in Figure 3.1. A base station is located in each cell and is assumed to fully cover all nodes residing within its cell, that is, all nodes located in a cell can establish a direct wireless link with the corresponding base station.

The performance of a cellular network, expressed for instance in terms of number of served clients and QoS parameters, depends greatly on the number of "active" nodes (i.e., customers making a call) residing in each cell, and how these numbers evolve over time. The number of active nodes in the various cells and their evolution over time depend on many factors, including node mobility. In particular, a call initiated in a certain cell C_i can be transferred to an adjacent cell C_j as the active node moves from C_i to C_j (see Figure 3.1). The process of transferring an active call between adjacent cells is called *cell handoff*, and it is a very critical operation in cellular networks, since handoff should be completed without interrupting the call and with minimal impact on QoS parameters. Another parameter influencing cellular network performance is the *cell residence time*, that is, the time that an active node is expected to spend in a certain cell, which is important for optimizing the allocation of radio resources in the cell. Cell handoff and residence time are not the only cellular network parameters influenced by node mobility, but they are sufficient to illustrate why cellular

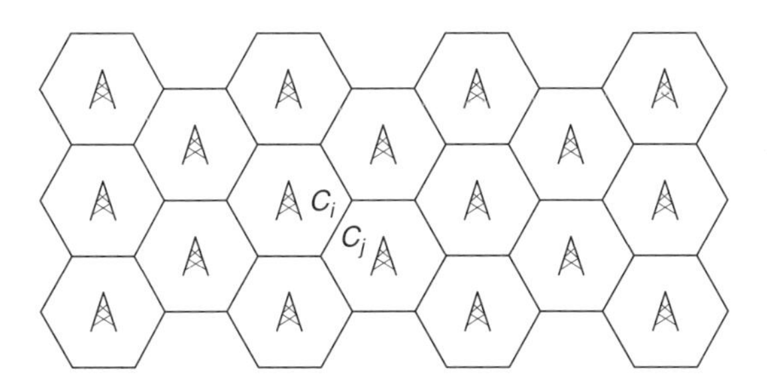

Figure 3.1 A cellular network architecture.

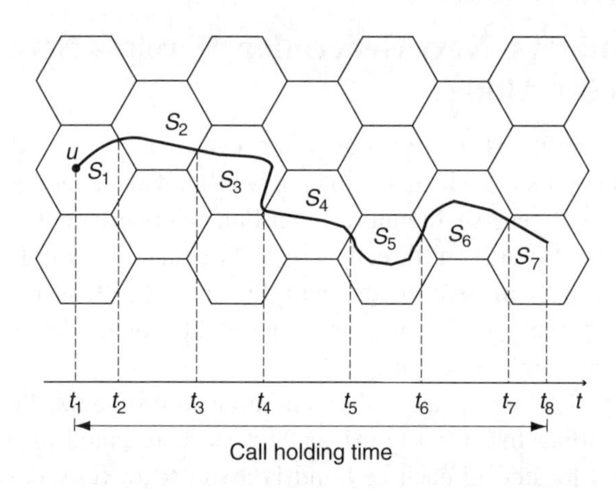

Figure 3.2 Node mobility in a cellular network.

network mobility models cannot be directly used to analyze the performance of next generation wireless networks.

The effect of node mobility on cell residence time and handoff probability can be modeled and evaluated using a *macro mobility model*, operating at the spatial granularity of a network cell: what is relevant from the point of view of a network designer is not the actual trajectory that a specific active node follows, but the instants of time at which a node trajectory crosses the border of a cell. This is illustrated in Figure 3.2, which is adapted from Kim and Choi (2010): the trajectory of an active node is divided into a number of segments $S_1, \ldots, S_k$, corresponding to intervals of times in which the node resides in the same cell. The first segment is determined by the time interval elapsing between time t_1 when the call is initiated and time t_2 at which the node exits from the initial cell. Segments $S_2, \ldots, S_{k-1}$ are determined by cell crossing times $t_2, \ldots, t_{k-1}$, while the last segment S_k is determined by the time interval elapsing between time t_{k-1} when the node enters the last cell and time t_k when the node terminates the call.

It is important to observe that, from the network designer's point of view, the relevant events related to a node mobility are the cell crossing times t_i, as they determine cell residence time and handoff probability. On the other hand, the details of the trajectory that a node follows within a specific cell – that is, the position of node u within a time interval (t', t''), with $t_i < t' < t'' < t_{i+1}$ for some $1 \leq i \leq k-1$ – has very little relevance from the *network* designer's point of view. In this case, the extreme case of a node moving within a single cell is seen as no mobility.

In the previous paragraph, we emphasized the word *network* when referring to a designer interested in characterizing macro-level node mobility, and stated that, from this designer's point of view, mobility of a node within a cell is not very relevant. Intra-cell mobility is instead of interest to the

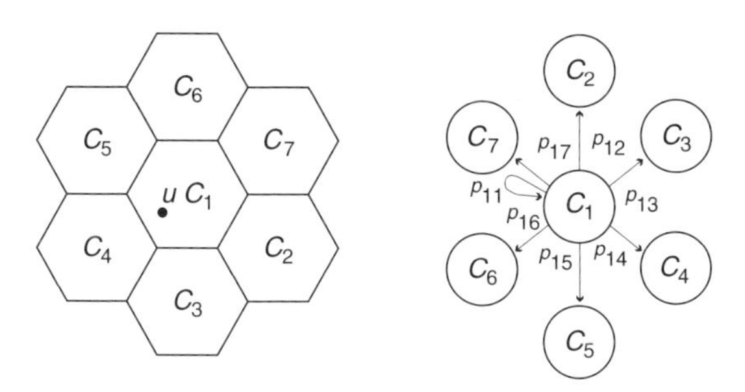

Figure 3.3 Portion of a cellular network (left) and portion of the corresponding Markov chain modeling node mobility between cells (right).

wireless communication engineer in the process of optimizing the quality of the radio transmission between the base station and the client node. However, as explained in Section 2.1, the effects of small-scale, intra-cell node mobility on link quality are typically considered in the definition of the radio channel model – more specifically in the definition of small-scale fading models.

Macro-level node mobility in cellular networks is typically modeled using discrete-time Markov chains – see Appendix A for a formal definition of Markov chains and their main properties – with states corresponding to cells, and transition probabilities corresponding to the probability of a node moving to another cell at the next time slot. In Figure 3.3, a portion of a cellular network is shown on the left hand side, with a node u located in cell C_1 at time t. Under the quite common assumption that nodes can move only to adjacent cells, the right hand side of Figure 3.3 shows the portion of Markov chain representing possible transitions from state C_1, and the respective probabilities. Thus, node u has probability p_{11} of remaining in cell C_1 at time $t + 1$, and probability $(1 - p_{11})$ of moving to another cell. In particular, node u moves to cell C_2 with probability p_{12}, to cell C_3 with probability p_{13}, and so on.

In the vast literature on cellular network mobility modeling, several Markov chain-based mobility models have been introduced, and their properties extensively studied. The interested reader is referred to Kim and Choi (2010) and references therein.

Cellular network mobility models as described above are not suitable for modeling mobility in next generation wireless networks for the following reasons:

1. *Different spatial granularity*: as commented above, cellular network mobility models are concerned with modeling inter-cell mobility. Even if the size of a cell has been reducing over the years, the spatial granularity of movements considered in cellular network mobility models is still in the order of hundreds of meters, or a few kilometers. In other words, movement of the nodes within the range of a few tens of

meters is typically not relevant in cellular network mobility models. On the other hand, given short communication ranges typical of next generation wireless networks, movement of nodes even over short ranges in the order of a few tens of meters is relevant in that context. Thus, while cellular network mobility models are concerned with *macro-level mobility*, models for next generation wireless networks must deal with *micro-level mobility*.

2. *Different network architecture*: in cellular networks, base station deployment is carefully planned and optimized, resulting in a quite regular tiling of the plane in non-overlapping cells. Given optimized base station deployment, the assumption underlying cellular network mobility models – a base station entirely covers its cell – turns out to be quite accurate. Furthermore, communication in cellular networks occurs only between a base station and a mobile station, in a single-hop fashion. Assumptions of full radio coverage in a cell and single-hop communication justify the interest in characterizing inter-cell mobility, since mobility within a single cell is not interesting from a network-level perspective: as long as a node moves within a cell C, the node establishes a wireless link with the base station located in C, independently of its specific position within the cell. On the other hand, next generation wireless networks are typically characterized by the presence of at most a light, relatively unplanned network infrastructure, and heavily rely on wireless multi-hop communications. Thus, the interest of a next generation wireless network designer lies in answering questions such as: "Are two specific network nodes u and v within each other's transmission range?," or "Assuming there exists a link between nodes u and v, what is the expected duration of that link?," and so on. These questions can be answered only by looking at micro-level mobility, where trajectories of single nodes are taken into account.

3.3 A Taxonomy of Existing Mobility Models

In the previous section, we explained why cellular network mobility models cannot be directly applied to model mobility in next generation wireless networks. This fact, coupled with the paramount importance of mobility when studying next generation wireless network performance, explains the many efforts made in recent years by the research community to define and study suitable next generation wireless network mobility models. Surveys of this body of research work can be found in Bai and Helmy (2006), Camp et al. (2002), and Musolesi and Mascolo (2009). In this book, we will provide a more thorough and systematic treatment of this body of knowledge, which

Table 3.1 Classification of representative mobility models according to different criteria

Model	Spatial scope	Application	Nature
Random walk	Micro	General purpose	Synthetic
Random waypoint	Micro	General purpose	Synthetic
RPGM	Micro	General purpose	Synthetic
LH	Micro	Application-specific	Trace-based
SUMO	Micro	Application-specific	Synthetic
SWIM	Micro	Application-specific	Synthetic
Model	**Correlation**	**Geography**	**Trajectory**
Random walk	Independent	Unconstrained	Trajectory-based
Random waypoint	Independent	Unconstrained	Trajectory-based
RPGM	Correlated	Unconstrained	Trajectory-based
LH	Independent	Unconstrained	Contact-based
SUMO	Correlated	Constrained	Trajectory-based
SWIM	Correlated	Unconstrained	Trajectory-based

we will attempt to characterize and present as a scientific discipline rather than a collection of loosely related mobility models and results.

As a first effort in this direction, in this section we propose a taxonomy of existing mobility models – from now on, by the term "mobility model" we mean "mobility model for next generation wireless networks." Instead of proposing a well-structured, hierarchical taxonomy with categories and sub-categories, we present and describe extensively a set of independent criteria according to which mobility models can be classified. The criteria that will be used to classify existing mobility models are listed in Table 3.1, along with a classification of some of the mobility models which we will describe in this book according to these criteria.

3.3.1 Spatial Scope

The first criterion used to classify mobility models is spatial scope, as already described in the previous section. As commented above, given the relatively short transmission range typical of nodes in a next generation wireless network, all mobility models proposed in the literature aim at modeling mobility in a relatively small spatial scope, that is, with a spatial granularity in the order of a few tens of meters. Accordingly, mobility models for next generation wireless networks can be considered as members of the class of micro-mobility models (see Table 3.1).

3.3.2 Application Scenario

A second criterion useful for classifying mobility models is whether a model is tailored to a specific application scenario or not. In the former case, we use the term *application-specific* mobility model, while in the latter we use the term *general-purpose* mobility model.

General-purpose mobility models are typically synthetic (for a definition of synthetic mobility model see Section 3.3.3) models defined through simple mathematical rules. Simplicity is the primary reason why general-purpose mobility models have been used so extensively in the next generation wireless networking literature. In fact, the simple mathematical rules used to define node mobility can be easily embedded into a wireless network simulator, thus significantly easing the complex task of simulating wireless network behavior. A further boost to usage of general-purpose mobility models is that an implementation of the most representative models in this class is included in recent releases of widespread wireless network simulators, such as Ns2 (Team 2011d), GloMoSim (Team 2011b), and GTNetS (Team 2011c). Another factor contributing to the success of general-purpose mobility models is that they are among the few mobility models that, due to their simplicity, allow derivation of theoretical, foundational results about mobile networks.

Representative examples of the class of general-purpose mobility models are the family of random walk models, which will be described in Chapter 4, and the random waypoint mobility model (Johnson and Maltz 1996), which is by far the most commonly used mobility model in ad hoc network simulation.

The major shortcoming of general-purpose mobility models comes from the primary reason for their widespread use, namely, their simplicity. Since these models are defined through simple mathematical rules, it is very difficult to account for the many factors influencing mobility in a real-world application scenario. For instance, general-purpose mobility models typically assume that nodes move in a spatial domain without any constraint regarding the trajectories they can follow – that is, they are unconstrained mobility models, see Section 3.3.5. Except for a few situations (e.g., people or animals moving in a vast, unobstructed space), movement is instead highly constrained in the real world: individuals move along sidewalks and pathways, cars move along roads, etc.

Given these shortcoming, several mobility models have been proposed with the explicit purpose of modeling a specific application scenario with a related network architecture. Mobility models specifically designed for WLAN environments, vehicular networks, and so on, have been introduced in the literature in recent years. Unfortunately, tailoring a mobility model to a specific application scenario often comes at the expense of significantly complicating the definition of the model, making the simulation and (even more) the analysis task much more challenging than in the case of general-purpose mobility models.

3.3.3 Nature

Mobility models can be classified also according to whether they are synthetically defined through simple mathematical rules, or rather they are stochastically defined to resemble mobility patterns observed in real-world data traces. In the former case, we use the term *synthetic* mobility model, while in the latter case we use the term *trace-based* mobility model.

In a synthetic mobility model, rules are defined to determine the initial position of a node, as well as its *trips*, where a trip is typically characterized by a starting point, an endpoint, a trajectory connecting the starting and endpoints, and a velocity. On the other hand, in trace-based mobility models, what is relevant are not the actual trajectories followed by the nodes, but rather their contact patterns, which should resemble as closely as possible those observed in representative traces collected in real-world experiments. In this respect, trace-based mobility models display some resemblance to cellular network mobility models, since in these models also the specific trajectory followed by a node is not of interest to the network designer. However, the spatial scope of trace-based mobility models is typically small, so differently from cellular network mobility models they belong to the class of micro-level mobility models.

3.3.4 Correlation

Another criterion according to which mobility models can be classified is whether the movement of a node is independent of or correlated with the movement of other nodes. In the former case, we use the term *independent* mobility model, while in the latter case we use the term *correlated* mobility model.

In an independent mobility model, the movement of a node is stochastically independent of the movement of any other node in the network. Typically, movement patterns of different nodes are not only *independent*, but also *stochastically equivalent*. In other words, the probabilistic rules governing node mobility are the same for all the nodes, resulting in the same stochastic behavior of all network nodes.

Anyone who has some experience in the simulation and analysis of relatively complex systems will understand the importance of the independence assumption when simulating or analyzing the performance of a wireless network: both simulation and analysis are significantly simplified if nodes move in a stochastically independent fashion. Unfortunately, the mobility of nodes is indeed highly correlated in my real-world situations. Think for instance of a typical vehicular network scenario: well-defined traffic rules strongly correlate the movement of a vehicle with those of surrounding vehicles. Another example is instances of group mobility, such as a scenario where a group of tourists follows a tourist guide. Correlated mobility models have been introduced to model situations in which the movement of a node is highly correlated with, if not completely dependent on, the movement of other nodes in the network.

3.3.5 Geography

A further criterion distinguishing mobility models is whether node mobility is geographically constrained or not. In the former case, we use the term *geographically constrained* mobility model, while in the latter case we use the term *unconstrained* mobility model.

In unconstrained mobility models, nodes are free to move everywhere within a predefined geographical domain representing the range of possible node movements. Typical representatives of the class of unconstrained mobility models are general-purpose, synthetic – hence, simple – mobility models. While simplicity is definitely a positive feature of a mobility model, unconstrained mobility models typically share with general-purpose, synthetic models also the major shortcoming of being scarcely representative of many real-world scenarios, in which node mobility is indeed highly geographically constrained. Examples of geographically constrained mobility abound: vehicles moving along roads, individuals moving along sidewalks and pathways, etc.

It is interesting to observe the relation between geographical constraints and correlation of node movements: while correlation of node movements might indeed be imposed by geographical constraints, a geographically constrained mobility model is not necessarily correlated. For instance, when considering mobility models for vehicular networks, vehicles can be constrained to move along roads, but traffic rules can be ignored – like considering vehicles as "transparent" entities to avoid accidents.

3.3.6 Trajectory

Finally, another criterion that can be used to classify mobility models is whether they generate explicit trajectories for each node, or rather generate traces of mutual contacts between nodes. In the former case, we use the term *trajectory-based* mobility model, while in the latter we use the term *contact-based* mobility model.

While node trajectories, coupled with a notion of transmission range, can be used to generate node contact patterns if needed, the main difference between the two classes of models is that trajectory-based mobility models explicitly model and give access to a notion of node trajectory, while in contact-based mobility models no notion of node trajectory is explicitly or implicitly used to generate the contact traces which are returned as an output of the model. Contact traces are typically obtained through stochastically modeling the occurrence/disappearance of links between pairs of nodes, using Markov chain-based models.

As the reader might have guessed, trace-based mobility models typically do not explicitly represent node trajectories, but rather their contact patterns; that is, they typically belong to the class of contact-based mobility models.

3.4 Mobility Models and Real-World Traces: The CRAWDAD Resource

Mobility models have the main goal of mimicking certain types of mobility (human, vehicular, animal, etc.), or at least of resembling their most distinguishing features. Thus, the ultimate way of assessing the "performance" of a mobility model is to compare the mobility traces it generates with those collected in real-world mobility experiments for the relevant mobility scenarios.

Mobility traces collected in such experiments are a fundamental resource for next generation wireless network designers. On the one hand, they can be used to validate and/or fine-tune artificial mobility models as described above. On the other hand, they can also be directly used in the network performance evaluation process, for example, by feeding wireless network simulators with the collected mobility traces. However, real-world mobility traces should be seen as complementing, not replacing, the role of artificial mobility models in the network performance evaluation process, for the following reasons:

1. *Mobility parameters flexibility*: when optimizing the design of a networking protocol or a mobile application, it is important to understand how the performance of the protocol/application varies under different mobility conditions. This type of evaluation can be done only using artificial mobility models, where mobility parameters such as node speed, average length of a trip, etc., can be changed. Real-world data traces can instead be used as fixed benchmarks upon which a protocol/application performance can be tested. If these data traces are made publicly available, other researchers can use the same traces to test their protocols, making it possible to really "benchmark" their design.
2. *Cost issues*: the process of generating a mobility trace is costly, in terms of both monetary and time resources. Thus, collection of data traces is usually possible with considerable funding and in established research projects of relatively long duration. To better understand why producing mobility traces is a time-consuming process, the reader has to consider

that generating a data trace typically requires a considerable amount of preparatory work, such as developing software for data collection, installing and testing the software on different types of mobile devices, and so on. Once these tasks have been accomplished, the data collection campaign itself can also be of relatively long duration: for instance, the trace collected during the reality mining experiment at MIT (Team 2006) covered a period of nine months.

3. *Privacy issues*: even if the above cost issues are resolved, collecting mobility traces requires tracing the (at least approximate) position of a mobile device for a relatively long period of time. If the mobile device is carried by a person, privacy issues arise. If, besides tracing positions, also *interactions* between people are traced, privacy concerns become even more important. A first action to lessen privacy issues is tracing the mobility of volunteers, who have to be carefully informed about the purposes of the experiment, the use of the data collected, etc. Furthermore, data anonymization techniques are typically used to conceal the identity of individuals participating in the mobility experiment. However, the fact remains that severe privacy issues are raised in any data collection campaign involving people, and that dealing with (or, at least, softening) these issues adds further complexity to the mobility data collection task.
4. *Data availability*: since collecting data traces is a costly and complex task, it is often not possible to generate data traces for the protocol/application under study. Thus, data traces generated as outcomes of other research projects should be used to benchmark the designed protocol/application. However, many mobility data traces are not publicly available and thus cannot be used for benchmarking purposes.

While many of the issues listed above are in a sense intrinsic to the mobility data collection task, significant efforts have been recently undertaken in the research community to make a large number of mobility traces publicly available. The most important initiative in this direction is the Community Resource for Archiving Data At Dartmouth (CRAWDAD) (Team 2011a), a project led by David Kotz of Dartmouth College, USA. The project started in 2004 with a data collection campaign of `syslog`, `SNMP`, and `tcpdump` data from the WLAN installed at Dartmouth College. Since then, the project has evolved to become a community resource for archiving mobility traces related to wireless networks.

The homepage of the CRAWDAD website is displayed in Figure 3.4. CRAWDAD resources amount not only to a large set of publicly available mobility traces – as of the beginning of 2012, 78 traces are available – but also to a set of tools for visualizing and processing these traces – 23 tools are available – and to browsing tools allowing navigation of the traces by keywords, etc.

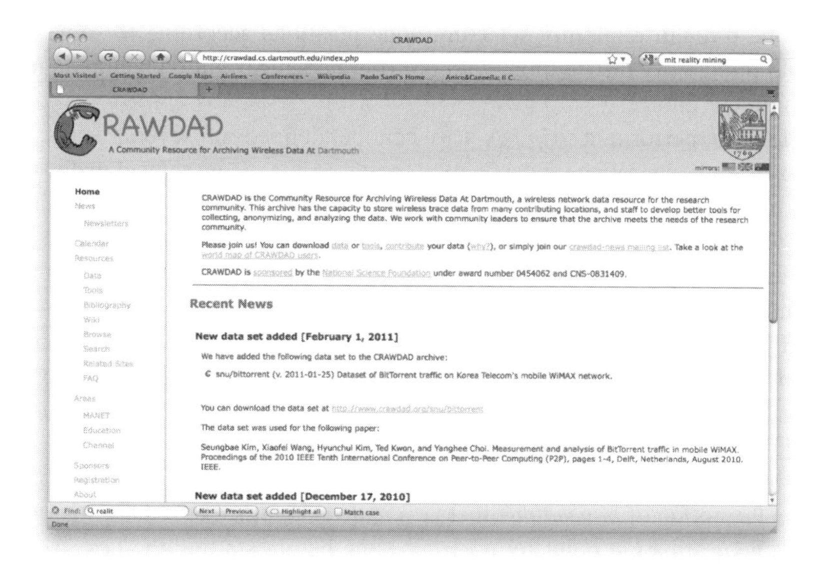

Figure 3.4 The CRAWDAD website.

In recent years, CRAWDAD has become a very valuable resource for next generation wireless network designers, as witnessed by the fact that – as of the beginning of 2011 – more than 500 scientific papers using CRAWDAD mobility traces have been published.

3.5 Basic Definitions

In this section, we provide the basic definitions and terminology that will be used to describe the various mobility models presented in this book. Most of these definitions apply to the class of synthetic mobility models which, as can be seen from Table 3.1, covers the vast majority of mobility models commonly used in the literature.

A synthetic mobility model can be formally defined through a set of rules specifying the following:

1. A *spatial domain* within which nodes move. The spatial domain is typically a region R of one-, two-, or three-dimensional space. A simple example of a spatial domain is the unit cube $[0, 1]^d$, with $d = 1, 2$, or 3. In the case of geographically constrained mobility, the formal definition of a spatial domain is more difficult, and informal definitions (e.g., a set of roads is defined, and nodes are constrained to move only along roads) of the spatial domain are used instead.
2. The *initial position of nodes*. The position of nodes at the beginning of the time period of interest should be determined. Typically, this is

done through randomly selecting node positions according to some spatial probability distribution, with uniform node distribution being the most common choice.

3. How to perform a *trip*. A trip can be defined in several ways: for example, specifying a destination point and a geographical path – the *trajectory* – connecting the current node position with the destination point. Note that both the destination point and the trajectory must be entirely contained in the spatial domain R. Alternatively, a direction of movement and the duration of the trip can be defined.
4. How to select *speed*. The speed of a trip is typically selected according to some random process. A common choice is to select speed uniformly at random in an interval $[v_{min}, v_{max}]$, where v_{min} and v_{max} represent the minimum and maximum possible speed, respectively.
5. The *transition from one trip to the next*. A common transition rule is based on the notion of *pause time*, that is, a node is assumed to remain stationary at the destination point for a certain time before starting the next trip.
6. The *border rule*. In some mobility models, it is possible that a node will hit the border of the spatial domain R during a trip. In such a case, since nodes are not allowed to exit R, a border rule must be defined to describe how to deal with this situation. Typical border rules are billiard-like reflection and toroidal wrapping.

A very important notion in mobility modeling is that of *stationary node spatial distribution*. Consider a certain synthetic mobility model $\mathcal{M}$, and let X_0 be a variable denoting the position of a node u at time 0. Note that X_0 is determined according to the rules for initial node positioning as defined in $\mathcal{M}$. In particular, if X_0 is randomly selected, X_0 is a random variable with values in R, with f_0 being the corresponding pdf. As node u starts moving, its position varies within R. Let $X(t)$ denote the position of u at time $t > 0$. If trips are randomly selected, $X(t)$ also can be considered as a random variable defined within the spatial domain R. Let f_t denote the pdf of random variable $X(t)$. Assuming for simplicity that R is a two-dimensional domain, function $f_t(x, y)$ corresponds to the probability density of finding node u exactly at position (x, y) at time t. Note that, given the definition of mobility model $\mathcal{M}$, the support of both $f_0(x, y)$ and $f_t(x, y)$ is contained in R.

Definition 3.1 *A mobility model $\mathcal{M}$ is said to be* stationary *if there exists a unique pdf $\hat{f}$ such that*

$$\lim_{t \to \infty} f_t = \hat{f},$$

that is, if, $\forall(x, y) \in R$, $\lim_{t \to \infty} f_t(x, y) = \hat{f}(x, y)$. In such a case, function $\hat{f}$ is said to be the stationary node spatial distribution *of $\mathcal{M}$.*

It is important to observe that the stationary node spatial distribution of a mobility model might be different from the initial node spatial distribution, that is, $\hat{f} \neq f_0$ in general.

References

Bai F and Helmy A 2006 A survey of mobility modeling and analysis in wireless ad hoc networks. *Wireless Ad Hoc and Sensor Networks*. Springer, Berlin.

Camp T, Boleng J and Davies V 2002 A survey of mobility models for ad hoc network research. *Wireless Communications and Mobile Computing* **2**, 483–502.

Johnson D and Maltz D 1996 Dynamic source routing in ad hoc wireless networks. *Mobile Computing*, pp. 153–181. Kluwer Academic, Dordrecht.

Kim K and Choi H 2010 A mobility model and performance analysis in wireless cellular network with general distribution and multi-cell model. *Wireless Personal Communications* **53**, 179–198.

Musolesi M and Mascolo C 2009 Mobility models for system evaluation. *State of the Art on Middleware for Network Eccentric and Mobile Applications*. Springer, Berlin.

Team C 2011a *http://crawdad.cs.dartmouth.edu/index.php*.

Team G 2011b *http://pcl.cs.ucla.edu/projects/glomosim/*.

Team G 2011c *http://www.ece.gatech.edu/research/labs/MANIACS/GTNetS/*.

Team N 2011d *http://www.isi.edu/nsnam/ns/*.

Team R 2006 *http://reality.media.mit.edu/*.

Part Two

"General-Purpose" Mobility Models

In this part of the book, we will consider mobility models targeted at modeling mobility in a short-range, multi-hop wireless network in general, that is, a MANET. On the positive side, models in this class are typically synthetic models that can be concisely described through a limited set of mathematical rules. Hence, they can be easily integrated into a wireless network simulator, and they are apt for use in theoretical analysis. On the negative side, models in this class do not account for application-specific aspects of mobility (e.g., presence of roads that constraint movement in a vehicular network) and thus cannot often be considered as realistic mobility models.

Usage of models in this class in the network performance evaluation process should be carefully considered based on the above observations: "general-purpose" mobility models can be used to gain an idea of the overall performance trend – through theoretical analysis and/or simulation – of a certain application/protocol in the presence of mobility, especially when the network size grows large. If a more accurate performance evaluation for specific application scenarios is sought, relevant application-specific mobility models as described later in the book should be used instead.

4

Random Walk Models

Random walks are a large class of mobility models used in various scientific disciplines to study natural and human phenomena such as the path traced by a molecule as it travels through a liquid, the trajectory of a foraging animal, the financial status of a gambler, the time-varying price of a stock or share, and so on.

The term random walk was first introduced by the mathematician Karl Pearson in 1905 (Pearson 1905), and refers to a movement trajectory consisting of successive random steps. As in the taxonomy introduced in Chapter 3, random walks belong to the class of synthetic, entity-based mobility models. In particular, nodes in a network with random walk mobility move independently of each other. Thus, a mobile network with n nodes is modeled using n independent and stochastically equivalent random walks.

As mentioned above, several types of random walks have been defined and studied in various scientific disciplines starting at the end of the nineteenth century. An exhaustive coverage of the theory of random walks is well beyond the scope of this book, so the interested reader is referred to several existing books on random walks, such as Lawler and Limic (2010) and Rudnick and Gaspari (2010). In this chapter, we will present examples of random walks relevant to next generation wireless networks.

Random walks can be classified as either *continuous* or *discrete* depending on whether the spatial domain R is a continuous subregion or a discrete set of points in d-dimensional space. Alternatively, the notion of continuous or discrete random walk might refer to the time instead of the spatial domain. In particular, in time-discrete random walks the position of a node is defined at discrete times $t, t+1, t+2, \ldots$, where rules are defined to determine $X(t+1)$ – the position of the node at time $t+1$ – based on $X(t)$. On the contrary, continuous-time random walks take their steps at random times, hence the position of a node $X(t)$ is defined for the continuous times $t \geq 0$.

Mobility Models for Next Generation Wireless Networks: Ad Hoc, Vehicular and Mesh Networks, First Edition. Paolo Santi.

Finally, random walks are defined as d-dimensional, with d typically 1, 2, or 3, according to the dimensionality of the underlying spatial domain.

In this chapter, we first present relevant discrete random walks such as those defined on grids and graphs. We will then proceed to present continuous random walks including Brownian motion and Lévy flights. Finally, we will introduce important properties of random walk models that have been studied in the literature.

4.1 Discrete Random Walks

In discrete random walks, the spatial domain of movement is composed of a set of points in d-dimensional space, representing all possible positions of a node in the movement domain. Time is also discretized, that is, movement of a node is modeled as a sequence of fixed-duration time steps, with $X(t), X(t+1), X(t+2), \ldots$ representing the sequence of node positions at time steps $t, t+1, t+2, \ldots$.

Discrete random walks are typically modeled as memory-less stochastic processes, that is, the value of $X(t+1)$ is randomly chosen based on the value of $X(t)$ only: the history of past node positions before time t is not relevant in determining $X(t+1)$. Transition from position $X(t)$ to the next position $X(t+1)$ is typically governed by a neighborhood relationship defined on the countable set $R = \{p_1, \ldots, p_m, \ldots\}$ of the possible positions of a node in the spatial domain. In particular, for each $p_i \in R$, a set $N(p_i) \subseteq R$ of *neighbors* is defined. The transition rule then dictates that, if $X(t) = p_i$, the next position $X(t+1)$ is chosen uniformly at random in $N(p_i)$.

Discrete random walks are equivalent to a discrete-time Markov chain, with the states of the chain representing the possible positions p_i of the node in R, and transition probabilities defined as follows:

$$P(p_i \to p_j) = \begin{cases} \dfrac{1}{|N(p_i)|} & \text{if } p_j \in N(p_i) \\ 0 & \text{otherwise} \end{cases}.$$

An elementary example of a discrete, one-dimensional random walk is the random walk on the integer number line $\mathbb{Z}$, defined as follows:

$$X(0) = 0,$$

and

$$X(t+1) = \begin{cases} X(t) + 1 & \text{with probability } \frac{1}{2} \\ X(t) - 1 & \text{with probability } \frac{1}{2} \end{cases}.$$

An example of random walks on the integer number line is reported in Figure 4.1, with the x-axis reporting time steps and the y-axis the current position on the integer line.

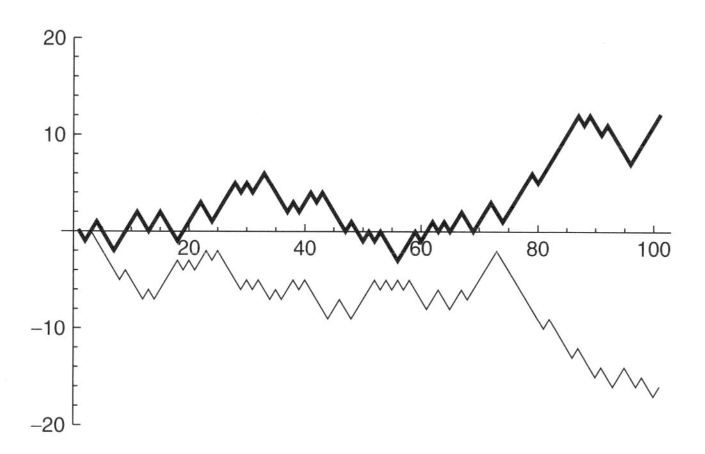

Figure 4.1 Two random walks on the integer number line starting at 0.

4.1.1 Random Walks on Grids

A popular family of random walks, which is useful for multi-hop wireless network performance evaluation, is that of random walks defined on grids, and in particular on two-dimensional grids. A random walk on a two-dimensional grid is defined as follows:

1. The spatial domain is $R = \{(x, y) \in \mathbb{Z}^2\}$ – the *unbounded grid* – or $R = \{(x, y) | (x, y \in \mathbb{Z}) \wedge (|x|, |y| \leq k)\}$ for some integer $k > 0$ – the *bounded grid*.
2. The neighbor set is defined as follows: $\forall (x, y) \in R, N(x, y) = \{(x + 1, y), (x, y + 1), (x - 1, y), (x, y - 1)\}$. In case R is bounded, the neighbor set is defined using summation and subtraction modulo k.

Note that the definition of neighbor for bounded grids implicitly defines a toroidal wrap-around border rule. An example of a random walk on a two-dimensional unbounded grid is shown in Figure 4.2.

The random walk on the two-dimensional grid is also called the *drunkard model* in the literature, due to the following example. Consider a drunkard living in a typical US city with a Manhattan-like map. Assuming square blocks, the coordinates of the road intersections on the map correspond to integer pairs (x, y). The drunkard starts from a random intersection on the grid (the bar where he got drunk); at each intersection, he chooses the next road uniformly at random (including the road he came from). Then, the trajectory followed by the drunkard can be modeled as a random walk on a two-dimensional grid.

As we will see later in this chapter, questions such as "Will the drunkard ever go back to the bar?," or "Given a position $(\bar{x}, \bar{y})$ corresponding to the

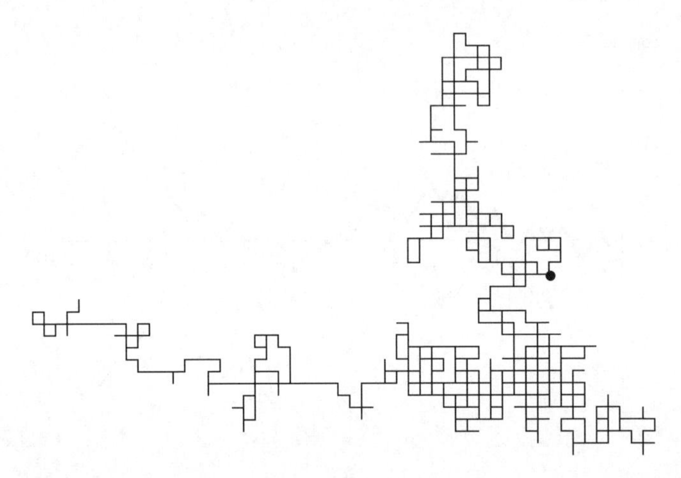

Figure 4.2 A random walk on the two-dimensional, unbounded grid starting at (0, 0) (point in bold).

drunkard's home, will the drunkard ever reach home starting from the bar?," etc., have been studied in the literature.

Random walks on grids can be generalized in several ways, for instance, to somehow account for node velocity. For example, Clementi et al. (2009) define a random walk on a bounded grid of side $k = \lfloor\sqrt{n}\rfloor$, where n is the number of mobile nodes, and the spatial domain R and neighbor set are defined as follows:

$$R = \{(\epsilon \cdot x, \epsilon \cdot y)\},$$

with $\epsilon > 0$ an arbitrary real constant called the *resolution coefficient*, x, $y \in \mathbb{N}$, and $\epsilon \cdot x, \epsilon \cdot y \leq k$; and

$$N((a = \epsilon \cdot x, b = \epsilon \cdot y)) = \{(a', b') \in R | \|(a, b) - (a', b')\| \leq r\},$$

where r is a parameter called the *move radius*. In words, the neighbors of a location (a, b) are the set of points in R within Euclidean distance r from (a, b). Thus, the move radius implicitly accounts for node velocity, with larger velocities corresponding to relatively larger move radii. Note that, since the spatial domain R is bounded, neighbor sets are indeed defined using distance in the toroidal two-dimensional space.

The above model has been used (Clementi et al. 2009) to study the process of information spreading (known also as *flooding* in the wireless networking literature) in MANETs.

4.1.2 Random Walks on Graphs

Random walks on graphs are a class of random walks in which the set of possible locations of a mobile entity corresponds to nodes in a graph, rather than to points in a spatial domain. Given a graph $G = (V, E)$, a random walk on G is defined by randomly selecting a starting node $u \in V$, and by moving to a randomly selected neighbor graph node at each time step. Using the terminology introduced at the end of Chapter 3, the (spatial) domain R and neighbor set are defined as follows:

$$R = V,$$

and

$$\forall u \in V, \ N(u) = \{v \in V | (u, v) \in E\}.$$

The graph underlying the random walk can be either directed – in which case movement is allowed only toward outgoing neighbors of a graph node – or undirected.

Note that random walks on graphs can be used to model geographical mobility if the graph underlying the random walk is generated according to a reference geography. For instance, consider the map of a city with a relatively irregular road system, and generate a graph $G = (V, E)$ as follows: a node is added in V for each intersection on the map; an edge (u, v) is added in the graph if there exists a road segment connecting intersection u with intersection v. With this definition of the underlying graph, the movement of a drunkard in the city can be modeled as a random walk on the graph. Questions about the drunkard's movement similar to the ones mentioned above have been studied also when random walks take place on a graph.

4.2 Continuous Random Walks

In continuous random walks, the spatial domain of movement is a continuous subregion of d-dimensional space. Similarly to discrete random walks, continuous random walks are typically also time continuous, meaning that the position of a node is modeled not only at specific time instants t, $t + 1, t + 2, \ldots$, but on the continuous time line.

In this section, we present two important instances of continuous random walks, namely *Brownian motion* and *Lévy flight*.

4.2.1 Brownian Motion

Brownian motion refers to the random movement observed in particles suspended in a liquid or gas. This kind of movement was first studied by Robert Brown in 1827, when he examined the motion of pollen grains in water.

In mathematical terms, Brownian motion can be described by a stochastic process called the Wiener process. While the formal mathematical definition of a Wiener process is not straightforward and is beyond the scope of this

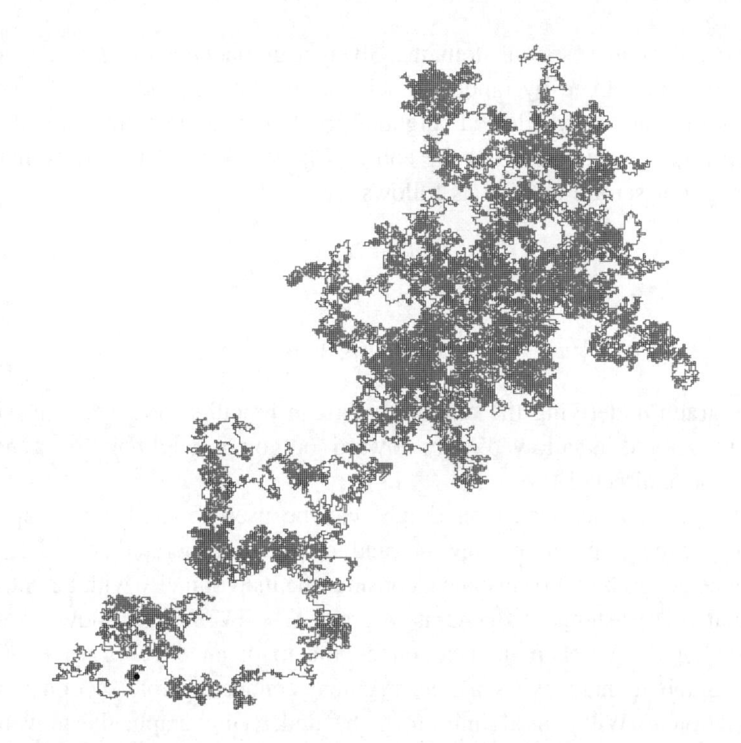

Figure 4.3 An example of two-dimensional Brownian motion.

book, intuitively a Wiener process is the scaling limit of a random walk on the d-dimensional grid as the step of the grid becomes smaller.

More specifically, consider the following random walk on a grid – for simplicity, we assume a one-dimensional grid, but the intuition applies to higher dimensions:

$$R = \{\epsilon \cdot x | x \in \mathbb{Z}\},$$

where $\epsilon > 0$ is a real number corresponding to the grid step; and

$$N(\epsilon \cdot x) = \{\epsilon \cdot (x - 1), \epsilon \cdot (x + 1)\}.$$

Denoting by $RW(\epsilon)$ the above-defined random walk, we have that the Wiener process is a stochastic process corresponding to $\lim_{\epsilon \to 0} RW(\epsilon)$. Note that in the Wiener process both space and *time* are scaled down by a factor $\epsilon > 0$. Thus, the limiting process as $\epsilon \to 0$ (i.e., Brownian motion) is continuous in both the spatial and time domain. An example of 100 000 steps of two-dimensional Brownian motion is shown in Figure 4.3.

4.2.2 Lévy flight

A Lévy flight – named after the French mathematician Paul Pierre Lévy – is a random walk in which the length of steps is distributed according to a

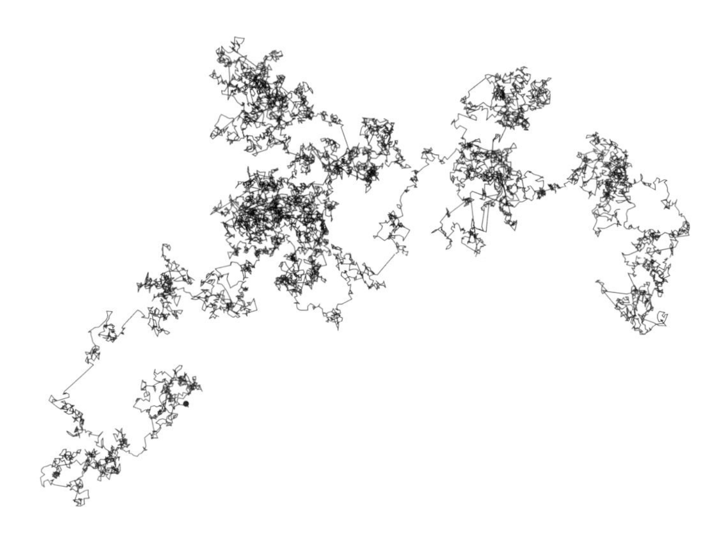

Figure 4.4 An example of two-dimensional Lévy flight.

heavy-tailed probability distribution. More specifically, step lengths are chosen according to a power law of the form $y = x^{-\alpha}$, with $1 < \alpha < 3$.

The main difference between Brownian motion and Lévy flight is that in the latter model, although on average the step lengths are relatively short, long-distance steps occur occasionally. Thus, Lévy flights can be used to model erratic behavior in a close neighborhood with occasional long-distance travel. Such a movement pattern is displayed for instance by animals during foraging and hunting.

A Lévy flight is formally defined as follows (again, we present the one-dimensional case for simplicity): the spatial domain R is the real-number line; when a node is located at position $x \in R$, the next position is either $x + \delta$ or $x - \delta$ with equal probability, where δ is a random variable with a power-law distribution, that is,

$$Prob(\delta = y) = \begin{cases} 0 & \text{if } y < \delta_{min} \\ \dfrac{\alpha - 1}{\delta_{min}} \cdot \left(\dfrac{y}{\delta_{min}}\right)^{-\alpha} & \text{otherwise} \end{cases},$$

where δ_{min} is the minimum step size. An example of 1000 steps of a two-dimensional Lévy flight is shown in Figure 4.4: as can be seen, Lévy flights alternate local movement patterns similar to Brownian motion with occasional long-distance trips.

4.3 Other Random Walk Models

Several variations of random walk models have been introduced in the literature. Here, we browse through some of them, referring the reader to existing books on random walks (Lawler and Limic 2010; Rudnick and Gaspari 2010) for a more detailed description of these models.

In Gaussian random walks, the step size is chosen according to a normal distribution. One-dimensional Gaussian random walks are used for instance to model real-world time series data such as those observed in financial markets. The definition of Gaussian random walk is very similar to that of Lévy flight, the only difference being in the probability distribution used to determine the step length, which has a thin tail in the former case and a fat tail in Lévy flight. Thus, relatively long step sizes are seldom observed in Gaussian random walks, while they occur relatively frequently (see Figure 4.4) in Lévy flights.

All random walk models considered so far assume that the direction and/or length of steps are independent random variables. A variant of random walk models, called *random walks with correlated steps*, assumes instead that consecutive steps are correlated. Considering for instance a random walk with fixed step size (e.g., the drunkard model), and denoting by $D(t)$ the direction (North, South, East, West) taken by the node at time t, a random walk with correlated steps can be defined as follows:

$$P(D(t+1) = Dir) = \begin{cases} p_1 & \text{if } D(t) = Dir \\ p_2 & \text{otherwise} \end{cases},$$

with $p_1 > p_2$, $3p_2 + p_1 = 1$, and $Dir \in \{North, South, East, West\}$.

More complex random walk models include the *self-avoiding walk*, in which a node moves in a grid with the constraint that the same point cannot be visited more than once, and the *loop-erased random walk* on graphs, which is obtained from a random walk on a graph by erasing all loops – thus obtaining a simple path in the graph.

4.4 Theoretical Properties of Random Walk Models

The properties of random walk mobility models have been extensively studied in the literature. We briefly review some of them here.

4.4.1 Stationary Node Spatial Distribution

A first property of interest is the stationary node spatial distribution, which we formally defined in 3. Most random walks are characterized by a *uniform* stationary node spatial distribution, independently of how the initial location of a node is selected. In other words, after a large enough number of mobility steps, the probability of finding a node at any location in R is constant, and defined as follows:

$$P(X(t) = p_i) = \frac{1}{|R|}$$

if R is a discrete set of points, or

$$P(X(t) \in A) = \frac{vol(A \cap R)}{vol(R)}$$

if R and A are continuous subregions of the d-dimensional space, where $vol(S)$ denotes the d-dimensional volume of region S.

4.4.2 The Level-Crossing Phenomenon

An important property of discrete random walks is the *level-crossing phenomenon*, also known as *recurrence* or *the gambler's ruin*. Considering a simple one-dimensional random walk on the integer line, the level-crossing phenomenon refers to the fact that, fixing any point p on the integer line, the random walk will cross p an infinite number of times, independently of the initial location of the walk. In other words, the random walk will eventually cover *all* the integer line. To understand why this property is also called the gambler's ruin, consider a gambler with some initial amount of money p. This is the gambler's starting position on the integer line. Assume that, at each play, the gambler wins a token with probability 1/2, or loses a token with the same probability 1/2 – this is called a *fair* game. If the bank has an infinite amount of money, the above game can be modeled as an infinite random walk on the integer line. Thus, the level-crossing phenomenon states that, independently of the initial amount of money p that the gambler has, the game at some point will reach a situation in which the gambler has no money; at that point the game is over.

The level-crossing phenomenon can be generalized to random walks on two-, three-, or higher dimensional random walks. We now go back to the drunkard example, and use the level-crossing phenomenon to answer the following questions: "Will the drunkard ever go back to the bar?," and "Given a position $(\bar{x}, \bar{y})$ corresponding to the drunkard's home, will the drunkard ever reach home starting from the bar?" It turns out that the level-crossing phenomenon occurs also in two-dimensional random walks, implying that both questions have a positive answer. That is, with an infinitely long random walk, the drunkard is guaranteed to reach both the bar again and any other location in the city, including home.

It is interesting to observe that, denoting by $P(d)$ the probability that an infinitely long random walk on a d-dimensional grid starting at the origin returns to the origin, we have $P(1) = P(2) = 1$, but $P(d) < 1$ for $d \geq 3$. For instance, $P(3) \approx 0.34$. Thus, the level-crossing phenomenon no longer occurs for three- and higher dimensional random walks.

4.4.3 Hitting Time, Return Time, and Cover Time

A set of notions is defined for discrete random walks.

Definition 4.1 *Let u, v be points in R. The* hitting time *from u to v is the expected number of mobility steps needed to reach v starting from node u.*

Definition 4.2 *Let u be a point in R. The* return time *is the expected number of mobility steps needed to return to node u, starting from node u.*

Definition 4.3 *Let u be a point in R. The* cover time starting from *u, denoted C_u, is the expected number of mobility steps needed to cover each location in R starting from u. The* cover time *is defined as*

$$\max_{u \in R}\{C_u\}.$$

We now report some known results about random walks on two-dimensional grids. For this class of random walks, it is known that the average hitting time between two nodes u and v (i.e., the expected value of the hitting time when u and v are chosen uniformly at random in R) is $\Theta(m \log m)$, where $m = \sqrt{l}$ for some $l > 0$, the number of locations in R. Returning to the drunkard example, this results implies that, if the bar and the home are randomly located in the city, the expected number of intersections that the drunkard will traverse on his way home (which we know exists with probability 1 due to the level-crossing phenomenon) is $\Theta(m \log m)$. On the other hand, the cover time for the same class of random walks is $\Theta(m \log^2 m)$.

References

Clementi A, Monti A, Pasquale F and Silvestri R 2009 Information spreading in stationary Markovian evolving graphs. *Proceedings of the IEEE International Symposium on Parallel and Distributed Processing*, pp. 1–12.

Lawler GF and Limic V 2010 *Random Walk: A Modern Introduction*. Cambridge University Press, Cambridge.

Pearson K 1905 The problem of the random walk. *Nature* **72**, 294.

Rudnick J and Gaspari G 2010 *Elements of the Random Walk: An Introduction for Advanced Students and Researchers*. Cambridge University Press, Cambridge.

5

The Random Waypoint Model

We have seen in the previous chapter that a major shortcoming of random walk models is the lack of a fundamental aspect of intentional mobility, namely, the notion of *trajectory* of a movement. Indeed, random walks are aimed at mimicking completely random movement patterns, and as such they can be considered as relatively accurate in describing mobility patterns where mobile entities display such highly random behavior.

Are there many next generation wireless network application scenarios where node mobility can be classified as "highly random?" Unfortunately (or luckily, depending on the viewpoint), the answer is *no*: apart from a few wireless sensor network scenarios (e.g., animal tracking in open environments, in which mobility can be faithfully modeled by Lévy flights), in the vast majority of next generation wireless network scenarios node mobility is not random, but obeys some (often strict) rules.

The random waypoint model (RWP), first defined by Johnson and Maltz (1996) to study the performance of a routing protocol for MANETs, can be considered as a first attempt to define a simple, synthetic mobility model aimed at modeling intentional human movement. Thus, from the viewpoint of modeling accuracy, and despite its several shortcomings as described in this chapter, the RWP model can be considered a breakthrough in next generation wireless network mobility modeling compared to random walks. This, combined with the popularity of the Dynamic Source Routing (DSR) protocol for MANETs introduced in Johnson and Maltz (1996), now part of the IEEE 802.11 standard, explains the success of this mobility model. The RWP model is in fact by far the most widely used mobility model in next generation wireless network simulation, and it is provided as a default mobility

Mobility Models for Next Generation Wireless Networks: Ad Hoc, Vehicular and Mesh Networks, First Edition. Paolo Santi.

model in popular wireless network simulators such as Ns2 (Ns2Team 2011), GloMoSim (Team 2011a), GTNetS (Team 2011b), etc.

The popularity of the RWP model has partly declined recently, mainly due to some research work highlighting possible inaccuracy in simulation results that might occur if the RWP model is not used properly; one of the case studies in Part Seven of this book will be devoted to this topic. Another factor negatively impacting RWP model popularity is the "specialization" trend observed in next generation wireless network research in recent years: as commented in Section 1.2, the concept of "general-purpose" MANET for which the RWP model was originally defined is being replaced by narrower scoped network architectures and scenarios such as vehicular networks, opportunistic networks, and so on. Thus, mobility models specifically defined for the scenario at hand, like the ones presented later in this book, have gained popularity in recent years.

Considering its past and present popularity, the RWP mobility model is by far the most thoroughly studied mobility model for next generation wireless networks, and several of its fundamental properties have been characterized in the literature. In this chapter, after a formal definition of the RWP mobility model, we will present a characterization of the two most important properties of RWP mobile networks, namely, the stationary node spatial distribution and average nodal speed. For both these properties, we will report not only the characterization, but also some details about their derivation. This should show the reader how significant examples of tools from geometric probability and statistics are used within the field of mobility modeling. We will then conclude by presenting variations of the basic RWP model which have been proposed.

5.1 The RWP Model

Similar to random walk models, the RWP model belongs to the class of synthetic, entity-based mobility models. In particular, this means that the mobility of different network nodes is modeled by *independent* stochastic processes. Hence, a mobile network with n nodes is modeled by means of n replicas of the same stochastic process governing the mobility of a single node. In the following, we describe such a stochastic process.

Informally speaking, the RWP mobility model in a d-dimensional spatial domain R is defined by initially selecting the position of the node uniformly at random in R. After a predefined *pause time* t_p, a parameter of the model, the node selects a destination – called *waypoint* – uniformly at random in R, and starts moving along a straight-line trajectory toward the waypoint with constant speed v chosen uniformly at random in an interval $[v_{min}, v_{max}]$, where v_{min} and v_{max} are two further parameters of the model. Once the waypoint is reached, the node pauses at the waypoint for time t_p. Then, it starts a new trip according to the same rules: the next waypoint is chosen uniformly at random in R; the node moves along a straight-line trajectory toward the waypoint with random speed, etc.

More formally, a RWP mobile node is defined by the following stochastic process:

$$\{D_i, T_i, V_i\}, \quad \text{for } i = 1, 2, \ldots,$$

where D_i is a d-dimensional random variable corresponding to the coordinates of the ith waypoint, T_i is the pause time at the ith waypoint, and V_i is a random variable corresponding to the node speed during the trip toward the ith waypoint. In the original RWP definition, D_i is uniformly distributed on R, T_i is a constant equal to t_p, and V_i is chosen uniformly at random in $[v_{min}, v_{max}]$. The position of the initial node position D_0 is also selected uniformly at random in R.

An example of RWP mobility in the unit square $R = [0, 1]^2$ is shown in Figure 5.1: the starting point is the black dot; the next waypoint is chosen uniformly at random in R, with the node moving along a linear trajectory. The figure reports the trajectory followed by the node during 20 mobility steps.

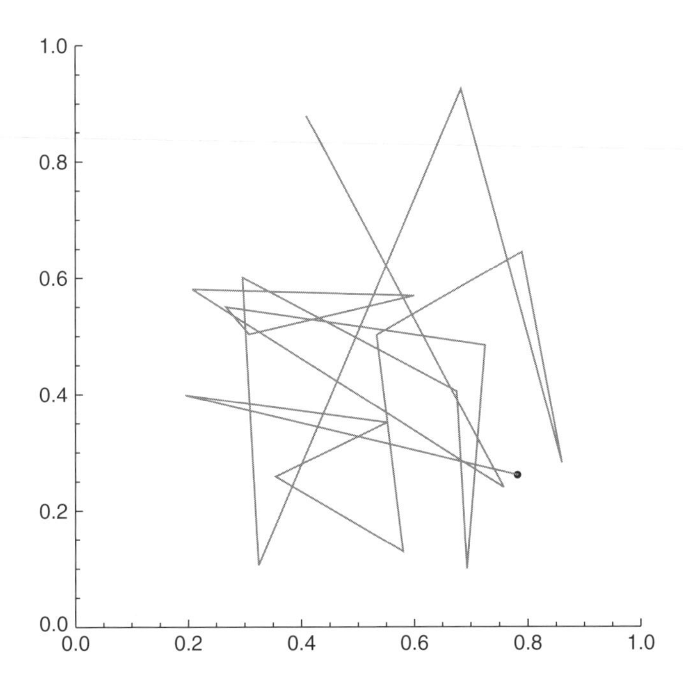

Figure 5.1 An example of two-dimensional RWP mobility in the unit square.

Two observations about RWP mobility are in order:

1. If the spatial domain R is a convex region, then the trajectory of a RWP mobile node never crosses the border of R. This means that no border rule needs to be specified for RWP mobility.
2. Careful observation of Figure 5.1 reveals that the trajectories of a RWP mobile node moving in a bounded region are more likely to cross the center than the border of the movement domain R. In other words, as we will see in the next section, the stationary node spatial distribution of the RWP model in a bounded domain *is not uniform*: a RWP mobile node is relatively more likely to be positioned in the center than on the border of R. The fact that node spatial density on the border is relatively sparse is known as the *border effect*, since it is caused by the fact that nodes move within a *finite* spatial domain. If the spatial movement domain were unbounded or toroidal (think about nodes moving on the surface of the Earth), then the node spatial distribution of the RWP model would be uniform, and no border effect would occur. Finally, it is interesting to observe that the border effect occurs with bounded RWP mobility even if a mobile node never crosses the border of R; thus, the border effect is not necessarily caused by trajectories hitting the border and definition of a border rule.

5.2 The Node Spatial Distribution of the RWP Model

The stationary node spatial distribution is the first fundamental property of RWP mobility to have been investigated in the literature. The above-described border effect was first noticed and quantitatively measured in some simulation-based studies (e.g., Bettstetter and Krause (2001), Blough et al. (2002)); then, the border effect was formally characterized through derivation of the stationary node spatial distribution.

Before proceeding further, the reader might question why characterizing the stationary node spatial distribution of a mobility model is considered a very important problem within the wireless networking research community. The answer to this question lies in the fact that a stationary node spatial distribution which is different from the initial distribution of nodes (which is typically random uniform) can severely impair simulation accuracy. The impact of the RWP stationary node spatial distribution on simulation accuracy will be discussed extensively in Part Seven of this book.

The stationary node spatial distribution of a RWP mobile node moving in the unit square was first closely approximated by Bettstetter et al. (2003). Later on, Hyytia et al. (2006) derived the exact node spatial distribution for arbitrary convex two-dimensional shapes of the spatial domain R and for arbitrary waypoint distribution. In this section, we present their approximate derivation, which is more intuitive. In what follows, the spatial domain is assumed to be the unit square $R = [0, 1]^2$.

Consider a RWP mobile node defined by the stochastic process $\{D_i, t_p, V_i\}$ as defined in the previous section, and let $d_i = (x_i, y_i)$ denote the ith realization of random variable D_i. A first important result proven by Bettstetter et al. (2003) concerns the *mean ergodicity* of the sequence of random variables $\{L_i\}$, where $L_i = \|d_i - d_{i-1}\|$, that is, the sequence of random trajectory lengths. Ergodicity is one of the most important notions in statistics, and states that a stochastic process $\{X_1, X_2, \ldots\}$ is ergodic when, for large enough k, taking k samples from a single random variable (e.g., X_1) is statistically equivalent to taking a single sample from k of the X_i variables. Thus, the mean ergodicity result stated by Bettstetter et al. (2003) implies that the mean of a set of samples taken from random variables $\{L_1, L_2, \ldots\}$ (one sample for each random variable) is the same as that obtained by repeatedly sampling from a single random variable, say L_1. Note that this result is not immediate, since random variables L_i and L_{i+1} *are not* independent: in fact, two consecutive trajectories share an endpoint. However, random variables L_i and L_{i+2} *are* indeed independent, since they do not share endpoints. Sequence $\{L_i\}$ can then be divided into two sub-sequences $\{L_{2k}\}$ and $\{L_{2k+1}\}$ of even and odd trips, respectively, with the property that mean ergodicity holds for both sub-sequences. Due to linearity of the mean, the mean ergodicity of the entire sequence $\{L_i\}$ then follows.

Why is the mean ergodicity result so important? To understand this result, let us start by investigating the effect of the pause time on the node spatial distribution. Let us consider the position (x_t, y_t) of a mobile node u at time t, for a sufficiently large t. According to the RWP model definition, node u can be in one of two possible states at any instant of time: *pausing*, if it is resting at a waypoint; or *traveling*, if it is traveling toward the next waypoint. Let S_t denote the random state of node u at time t, with S_t taking values P (pause) or T (travel). The node spatial distribution of the RWP model, that is, the pdf of (x_t, y_t) as $t \to \infty$, can be defined as follows:

$$f_{RWP}(x, y) = f_P(x, y) \cdot Prob(S = P) + f_T(x, y) \cdot Prob(S = T),$$

where $f_P(x, y) = U(x, y)$ is the pdf of waypoints (i.e., the uniform distribution in R), $f_T(x, y)$ is the pdf of a node when traveling between consecutive waypoints, and $Prob(S = P) = \lim_{t\to\infty} Prob(S_t = P)$ (respectively, $Prob(S = T)$) is the stationary probability of finding the node in state P (respectively, T). Thus, in order to obtain f_{RWP}, we need to derive the node spatial distribution f_T of a node moving along a trajectory whose endpoints are chosen uniformly at random in R, which is called the *mobility component* of the distribution. The mean ergodicity property implies that this problem is equivalent to that of computing the intersection between a segment with

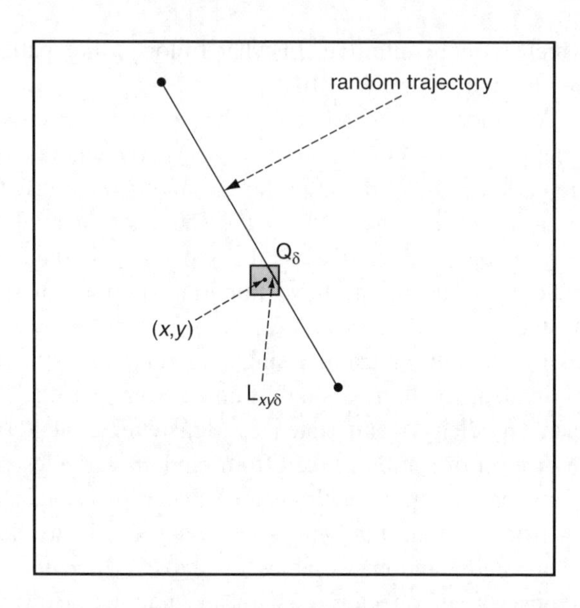

Figure 5.2 The derivation of the mobility component of distribution f_{RWP}.

uniformly chosen endpoints and a small area, say, a square of side δ with $0 < \delta \ll 1$ centered at a certain position (x, y). The latter problem can then be solved using tools from geometrical probability.

Figure 5.2 depicts the situation. The goal is to compute the probability mass lying in a small square Q_δ of side $\delta > 0$ centered at an arbitrary position $(x, y) \in R$, that is, the probability $P_t(x, y, \delta)$ of finding node u in Q_δ at time t, for $t \to \infty$. We first observe that, if δ is sufficiently small, the pdf f_T can be considered to be constant in Q_δ. Thus, we can write

$$P(x, y, \delta) \approx \delta^2 \cdot f_T(x, y)$$

and

$$f_T(x, y) = \lim_{\delta \to 0} \frac{P(x, y, \delta)}{\delta^2},$$

where $P(x, y, \delta) = \lim_{t \to \infty} P_t(x, y, \delta)$.

We are then left with the business of evaluating $P(x, y, \delta)$. To this end, we first observe that in the RWP model, node velocity during a specific trip, although randomly selected at the beginning of the trip, is *constant*; thus, taking any trajectory in the sequence of trips $\{D_i, t_p, V_i\}$, and taking two equal length portions (segments) s_1, s_2 within this trajectory, we have that the amount of time that the node spends in s_1 is the same as that spent in s_2. This implies that space and time are equivalent quantities in the RWP model, and $P(x, y, \delta)$ can be expressed as the ratio between the expected *length* of the intersection $L_{xy\delta}$ of a random trajectory with square Q_δ and the expected length of a random trajectory. In other words, we can write

$$P(x, y, \delta) = \frac{E[L_{xy\delta}]}{E[L]}.$$

Quantity $E[L]$ represents the expected length of a random segment with uniform endpoints in the unit square, which is known from the geometric probability literature to be equal to 0.521 405 (Santaló 2002).

Deriving $E[L_{xy\delta}]$ is more complicated. Let us fix the position $d_{i-1} = s$ of the trajectory starting point (waypoint $i - 1$). It can immediate be seen that $L_{xy\delta} \neq \emptyset$ if and only if the endpoint of the trajectory (waypoint $d_i = e$) lies within a subregion of R, represented as the shaded area in Figure 5.3 and denoted $A(x, y, \delta)$. Under the simplifying assumption that the length $l_{xy\delta}$ of segment $L_{xy\delta}$ is expressed as (here it comes as the approximation in Bettstetter et al.'s approach)

$$l_{xy\delta} = \begin{cases} c \cdot \delta & \text{if } e \in A(x, y, \delta) \\ 0 & \text{otherwise} \end{cases},$$

we can write

$$E[L_{xy\delta}] = \int_{s \in R} \left(\int_{e \in A(x,y,\delta)} c \cdot \delta \, de \right) ds = c \cdot \delta \int_{s \in R} A(x, y, \delta) \, ds,$$

where the notation $A(x, y, \delta)$ in the right hand side integral above is slightly abused to denote the area of the corresponding region.

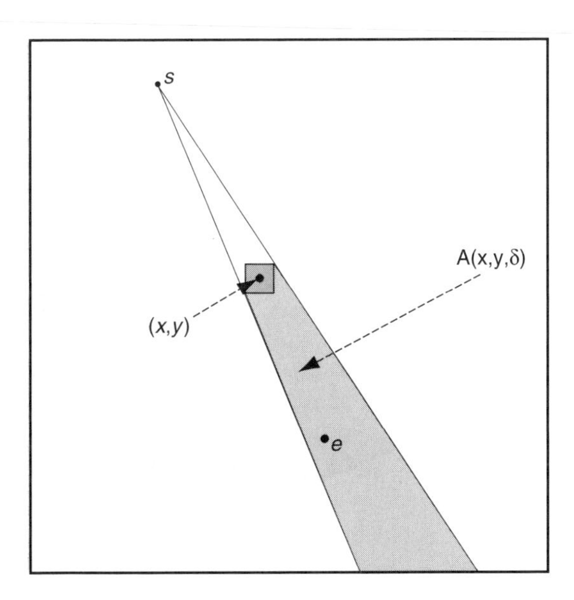

Figure 5.3 Derivation of $E[L_{xy\delta}]$.

The integral on the right hand side of the formula at the end of the previous page can be exactly computed by properly partitioning R based on the shape of the $A(x, y, \delta)$ regions. The derivation is quite lengthy and is not reported here – the interested reader is referred to Bettstetter et al. (2003). The resulting expression for f_T in the region $(0 \leq x \leq 0.5) \cup (0 \leq y \leq x)$ – values of f_T in the other regions of R are obtained by symmetry – is

$$f_T(x, y) = 6y + \frac{3}{4}\left(1 - 2x + 2x^2\right)\left(\frac{y}{y-1} + \frac{y^2}{x(x-1)}\right) + \frac{3y}{2}\left[(2x-1)(y+1)\ln\left(\frac{1-x}{x}\right) + (1 - 2x + 2x^2 + y)\ln\left(\frac{1-y}{y}\right)\right].$$

Density function f_T is shown in Figure 5.4. As can be seen in the figure, the mobility component of the RWP model spatial density function is bell-shaped, with a relatively higher node density in the center of the spatial domain R.

The last step in the derivation of the RWP model node spatial distribution is to compute probabilities $Prob(S = P)$ and $Prob(S = T)$. These probabilities, under the assumption that node speed is kept constant at $v > 0$ (i.e., $v_{min} = v_{max} = v > 0$), can be easily computed as follows:

$$Prob(S = P) = \frac{t_p}{t_p + \left(\frac{E[L]}{v}\right)} \quad \text{and} \quad Prob(S = T) = 1 - Prob(S = P).$$

The resulting distribution f_{RWP} when $t_p = 50$ and $v = 0.01$ (corresponding to 10 m/s if the unit side is considered to be 1 km) is displayed in

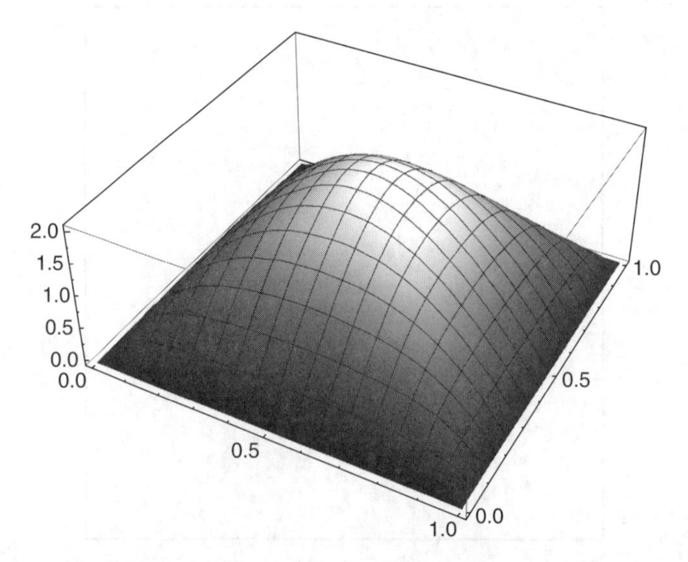

Figure 5.4 The mobility component of the RWP model spatial density function.

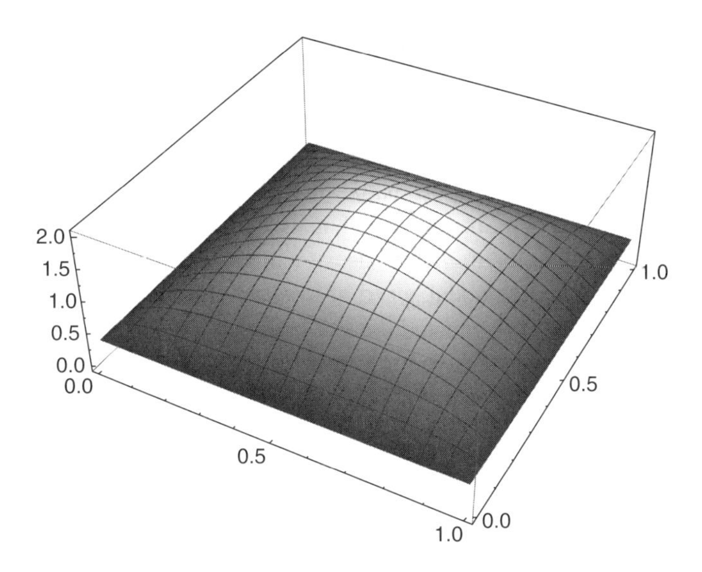

Figure 5.5 The RWP node spatial distribution with $t_p = 50$ and $v = 0.01$.

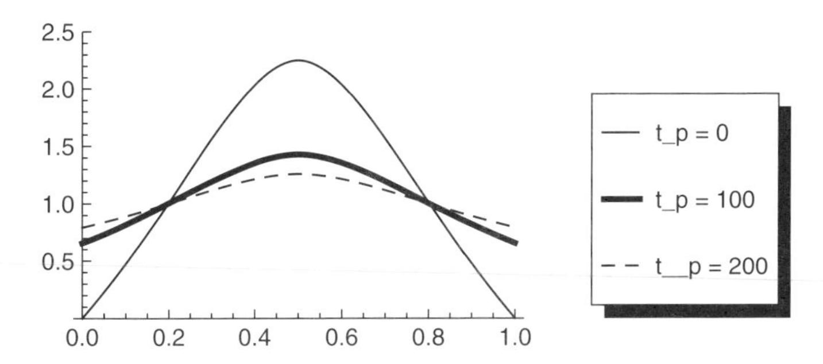

Figure 5.6 Diagonal cross-section of the RWP node spatial distribution with $v = 0.01$ and different values of pause time t_p.

Figure 5.5. It is interesting to observe that, in accordance with intuition, a relatively longer pause time results in a relatively flatter (i.e., more uniform) node spatial distribution. This is shown in Figure 5.6, which displays the diagonal cross-section of the distribution for different values of pause time.

5.3 The Average Nodal Speed of the RWP Model

A second property of the RWP model that has been studied in the literature is the *stationary average nodal speed*, which is formally defined as follows.

Assume that n nodes move independently within a spatial domain R according to the RWP mobility model, and denote by $v_i(t)$ the instantaneous speed of the ith node at time t. The stationary average nodal speed v_{RWP} is defined as

$$v_{RWP} = \lim_{t\to\infty} \frac{\sum_{i=1}^{n} v_i(t)}{n}.$$

Similar to the stationary node spatial distribution, interest in the stationary average nodal speed lies in the fact that its characterization allows correct setting of the simulation parameters, as we will see in Part Seven of this book.

The stationary average nodal speed of the RWP model was first studied by Yoon et al. (2003), who evaluated the stationary average nodal speed through simulation, and formally derived the value of v_{RWP}.

The value of v_{RWP} was derived under the following assumptions:

1. Nodes move in an unlimited, arbitrarily large area; given the current node location (x, y), the next waypoint is chosen uniformly at random in a circle of radius D_{max} centered at (x, y).
2. The pause time is zero.
3. The node velocity is chosen uniformly at random from $[v_{min}, v_{max}]$.

While the second and third assumptions are indeed part of the RWP model definition, the first assumption apparently changes considerably the features of RWP mobility. However, Yoon et al. (2003) prove that assumption 1 above, which is made to simplify the analysis, has no effect on the value of v_{RWP}.

To derive v_{RWP}, Yoon et al. describe the RWP mobility model as a stochastic process $\{V_i, D_i, S_i\}$, where V_i is the random variable denoting the velocity during trip i, D_i is the random variable denoting distance traveled during trip i, and S_i is the random variable denoting travel time during trip i. Setting $\sum_{i=1}^{n} v_i(t)/n = V(t)$, then v_{RWP} can be expressed as follows:

$$\begin{aligned} v_{RWP} &= \lim_{t\to\infty} V(t) \\ &= \lim_{T\to\infty} \frac{1}{T} \int_{[0,T]} V(t)\, dt = \lim_{T\to\infty} \frac{1/K(T) \cdot \sum_{k=1,\ldots,K(T)} d_k}{1/K(T) \cdot \sum_{k=1,\ldots,K(T)} s_k} = \frac{E[D_i]}{E[S_i]}, \end{aligned}$$

where $K(T)$ is the total number of trips undertaken within time T, including the (possibly incomplete) last one, and d_k (resp., s_k) is the travel distance (resp., time) of trip k.

Thus, similar to the derivation of the stationary node spatial distribution, the computation of v_{RWP} is reduced to the problem of computing the ratio of the expectations of two random variables, namely D_i and S_i. Yoon et al. (2003) show that

$$E[D_i] = \frac{2}{3} D_{max} \text{ and } E[S_i] = \frac{2D_{max}}{3(v_{max} - v_{min})} \cdot \ln\left(\frac{v_{max}}{v_{min}}\right),$$

yielding

$$v_{RWP} = \frac{v_{max} - v_{min}}{\ln\left(\frac{v_{max}}{v_{min}}\right)}.$$

What are the implications of the above characterization of v_{RWP}? A first important observation is that v_{RWP} is at most as large as the *initial* average nodal speed $v_0 = (v_{min} + v_{max})/2$. To see why this inequality holds, it is sufficient to study the following function:

$$g(\alpha) = \frac{v_{RWP}}{v_0} = \frac{\frac{(v_{max}-v_{min})}{\ln\left(\frac{v_{max}}{v_{min}}\right)}}{\frac{(v_{min}+v_{max})}{2}} = \frac{2(\alpha - 1)}{(\alpha + 1)\ln\alpha},$$

where $\alpha = v_{max}/v_{min} > 1$. The behavior of $g(\alpha)$ for increasing values of $\alpha > 1$ is displayed in Figure 5.7. As can be seen from the plot, $g(\alpha)$ is always below 1, implying that the stationary average nodal speed is at most as large as the initial speed v_0. This phenomenon is known as the *speed decay phenomenon*, and Yoon et al. (2003) also observed it experimentally. Furthermore, $\lim_{\alpha\to\infty} g(\alpha) = 0$, implying that when the maximal velocity v_{max} becomes asymptotically larger than the minimal velocity v_{min}, the stationary average nodal speed becomes asymptotically smaller than the initial speed. Thus, in this case the speed characteristics are substantially different between the beginning and steady state of the mobility process.

Another interesting observation is that $\lim_{\alpha\to 1} g(\alpha) = 1$, that is, when the interval within which random speeds are taken shrinks, the asymptotic average nodal speed becomes indistinguishable from the initial average speed. This can be seen from Table 5.1, which gives the values of v_{RWP} for different widths of the speed interval, given the same initial average speed $v_0 = 10$ m/s.

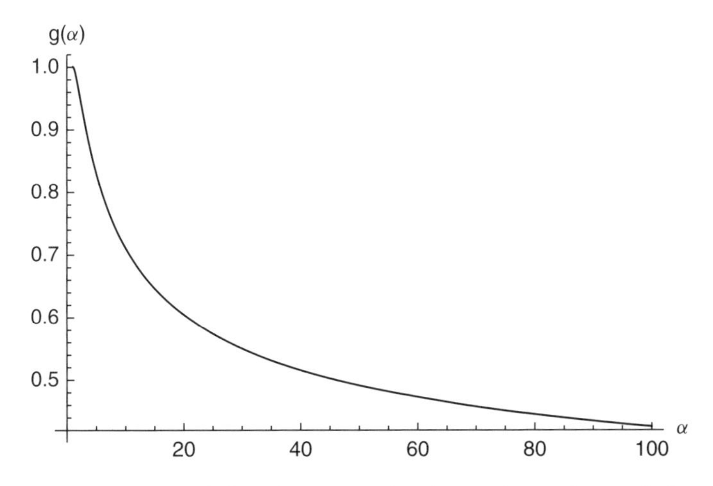

Figure 5.7 Behavior of function $g(\alpha)$ for increasing values of the ratio α between maximal and minimal node speed.

Table 5.1 Values of v_{RWP} for different widths of the speed interval (in m/s)

Speed interval	v_0	v_{RWP}	Speed interval	v_0	v_{RWP}
(1,19)	10	6.113	(6,14)	10	9.442
(2,18)	10	7.282	(7,13)	10	9.692
(3,17)	10	8.071	(8,12)	10	9.865
(4,16)	10	8.656	(9,11)	10	9.966
(5,15)	10	9.102			

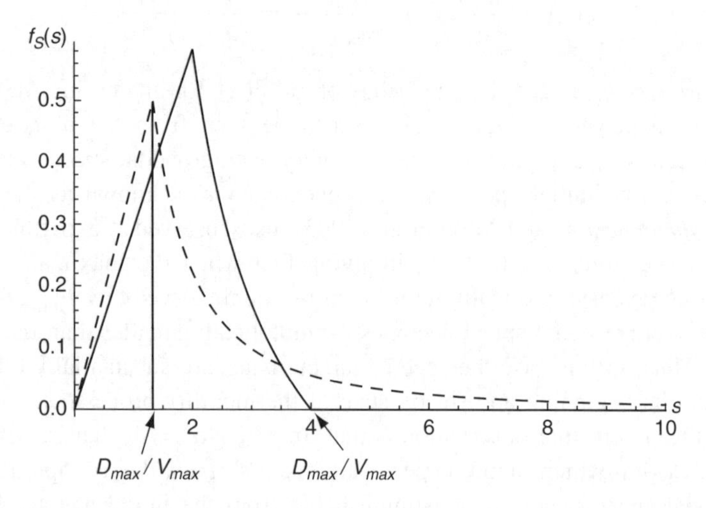

Figure 5.8 Shape of the pdf $f_S(s)$ of random variable S_i for different speed values: $v_{max} = 10$ m/s, $v_{min} = 5$ m/s (solid curve), and $v_{max} = 15$ m/s, $v_{min} = 0$ m/s (dashed curve).

A final, fundamental observation made by Yoon et al. (2003) is that the pdf $f_S(s)$ of the random variable S_i becomes heavy tailed when $v_{min} \to 0$. The shape of $f_S(s)$ is shown in Figure 5.8: the function increases linearly up to $s = D_{max}/v_{max}$; then, it starts decreasing till it reaches 0 when $s = D_{max}/v_{min}$. However, if $v_{min} = 0$ the right hand part of function $f_S(s)$ never reaches 0, and the distribution is heavy tailed. In this case, $E[S_i]$ becomes infinite, and $v_{RWP} \to 0$. Thus, if v_{min} is set to 0, the stationary regime of a RWP mobile network actually coincides with a static network ($v_{RWP} = 0$), and is reached only after infinite time. It is clear then that setting $v_{min} = 0$, which is the default setting in popular wireless networks, might severely impact simulation accuracy and representativeness.

The intuitive explanation of the above result is the following: if $v_{min} = 0$, for a fixed arbitrarily small $\epsilon > 0$, and the probability that a node chooses a speed in the range $[0, \epsilon]$ – *near zero* velocity – is a *constant*, then, by the law of large numbers, with infinite trials (trips) a near-zero velocity will eventually be chosen by *every* node in the network. When a near-zero velocity is chosen, the time needed to reach the destination, independently of its distance, approaches infinity. Thus, eventually each node in the network will get stuck in a trip with near-zero velocity, and the network will become virtually static.

5.4 Variants of the RWP Model

Several variants of the RWP model have been proposed in the literature, which we describe briefly in this section.

A variant of the RWP model, which is indeed a generalization, has been proposed by Bettstetter et al. (2003), where pause times are selected at each waypoint according to some probability distribution $f_T(t)$ with bounded support. Several other variants of the basic model have been proposed by Hyytia et al. (2006). In particular, these authors assume that nodes move in a unit disk, and consider a variant of the RWP model in which the position of waypoints is chosen according to an arbitrary, rotationally symmetric distribution

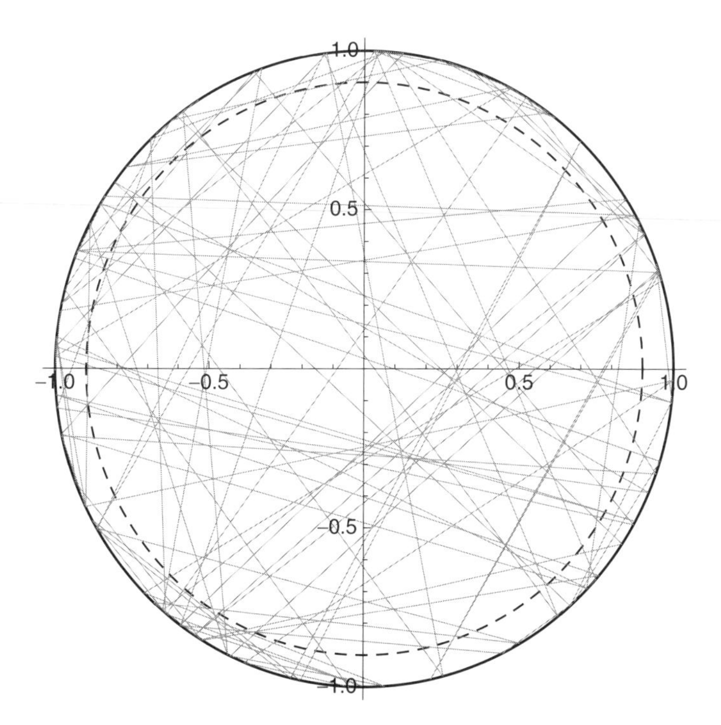

Figure 5.9 A hundred steps of RWPB mobility: the density of gray lines, representing trajectories, is relatively higher on the border (external annulus) than in the center of the disk.

with support in the unit disk. An interesting sub-case of this variant is when waypoints are chosen uniformly at random on the *border* of the disk, which is named the RWPB (where B stands for *border*) in Hyytia et al. (2006). In other words, waypoints are chosen uniformly at random among the set of points at distance 1 from the center of the disk. These authors show that the stationary node spatial distribution in the RWPB model is relatively more concentrated on the border of the disk. This can be seen from Figure 5.9: the density of lines is relatively higher on the border (external annulus) than in the center of the square.

References

Bettstetter C and Krause O 2001 On border effects in modeling and simulation of wireless ad hoc networks. *Proceedings of the IEEE International Conference on Mobile and Wireless Communication Networks (MWCN)*, pp. 20–27.

Bettstetter C, Resta G and Santi P 2003 The node distribution of the random waypoint mobility model for wireless ad hoc networks. *IEEE Transactions on Mobile Computing* **2**, 257–269.

Blough D, Resta G and Santi P 2002 A statistical analysis of the long-run node spatial distribution in mobile ad hoc networks. *Proceedings of the ACM Workshop on Modeling, Analysis and Simulation of Wireless and Mobile Systems (MSWiM)*, pp. 30–37.

Hyytia E, Lassila P and Virtamo J 2006 Spatial node distribution of the random waypoint mobility model with applications. *IEEE Transactions on Mobile Computing* **5**, 680–694.

Johnson D and Maltz D 1996 Dynamic source routing in ad hoc wireless networks. *Mobile Computing*, pp. 153–181. Kluwer Academic, Dordrecht.

Ns2Team 2011 *http://www.isi.edu/nsnam/ns/*.

Santaló LA 2002 *Integral Geometry and Geometric Probability*. Cambridge University Press, Cambridge.

Team G 2011a *http://pcl.cs.ucla.edu/projects/glomosim/*.

Team G 2011b *http://www.ece.gatech.edu/research/labs/MANIACS/GTNetS/*.

Yoon J, Liu M and Noble B 2003 Random waypoint considered harmful. *Proceedings of IEEE Infocom*, pp. 1312–1321.

6

Group Mobility and Other Synthetic Mobility Models

In the previous chapters, we introduced synthetic mobility models for MANETs sharing the property of being entity-based: that is, mobility of a network composed of n nodes is modeled as n independent stochastic processes with the same statistical properties, each modeling the movement of a single node. In many situations, though, movement of mobile entities displays spatial correlation, that is, the movement of an entity is influenced by the movement patterns of other entities in the surroundings. A typical such example is movement of vehicles along a road: if vehicle A follows vehicle B on a single-lane road where overtaking is forbidden, the velocity of A is clearly upper bounded by the velocity of the preceding vehicle B.

Among mobility models taking spatial correlation of movement into account, an important class is that of *group mobility* models, whose purpose is to model situations where the movement patterns of the members of a subset of the network entities (called a *group*) are highly correlated. Examples of group mobility in real-world scenarios are soldiers moving on the battlefield, disaster recovery and law enforcement operations, movement of tourist groups in a museum or a city, etc.

It is important to observe that group mobility can be considered as an instance of the more general class of spatially correlated mobility models. In fact, in the former class of models nodes are *explicitly* and *statically* partitioned into groups at the beginning, and movements of entities belonging to different groups are independent of each other. On the other hand, in spatially correlated mobility models explicit and static partitioning of mobile entities into separate groups is not mandatory. For instance, referring back to a vehicular movement scenario, the movement of vehicle A is influenced by

Mobility Models for Next Generation Wireless Networks: Ad Hoc, Vehicular and Mesh Networks, First Edition. Paolo Santi.

that of vehicle B; however, if vehicle A makes a turn at the next intersection, the dependence of its movement pattern on that of vehicle B is lost, while dependence from movement of another preceding vehicle C might occur. Thus, in this scenario vehicles cannot be explicitly and statically partitioned into separate groups, with the property that movement of a vehicle is only influenced by movement of other vehicles in the same group. Rather, very small groups are *implicitly* and *dynamically* formed (and destroyed) as the vehicles move along the roads.

In the first part of this chapter, we will present the most representative group mobility model for MANETs introduced in the literature, namely the reference point group mobility (RPGM) model defined in Hong et al. (1999). As we will see, many other group mobility models proposed in the literature can be considered as specific instances of the RPGM model, which can be considered as a very versatile group mobility model for MANETs.

In the second part of this chapter, we will present synthetic mobility models that have been proposed to improve a shortcoming of existing entity-based mobility models such as random walk and RWP: namely, that mobile nodes under these models display sudden changes in trajectories and/or speed. More specifically, we will present the smooth mobility model and the Gauss-Markov mobility model.

6.1 The RPGM Model

The RPGM model has been defined in Hong et al. (1999) to overcome a shortcoming of mobility models for MANETs at that time (i.e., their inability to model spatially correlated movements and in particular group mobility). Given its versatility in modeling different types of group mobility, the model has since then been established as the most representative group mobility model for MANETs.

In the RPGM model, the n mobile nodes $N = \{u_1, \ldots, u_n\}$ are first partitioned into groups $G_1, \ldots, G_k$, for some $k < n$. Members of group G_i are denoted $u_{i,1}, \ldots, u_{i,c_i}$ in the following, where c_i is the number of nodes in the ith group. An important observation is that the RPGM is a discrete-time mobility model, where movement of a node $u_{i,j}$ is modeled as a sequence of node positions $u_{i,j}(t), u_{i,j}(t+1), \ldots$ at discrete, equal duration time ticks $t, t+1, \ldots$.

Partitioning of nodes into groups obeys no specific rule in RPGM; this task is then completely left to the network designer, which contributes to the versatility of the RPGM model. Note in particular that singleton groups (i.e., groups composed of a single node) can be defined, thus allowing modeling situations in which individual, independent mobility and group-based mobility to coexist.

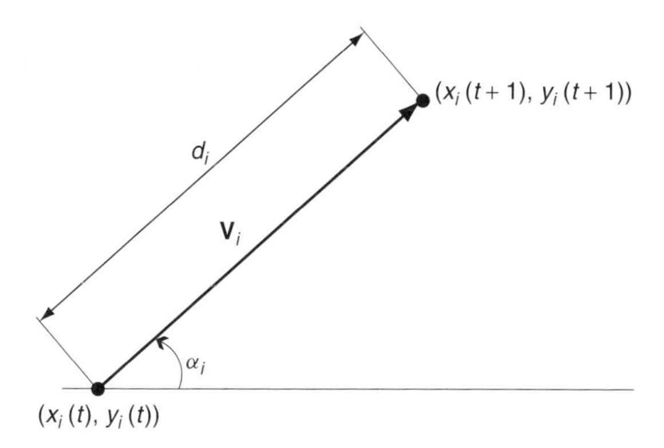

Figure 6.1 Movement vector of the logical "center" in the RPGM model in consecutive time steps.

In the RPGM model, each group G_i is assigned to a logical "center," which determines the movement of the group. In particular, the movement of a group is modeled as a sequence of locations $l_i(t), l_i(t+1), \ldots$ (called *checkpoints*) of its logical "center" at time $t, t+1, \ldots$. The sequence of checkpoints implicitly defines a sequence of vectors $\mathbf{V}_i(t), \mathbf{V}_i(t+1), \ldots$ corresponding to the speed and direction of the logical "center" during time interval $[t, t+1), [t+1, t+2), \ldots,$ respectively. For instance, considering the spatial domain to be the unit square $R = [0, 1]^2$ and letting $l_i(t) = (x_i(t), y_i(t))$ and $l_i(t+1) = (x_i(t+1), y_i(t+1))$, we have (see also Figure 6.1)

$$\mathbf{V}_i(t), \mathbf{V}_i(t+1), \ldots V_i(t) = (d_i(t), \alpha_i(t))$$

where

$$d_i(t) = \sqrt{\left(x_i(t+1) - x_i(t)\right)^2 + \left(y_i(t+1) - y_i(t)\right)^2}$$

is the traveled distance during the interval $(t, t+1)$ and

$$\alpha_i(t) = \arctan\left(\frac{y_i(t+1) - y_i(t)}{x_i(t+1) - x_i(t)}\right)$$

is the travel direction.

Given a sequence of group mobility vectors $\mathbf{V}_i(t), \mathbf{V}_i(t+1), \ldots$ for group G_i, the RPGM model dictates that mobility of a node $u_{i,j} \in G_i$ is modeled as follows. The initial position of $u_{i,j}$ is chosen uniformly at random in a disk of a certain radius d_{max} (a model parameter) centered at the initial position $l_i(0)$ of the logical "center" of G_i. Given the position $u_{i,j}(t) = (x_{i,j}(t), y_{i,j}(t))$ of node $u_{i,j}$ at time t, its position at time $t+1$ is determined as follows:

$$\begin{aligned} u_{i,j}(t+1) &= (x_{i,j}(t+1), y_{i,j}(t+1)) \\ &= (x_{i,j}(t) + \mathbf{V}_{i,j}(t)[1]\cos \mathbf{V}_{i,j}(t)[2], y_{i,j}(t) \\ &\quad + \mathbf{V}_{i,j}(t)[1]\sin \mathbf{V}_{i,j}(t)[2]), \end{aligned}$$

where $\mathbf{V}[h]$ denotes the hth component of vector $\mathbf{V}$, and $\mathbf{V}_{i,j}(t)$ is a random perturbation of vector $\bar{\mathbf{V}}_i(t)$ defined as follows:

$$\bar{\mathbf{V}}_i(t)V_{i,j}(t) = \mathbf{V}_i(t) + \mathbf{R}_{i,j}(t),$$

where $\mathbf{R}_{i,j}(t)$ is a vector with distance component uniformly distributed in $[0, d_{max}]$ and direction component uniformly distributed in $[0, 2\pi]$. An example of how node positions are generated given mobility of the group logical "center" is shown in Figure 6.2.

Note that the RPGM model contains no rules about how the movement of groups is modeled, that is, about how vector $\mathbf{V}_i(t+1)$ is generated starting from $\mathbf{V}_i(t)$. In this respect, the RPGM can be considered as a two-tier mobility model, where mobility is independently modeled at the group tier (mobility of logical "centers") and at the mobile entity tier. Group tier mobility in RPGM is left unspecified, but can be modeled through existing entity-based mobility models such as the RWP model, derived from scenario data files etc. On the other hand, RPGM defines rules according to which mobility at the lower tier of the single mobile entity is derived for a given mobility pattern at the group tier.

Summarizing, the RPGM model is a highly versatile mobility model giving the freedom to arbitrarily partition mobile nodes into groups, and to use

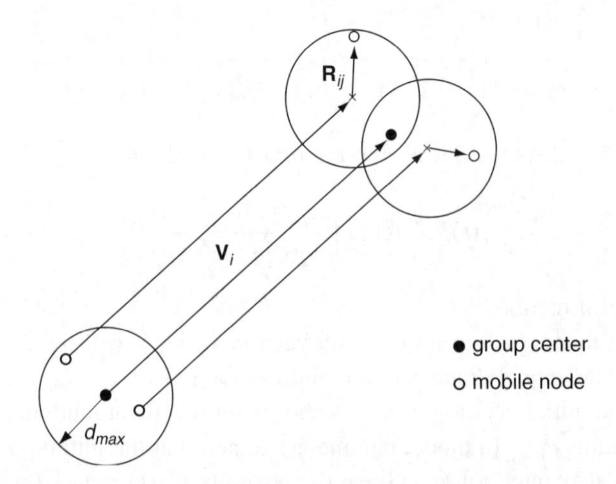

Figure 6.2 Movement of logical "center" and nodes in a group.

arbitrary mobility models or traces to generate group mobility. In the following, we present a few examples (mostly taken from Hong et al. (1999)) showing how different types of group mobility can be easily reproduced in RPGM by properly defining node partitioning and rules for group mobility.

6.1.1 RPGM with RWP Group Mobility

In this version of RPGM mobility, movement of group logical "centers" is governed by the RWP model. Indeed, some adjustments are needed in order to adapt RWP mobility – which is based on a continuous time model – to the discrete-time model used in RPGM, as explained next.

Consider the mobility of a single group G_i. First, the initial position of the G_i logical "center" is generated by choosing a location uniformly at random in the spatial domain R. Then, a sequence of waypoints $W = \{w_1, w_2, \dots\}$ for the logical centers and corresponding traveling speed $S = \{s_1, s_2, \dots\}$ are generated as dictated by the RWP model. For simplicity, assume that speed is expressed in units of length per time tick, so that each element in S can be interpreted as the space traveled by the logical "center" in a time tick during the ith trip from w_i to w_{i+1}. The next step is to generate a sequence of checkpoints $C = \{c_1, c_2, \dots, \}$ for the logical "center" of G_i starting from sequences W and S. Let d_i denote the distance between waypoints w_i and w_{i+1}; the straight-line trajectory between w_i and w_{i+1} is broken down into $m_i = \lceil d_i/s_i \rceil$ segments, with the first $m_i - 1$ segments having length s_i, and the last segment having length $d_i - (m_i - 1)s_i$. Checkpoints for the ith trip from w_i to w_{i+1} are then generated by taking the endpoints of the above-defined segments (see Figure 6.3).

Once the sequence of checkpoints is generated, an arbitrary number of mobile nodes belonging to group G_i can be defined, with mobility rules as dictated by the RPGM model. An example of RPGM with RWP group mobility for a group composed of three nodes is shown in Figure 6.4.

6.1.2 In-Place Mobility Model

This version of RPGM aims at modeling a situation in which the spatial domain R is divided into a number of non-overlapping regions, with a different group moving in each region. This mobility pattern is representative of, for instance, a battlefield situation, where different platoons are carrying out operations such as a landmine search in different areas.

A simple way of generating in-place mobility is to partition the spatial domain into relatively large square cells, and to realize RPGM with RWP group mobility separately in each subregion of the domain. An example of the resulting mobility pattern when R is divided into four squares of side 0.5, and groups of different size move in each square, is displayed in Figure 6.5.

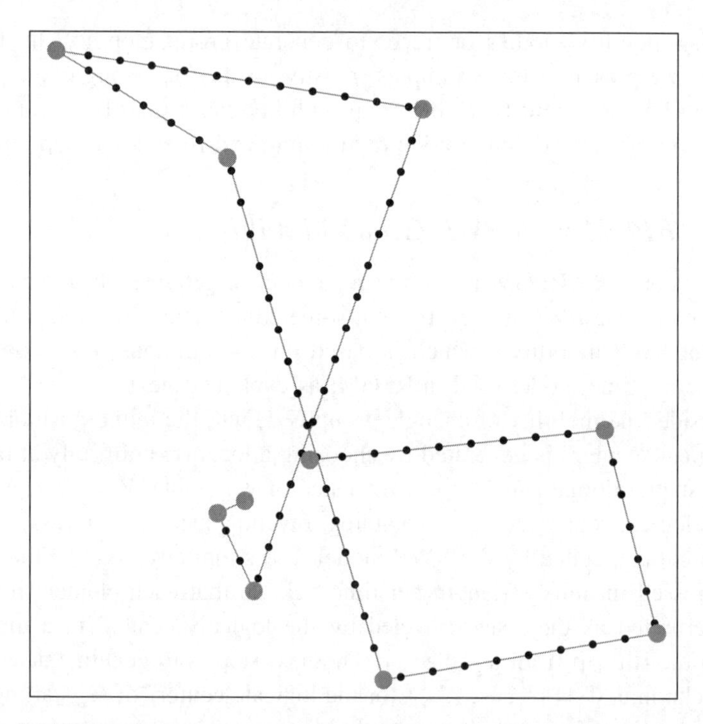

Figure 6.3 Transforming a sequence of RWP waypoints (large gray dots) into a sequence of RPGM checkpoints (black dots). The underlying RWP trajectories are represented by gray lines.

6.1.3 Convention Mobility Model

This version of RPGM models a convention scenario, more specifically the interactions between exhibitors and attendees. Similar to the in-place mobility model, the spatial domain can be thought of as divided into non-overlapping regions, representing different exhibition areas. Two types of groups are modeled in this scenario: *exhibitor* and *attendee* groups. Exhibitor groups are mapped to different exhibition areas, with members of the group moving within the corresponding exhibition area. On the other hand, groups of attendees roam freely between the various exhibition areas.

The convention mobility models can be realized in RPGM by defining different mobility patterns for the two types of groups: for instance, the logical "center" of an exhibitor group can be made to move within the inner part of the corresponding exhibition area; by suitably defining the mobility radius d_{max}, this ensures that nodes in the exhibitor group cannot exit the corresponding exhibition area. On the other hand, mobility of attendee groups can be realized by, for example, using RPGM with RWP group mobility over the entire spatial domain R. An example of the convention mobility model when R is divided into four non-overlapping squares is shown in Figure 6.6.

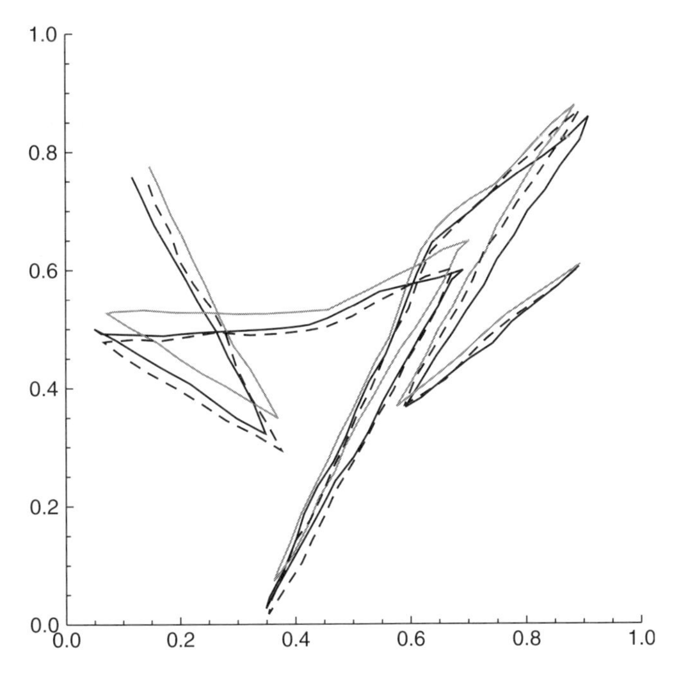

Figure 6.4 Example of RPGM with RWP group mobility for a group of three nodes.

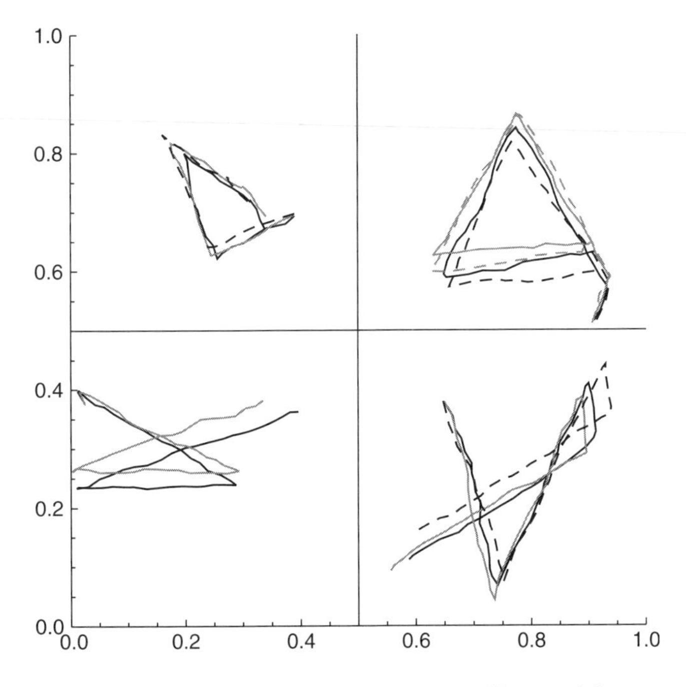

Figure 6.5 Example of in-place mobility model.

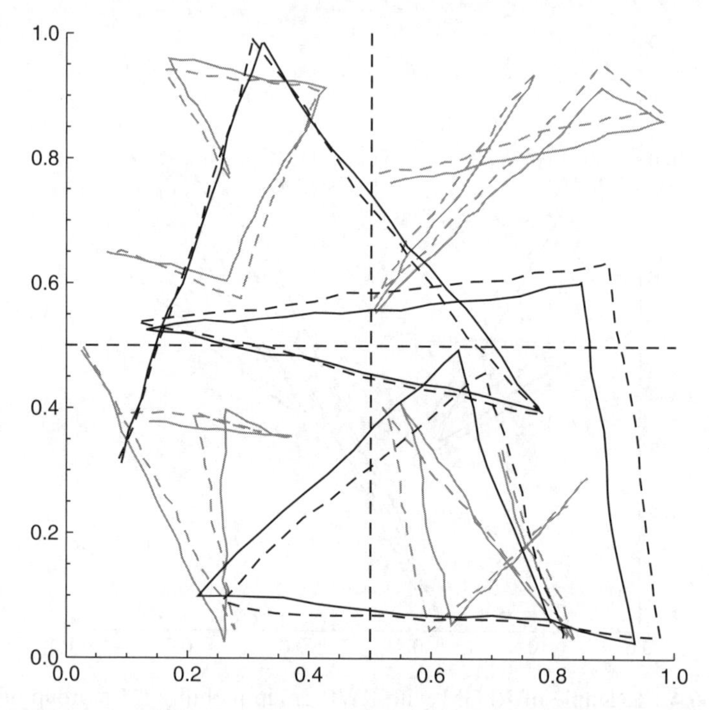

Figure 6.6 Example of convention mobility model with four exhibitor groups and one attendee group. Trajectories of nodes in exhibitor groups are in gray; trajectories of nodes in the attendee group are in black.

6.1.4 RPGM and Other Group Mobility Models

It has been observed (Camp et al. 2002) that many other group mobility models can be considered as instantiations of the RPGM model. This is the case, for instance, with the three models defined in Sanchez and Manzoni (1999), which we briefly describe in this section.

Sanchez and Manzoni (1999) define three group mobility models. The first model is called the *column mobility model* and is aimed at modeling mobile nodes (e.g., robots) performing a search and scanning activity. In this scenario, the group of nodes moves along a predefined grid in alternating directions, so as to systematically scan a certain area. The column mobility model can be easily realized through RPGM by arranging the checkpoints of the logical "center" of the group into a regular grid.

The second model is the *pursue mobility model*, whose purpose is to emulate situations where several nodes attempt to capture a single mobile node ahead, as in a target tracking or law enforcement scenario. To realize the pursue mobility model in RPGM, it is sufficient to define a singleton group formed by the evader node, and to use RWP, for example, to model mobility of the evader node. A second group is then composed of the pursuing nodes; the logical "center" of this group can be made to mimic movement of the evader node, with some additional random variations around this movement pattern.

The third model is the *nomadic community mobility model*, aimed at modeling mobility scenarios where a group of nodes freely moves in an environment: the reference point of a group moves randomly according to some mobility model (e.g., RWP), and group members move around the reference point with some small random deviation. This model is essentially equivalent to the RPGM model with RWP group mobility described in Section 6.1.1.

6.2 Other Synthetic Mobility Models

In the previous section, we presented mobility models aimed at modeling a form of spatially correlated mobility named group mobility. In this section, we present mobility models that have been proposed in the literature to overcome shortcomings in popular entity-based mobility models such as random walk and RWP, namely, that direction and speed of travel can change abruptly in these models. The purpose of the models presented in the following is to introduce a degree of *temporal* correlation in the mobility pattern of nodes, with the value of mobility parameters – speed and direction – at time $t+1$ being correlated with the respective values at time t. In particular, we present the smooth mobility model and the Gauss–Markov mobility model. Notice that both models belong to the class of entity-based mobility models, hence movements of different nodes in the network are modeled as independent stochastic processes.

6.2.1 The Smooth Random Mobility Model

In the smooth random mobility model (Bettstetter 2001), the movement of a node is described through two independent stochastic processes, one governing node speed, and the other governing node direction. Although the model can be adapted to work in both continuous-time and discrete-time conditions, the model is formally defined in Bettstetter (2001) under the discrete-time assumption, which we also embrace.

In the smooth random mobility model, the status of a mobile node at time t is described by the triplet $(l_u(t), v_u(t), \phi_u(t))$, where $l_u(t)$ denotes current location within the spatial domain R, $v_u(t)$ denotes the current velocity, and $\phi_u(t)$ denotes the current direction. Since we are focusing on the movement of a single node, in the following we drop subscript u to simplify the notation.

Given $l(t)$, the location of the node at the next time step is computed based on the current speed and direction values as follows:

$$l(t+1) = (x(t+1), y(t+1)) = (x(t) + v(t)\cos\phi(t),$$
$$y(t) + v(t)\sin\phi(t)),$$

under the usual assumption that speed is expressed in length units per time tick. The next speed and direction values are generated according to the following random processes.

Let us first consider the speed process. Let v_{max} be the maximum possible speed of a mobile node, and $v_1, v_2, \ldots \leq v_{max}$ a set of preferred speeds. Also, assume that two values a_{max} and d_{max} are defined to represent the maximum acceleration and maximum deceleration, respectively. The initial speed value $v(0)$ is determined according to the following probability distribution:

$$P(v) = \begin{cases} p_i & \text{if } v = v_i \text{ for some } i \\ 0 & \text{if } v > v_{max} \text{ or } v < 0 \\ \frac{(1-\sum_i p_i)}{v_{max}} & \text{otherwise} \end{cases},$$

where the p_i are arbitrary probability values with the property that $\sum_i p_i < 1$.

At each time step, a speed change event occurs with a certain probability p_v. Note that if $p_v \ll 1$, this process approximates a continuous Poisson process of intensity p_v. When a speed change event occurs, a new *target speed* is chosen according to the same probability distribution. Suppose a speed change event occurs at time t, and let $v'(t)$ be the new randomly selected target speed. The next step is to select an acceleration (if $v'(t) > v(t)$) or deceleration (if $v'(t) < v(t)$) value governing the speed change phase. In both cases, acceleration/deceleration is chosen uniformly at random in either the $[0, a_{max}]$ or $[0, d_{max}]$ interval. Let $\bar{a}$ be the selected acceleration value – without loss of generality, we are assuming that $v'(t) > v(t)$. In the following $\frac{(v'(t)-v(t))}{\bar{a}}$ time steps, speed values are changed as follows:

$$v(t+1) = v(t) + \bar{a}.$$

Once the speed has reached the target value, it remains constant over time until a new speed change event occurs.

The direction process obeys similar rules. Initially, a movement direction is chosen uniformly at random in the $[0, 2\pi]$ interval. Then, a direction change event occurs with a certain probability p_ϕ, with $p_\phi \ll 1$. If a direction change event occurs, a new target direction $\phi'(t)$ is selected uniformly at random in $[0, 2\pi]$. Similar to speed change, directional change also occurs gradually: denoting by Δ the angular difference between $\phi(t)$ and $\phi'(t)$, and with δ the allowed angular change per time tick – which in Bettstetter (2001) is computed based on a target curving time t_c – we have

$$\phi(t+1) = \phi(t) + \delta,$$

for the next Δ/δ time steps. In the above formula, we have assumed without loss of generality that $\Delta > 0$.

An example of mobility trace generated with the smooth random mobility model is shown in Figure 6.7.

A few observations are in order before concluding this section. First, a border rule is needed in order to determine how to change mobility parameters when a node hits the border, which occurs with non-zero probability. No specific border rule is mentioned in Bettstetter (2001), so in principle the model can be used in combination with any border rule (e.g., mirror reflection). Second, Bettstetter (2001) presents improvements over the basic model aimed at correlating the stochastic process modeling speed and direction changes, which are considered to be independent in the basic model. This is an unrealistic aspect of the mobility model, since in practice the velocity of a node changes (typically, decreases) when a change in movement direction occurs. Two optional rules, called *stop–turn–and–go* and *slowdown of turning nodes* are defined in Bettstetter (2001) to account for this problem. The interested reader is referred to Bettstetter (2001) for details.

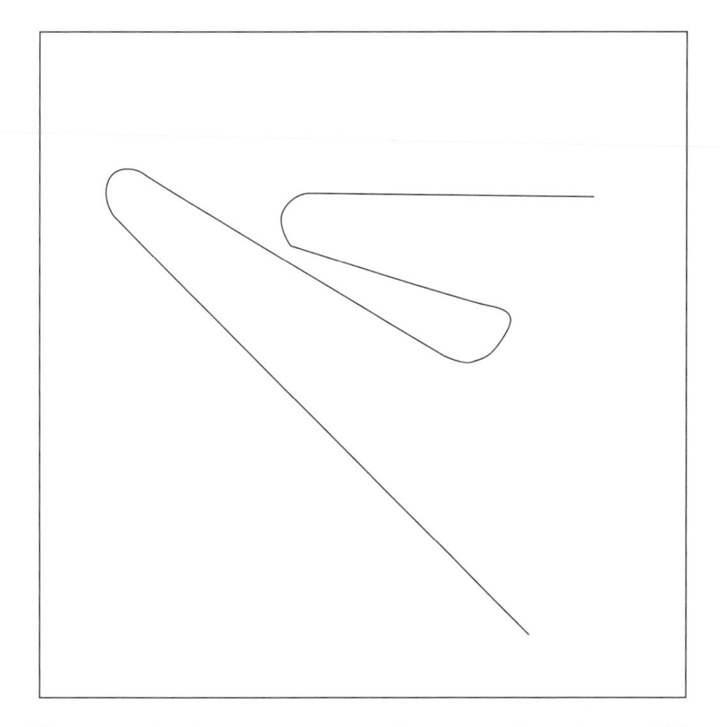

Figure 6.7 Example of trace generated with the smooth random mobility model.

6.2.2 Gauss–Markov Mobility Model

The Gauss–Markov mobility model was originally proposed in Liang and Haas (1999) to model mobility in personal communications service networks, a kind of network providing services similar to an advanced cellular network. However, the model aims at modeling mobility at a microscopic scale, so it can be used also to model MANET mobility. In the following, we mostly refer to the description of Gauss–Markov mobility as reported by Camp et al. (2002).

The Gauss–Markov mobility model is a discrete-time model. Initially, each mobile node is randomly located in the spatial domain R and assigned with a speed chosen uniformly at random in an interval $[v_{min}, v_{max}]$, and with a direction chosen uniformly at random in the interval $[0, 2\pi]$. Given speed and direction values at time t, the position of the node at the next time step is determined using trigonometric rules as in the smooth random mobility model. Speed and direction values for the next time step are then randomly updated according to the following rules. Let $v(t)$ and $d(t)$ denote the speed and direction values at time t, and let α be a parameter in the $[0, 1]$ interval allowing control of the degree of temporal correlation between speed and direction values. We have

$$v(t+1) = \alpha \cdot v(t) + (1-\alpha)\bar{v} + V \cdot \sqrt{(1-\alpha^2)}$$

and

$$d(t+1) = \alpha \cdot d(t) + (1-\alpha)\bar{d} + D \cdot \sqrt{(1-\alpha^2)}$$

where $\bar{v}$ (resp., $\bar{d}$) are constants representing the average speed (resp., direction of movement), and V (resp., D) is a normal random variable with zero mean and standard deviation σ, for some $\sigma > 0$. The facts that S and D have a Gaussian distribution, and that the speed and direction stochastic processes enjoy the memory-less property, explain the name given to this mobility model.

It is interesting to observe the role of parameter α in the definition of the mobility model. If $\alpha = 0$, we have speed and direction values at time step $t + 1$ which are completely uncorrelated with those at time t, and the model reduces to a form of Brownian motion. On the other hand, when $\alpha = 1$ we have maximum correlation between the speed and direction values at times t and $t + 1$, since $v(t + 1) = v(t)$ and $d(t + 1) = d(t)$. Intermediate values of parameter α correspond to intermediate degrees of temporal correlation in the speed and direction stochastic processes, which are actually the cases of interest for this mobility model.

Similar to the smooth random mobility model, the Gauss–Markov mobility model needs a border rule in case the spatial domain of movement R is bounded. Camp et al. (2002) suggest use of the mirror reflection border rule in combination with the Gauss–Markov mobility model.

References

Bettstetter C 2001 Smooth is better than sharp: a random mobility model for simulation of wireless networks. *Proceedings of the ACM International Conference on Modeling, Analysis and Simulation of Wireless Networks (MSWiM)*, pp. 19–27.

Camp T, Boleng J and Davies V 2002 A survey of mobility models for ad hoc network research. *Wireless Communications and Mobile Computing* **2**, 483–502.

Hong X, Gerla M, Pei G and Chiang CC 1999 A group mobility model for ad hoc wireless networks. *Proceedings of the ACM International Conference on Modeling, Analysis and Simulation of Wireless Networks (MSWiM)*, pp. 53–60.

Liang B and Haas Z 1999 Predictive distance-based mobility management for PCS networks. *Proceedings of IEEE Infocom*.

Sanchez M and Manzoni P 1999 A Java-based ad hoc networks simulator. *Proceedings of the SCS Western Multiconference – Web-based Simulation Track*.

References

7

Random Trip Models

In this chapter, we will present and analyze the properties of a class of mobility models introduced in Le Boudec and Vojnovic (2006) called *random trip models*. This class of mobility models is very general, and includes most of the mobility models described in the previous chapters such as random walks and random waypoint, as well as many mobility models which will be described in forthcoming chapters. The importance of the class of random trip models lies in the fact that, if a mobility model is proved to belong to this class, fundamental model properties such as the existence and characterization of a stationary regime can be established based on the framework developed for the general class of random trip mobility models.

In what follows, we will first formally define the class of random trip mobility models, and present theoretical results derived by Le Boudec and Vojnovic (2006) concerning the existence and characterization of the stationary regime for a random trip mobility model. We will then conclude the chapter with specific instances of mobility models belonging to this general class.

7.1 The Class of Random Trip Models

Informally, a *random trip* is a model of random, *independent* node movements. Note that the requirement of independent movements implies that group mobility models do not belong to the class of random trip models. Formally speaking, the random trip model is defined by the notions of *domain*, *phase*, *path*, and *trip*, and by the definition of a *default initialization rule*:

1. *Domain*: the domain $\mathcal{R}$ is a subset of $\mathbb{R}^d$, for some integer $d \geq 1$, and represents the mobility region.

Mobility Models for Next Generation Wireless Networks: Ad Hoc, Vehicular and Mesh Networks, First Edition. Paolo Santi.

2. *Phase*: a phase is one of the possible states of the mobile node. Formally, $\mathcal{I}$ represents the set of all possible phases in the model, where each element of $\mathcal{I}$ can be thought of as an integer describing one of the (countably many) possible states of a mobile node. For instance, a phase in a random trip model might indicate whether the mobile node is currently moving or pausing at a certain location.
3. *Path*: a path $P \in \mathcal{P}$ represents a possible trajectory of a mobile node in $\mathcal{R}$, where $\mathcal{P}$ is the set of all possible paths in $\mathcal{R}$. For instance, set $\mathcal{P}$ might be the set of all line segments with both endpoints in $\mathcal{R}$, where $\mathcal{R}$ is an arbitrary convex set. Formally, a path $P : [0, 1] \mapsto \mathcal{R}$ is a continuous mapping with a continuous derivative except possibly at a finite number of points. For any $P \in \mathcal{P}$, $P(0)$ is the starting point (origin) of the path, $P(1)$ is the ending point (destination) of the path, and $P(u)$, with $u \in (0, 1)$, represents the position of a node in $\mathcal{R}$ when a fraction u of its trajectory along path P is traversed.
4. *Trip*: a trip is defined by specifying a phase $I_n \in \mathcal{I}$, a path $P_n \in \mathcal{P}$, and a duration S_n, drawn according to some *trip selection rule*, specific to the model. For instance, in case only two phases (*pause* and *move*) are possible, the trip selection rule might dictate that a pause phase is always followed by a move phase, and that a move phase is always followed by a pause phase. Trip selection is performed at time instants $T_0 \leq 0 < T_1 < T_2 < \ldots$, where the $T_n \in \mathbb{R}^+$ are called *transition instants*. Given the phase I_n, the path P_n, and the duration S_n selected for the nth trip starting at time T_n, the next transition instant is $T_{n+1} = T_n + S_n$. The position of a mobile node at time t, with $T_n \leq t \leq T_{n+1}$, is a continuous random variable taking values in $\mathcal{R}$, defined as

$$X(t) = P_n\left(\frac{t - T_n}{S_n}\right).$$

Note that the trip selection rule must *always* satisfy the property that $P_{n+1}(0) = P_n(1)$, that is, the origin of the next path is the destination of the previous path. This is to avoid a mobile node being allowed to disappear from a point in $\mathcal{R}$ and reappear soon after at another, possibly far-away, point in $\mathcal{R}$. For a similar reason, the duration S_n of a trip is constrained to take strictly positive values, that is, $S_n \in \mathbb{R}^+$.
5. *Default initialization rule*: at time $t = 0$, representing the time at which observation of the mobile node starts, a phase, path, and remaining time till the next transition are drawn according to some *initialization rule*. The *default* initialization rule dictates that $T_0 = 0$ (i.e., the first transition instant occurs at time 0), and selects a phase, path, and trip duration according to the trip selection rule.

In order for a mobility model to belong to the class of random trip models, the following conditions on phase, path, and trip duration must be fulfilled. Note that a formal definition of these conditions is sometimes involved, requiring an understanding of non-trivial mathematical concepts. In what follows we have tried to simplify notation and presentation as much as possible, while only minimally sacrificing mathematical rigor. For a more rigorous definition of the conditions stated below, the reader is referred to Le Boudec and Vojnovic (2006).

7.1.1 Conditions on Phase and Path

The following conditions on phase and path must be satisfied:

1. *Markov property*: let $Y = (Y_n)$, where $Y_n = (I_n, P_n)$, be the stochastic process defined by observing the evolution of phase and path with time. Process Y must be a time-continuous, space-continuous Markov chain on $\mathcal{I} \times \mathcal{P}$. In particular, this means that process Y enjoys the *Markov property* (also called *memory-less* property): the state of the process at any time $t > T_n$ depends only on the state of the process at the last transition instant T_n, and not on the state of the process at previous transition instants $T_0, \ldots, T_{n-1}$.
2. *Harris recurrent*: the Markov chain Y must be *Harris recurrent*. The notion of Harris recurrence can be used to extend the notion of recurrence to Markov chains with continuous space state. When applied to the Markov chain $Y = (I_n, P_n)$ at hand, conditions for Harris recurrence can be stated as follows. Let $\Omega = \mathcal{I} \times \mathcal{P}$ be the state space of Y. The Markov chain is Harris recurrent if there exist sets $A, B \in \Omega$, a probability measure φ with support in B, and a number $\epsilon > 0$ such that:
   (a)
   $$P(\tau_A < \infty | Y_0 = y) > 0 \text{ for all } y \in \Omega,$$
   where $\tau_A = \inf\{n \geq 0 : y_n \in A\}$.
   (b)
   $$P(D < \infty | y_0 \in A) = 1 \text{ for each } y_0 \in A,$$
   where $D = \inf\{n \geq 1 : y_n \in A\}$.
   (c) If $y \in A$ and $C \subset B$, then
   $$P(y, C) \geq \epsilon \int_C \varphi(dx).$$

Informally speaking, condition (a) states that the probability of hitting a state in $A \subset \Omega$, starting from any state in Ω, is strictly positive. Condition (b) states that the probability of hitting set A in a finite time starting from a state $y_0 \in A$ is 1, that is, the set of states A is recurrent. Finally, condition (c) states that, starting from a state in A, the probability of making a transition into any subset $C \subset B$ is lower bounded by a quantity which is proportional to the probability measure of set C.
Harris recurrence of Markov chain Y ensures that the chain has a unique stationary measure π^0, defined by

$$\pi^0(A) = \int_\Omega P(y, A)\pi^0(dy).$$

3. *Positive Harris recurrence*: the chain Y is *positive* Harris recurrent, that is, the chain is Harris recurrent and the number of transitions between successive visits to the set A has a *finite expectation*.
   Positive Harris recurrence implies that the measure π^0 on Ω is such that $\pi^0(\Omega) < +\infty$, hence, it can be normalized to a probability measure, which represents the stationary probability distribution of the mobility model on the state space.

7.1.2 Conditions on Trip Duration

Random variables S_n describing trip duration must satisfy the following properties:

1. *Independence of trip duration distribution*: the distribution of trip duration S_n depends only on the current phase I_n and path P_n, and not on the history of past states $(I_1, P_1), \ldots, (I_{n-1}, P_{n-1})$ and past durations $S_1, \ldots, S_{n-1}$. Formally,

$$P(S_n \leq s | Y_n = y, Y_{n-1}, S_{n-1}, \ldots) = P(S_n \leq s | Y_n = y),$$

   for any $s \in \mathbb{R}$, $n \in \mathbb{Z}$, and $y \in \Omega$.
2. *Positive trip duration*: each trip takes a strictly positive time. Formally,

$$P(S_n > 0 | Y_n = y) = 1,$$

   for any $n \geq 1$ and $y \in \Omega$.
3. *A technical condition*: in order to state the convergence in distribution to a time-stationary distribution for sample paths initialized at $t = 0$ according to the default initialization rule (recall Section 7.1, bullet point 5), it is required that the Markov renewal process defined by (Y_n, S_n), with $n = 1, 2, \ldots,$ is *non-arithmetic*. Since this condition is quite technical, we refer the reader to Le Boudec and Vojnovic (2006) for the formal definition. Informally, this condition is verified in case the distribution of

random variable S_n has a density conditioned on a state set $Y^0 \subset \Omega$ of strictly positive measure π^0.

7.2 Stationarity of Random Trip Models

As mentioned at the beginning of this chapter, the importance of the class of random trip models lies in the fact that a formal characterization of necessary and sufficient conditions for the existence of a stationary regime has been derived for mobility models belonging to this class, as well as a characterization of the stationary state distribution in case a stationary regime exists. These results are summarized in the following theorem, derived by Le Boudec and Vojnovic (2006).

In what follows, the *state* of the mobile node at time t is described by the Markov process

$$\Phi(t) = (Y(t), S(t), U(t)),$$

where Y is the Markov chain describing the phase and path at time t, $S(t)$ is the duration of the trip which is being undertaken at time t, and $U(t) \in [0, 1]$ is the fraction of elapsed time on the current trip at time t.

Theorem 7.1 *For a random trip model:*

1. *There exists a time-stationary distribution π for Φ if and only if the expected trip duration $E(S_0)$ is finite. Whenever π exists, it is unique and defined as*

$$\pi(B) = \frac{E(\int_0^{T_1} 1_{\Phi(s)\in B}\, ds)}{E(S_0)},$$

for any $B \in \Omega \times \mathbb{R}^+ \times [0, 1]$. In the above formula, T_1 represents the first transition time after $t = 0$.

2. *If $E(S_0)$ is finite, then starting from the stationary distribution π^0 of Markov chain Y for almost any trip at time 0, the process $\Phi(t)$ converges in distribution to π, as $t \to +\infty$.*
3. *If $E(S_0)$ is infinite, then*

$$\lim_{t\to\infty} P(\Phi(t) \in A | Y_0 = y) = 0,$$

for any $y \in \Omega$, and any set $A \in \Omega \times \mathbb{R}^+ \times [0, 1]$ such that

$$E\left(\int_0^{T_1} 1_{\Phi(s)\in A}\, ds\right) < +\infty.$$

Informally speaking, the theorem states that there exists a stationary regime for a random trip model if and only if the expected duration of a trip is finite, and characterizes the stationary distribution (conditions 1 and 2); if the expected trip duration is not finite (condition 3), then the Markov process is *null-recurrent*, that is, the mean number of transitions between successive visits to a regeneration set is infinite, and there exists no stationary regime.

7.3 Examples of Random Trip Models

In this section, we present examples of mobility models belonging to the class of random trip models. We start with an informal proof of the fact that the random waypoint mobility model is a random trip model (for the formal proof, see Le Boudec and Vojnovic (2006)). We then describe a few other mobility models which are shown in Le Boudec and Vojnovic (2006) to belong to the class of random trip models.

7.3.1 Random Waypoint Model

Let us consider the classical random waypoint mobility model on a convex region (say, on the unit disk). In order to prove that the model is a random trip model, we first have to define the domain, path, trip duration, and default initialization rule, and then prove the relative properties – recall Section 7.1.

1. *Domain*: the domain of mobility is the unit disk $\mathcal{D} \subset \mathbb{R}^2$.
2. *Phase*: there are two possible phases, named $I_1 = \{pause\}$ and $I_2 = \{move\}$.
3. *Path*: set $\mathcal{P}$ is the set of all possible line segments with endpoints x_1, $x_2 \in \mathcal{D}$. Formally, a path $P \in \mathcal{P}$ is defined as $P : [0, 1] \mapsto \mathcal{D}$, where $P(u) = x_1(1 - u) + x_2 u$, where $x_1, x_2 \in \mathcal{D}$ are the starting and ending points of the path, respectively. Special cases are those paths P_p for which $x_1 = x_2$, corresponding to pauses at a specific location. Denoting by $\mathcal{P}_p \subset \mathcal{P}$ the set of pause paths as defined above, we have that the support of random variable P_n is in $\mathcal{P}_p$ if $I_n = \{pause\}$, while the support is $\mathcal{P} - \mathcal{P}_p$ if $I_n = \{move\}$.
4. *Trip selection rule*: the trip selection rule dictates that phases alternate between *pause* and *move* states. Formally, if $I_n = \{pause\}$, then $I_{n+1} = \{move\}$ and $I_{n+2} = \{pause\}$. Trip duration S_n is drawn as follows. If $I_n = \{pause\}$, then S_n is chosen according to the *pause time distribution* $F_{pause}(s)$. For instance, F_{pause} is the Dirac delta function centered at a certain pause time $s_p > 0$ in the standard random waypoint model. If $I_n = \{move\}$, the trip selection rule dictates that a point x_{n+1} is chosen uniformly at random in $\mathcal{D}$, and that the selected path is the line segment connecting points x_n and x_{n+1}. To determine trip duration S_n, a *speed*

V_n for the nth trip is chosen according to a distribution $F_{speed}(v)$, for example, uniformly at random in an interval $[v_{min}, v_{max}]$. Trip duration S_n is then defined as:

$$S_n = \frac{dist(x_n, x_{n+1})}{V_n},$$

where $dist()$ denotes Euclidean distance.

5. *Default initialization rule*: the default initialization rule dictates that a mobile node starts in *pause* time, at a location x_0 chosen uniformly at random in $\mathcal{D}$.

We now verify that conditions on phase, path, and trip duration are fulfilled.

1. *Markov property*: the stochastic process defined by $Y_n = (I_n, P_n)$ is clearly a Markov chain: the next phase I_{n+1} depends only on the previous phase I_n, and the distribution according to which path P_n is drawn depends only on I_n.
2. *Harris recurrence*: to prove Harris recurrence of Markov chain Y_n, it is sufficient to take set $A = \{pause, move\} \times \mathcal{D}$ as the recurrent set, to choose $\varphi = Unif \otimes Unif$ (product measure) as the probability measure on A, and to set ϵ to an arbitrary positive value – see Le Boudec and Vojnovic (2006) for a formal proof.
3. *Positive Harris recurrence*: to prove positive Harris recurrence, it is sufficient to observe that the recurrent set A is visited at each transition.
4. *Independence of trip duration distribution*: the distribution of trip duration depends only on the current phase I_n, so the property is satisfied.
5. *Positive trip duration*: if in *pause* phase, trip duration is set to $s_p > 0$; otherwise, it is a positive quantity determined by the ratio of two strictly positive numbers. In both cases, trip duration is strictly positive.
6. *Technical condition*: the technical condition is also satisfied, since random variable S_n has a density.

Since the expected duration of a trip in the random waypoint model is clearly finite, we have that, according to Theorem 7.1, the random waypoint model has a uniquely defined stationary regime. Note that the stationary regime as defined within the context of random trip models includes both the stationary node spatial distribution (given by random variable $X(t)$ defined as in Section 7.1) and the node speed. Thus, the framework of random trip models can be seen as a more formal way of characterizing the node spatial distribution and average nodal speed of the random waypoint model which were described in Chapter 5.

7.3.2 RWP Variants

Several variants of the random waypoint mobility model have been proved in Le Boudec and Vojnovic (2006) to belong to the class of random trip models:

- **The Swiss flag mobility model**: in this mobility model, the domain $\mathcal{R}$ is a non-convex set corresponding to the interior of a Swiss cross – see Figure 7.1. The set of phases is the same as in the original random waypoint model. The set of paths when in *move* phase is the set of shortest paths between any two points $x_1, x_2 \in \mathcal{R}$, where the path is entirely inside $\mathcal{R}$ – see Figure 7.1. The set of pause paths is defined as in the original random waypoint model. The trip selection rule chooses a new endpoint uniformly in $\mathcal{R}$, and the next path is the shortest path from the current to the next endpoint. In case there are several such paths, one of them is selected uniformly at random. The rules for trip duration selection are the same as in the original random waypoint model.
- **The restricted random waypoint model**: the model was originally introduced in Blazevic et al. (2004), but it is defined in a more general way in Le Boudec and Vojnovic (2006). In this model, the domain $\mathcal{R}$ is the convex closure of a set of convex sub-domains $R_1, \ldots, R_k$ – see Figure 7.2. Trip endpoints are chosen in the sub-domains R_i according to the following rule. Suppose the node starts from a point $x_0 \in R_i$. The number r of trip endpoints to be chosen in R_i is selected according to a probability distribution $F_i()$, specific to the sub-domain. For the next r trips, endpoints are

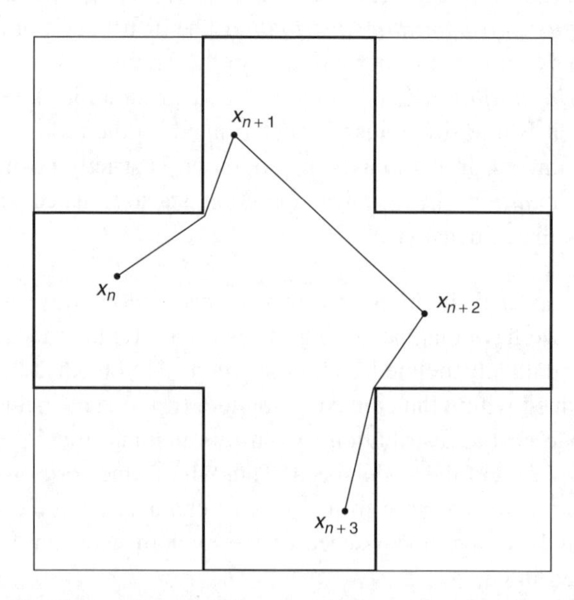

Figure 7.1 The Swiss flag mobility model. Depending on the position of endpoints, a path can be composed of either one or two linear segments.

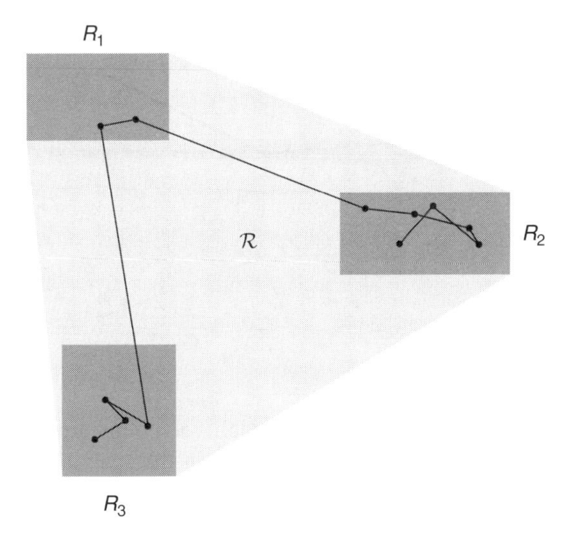

Figure 7.2 The restricted waypoint mobility model with three mobility sub-domains.

chosen uniformly at random in R_i. After the rth trip, a new sub-domain R_j, with j possibly equal to i, is chosen according to a probability distribution $F_{domain,i}()$. The destination for the $(r+1)$th trip is chosen uniformly at random in R_j, and the path from the last destination in sub-domain R_i to the first destination in R_j is a line segment. The rules for trip duration selection are the same as in the original random waypoint model.

- **The random waypoint on sphere**: this is a version of the random waypoint model in which the domain is the surface of the unit sphere in $\mathbb{R}^3$. The path between two endpoints x_1, x_2 (chosen uniformly at random on the surface of the sphere) is the shortest of the arcs on the circle of unit radius passing through x_1, x_2. If the two arcs have the same length, one of them is chosen uniformly at random. Rules for trip duration selection are as in the original random waypoint model.

7.3.3 Other Random Trip Mobility Models

These are as follows:

- **Space graph model**: in this model, originally introduced in Jardosh et al. (2003), there exists a set $R_1 \subset \mathcal{R}$ from which waypoints are chosen, corresponding to the set of possible destinations on a map. The mobility domain $\mathcal{R}$ corresponds to the union of R_1 with the edges of the planar graph connecting the points in R_1 – see Figure 7.3. The rules for trip selection are similar to that of the restricted waypoint mobility model with a single

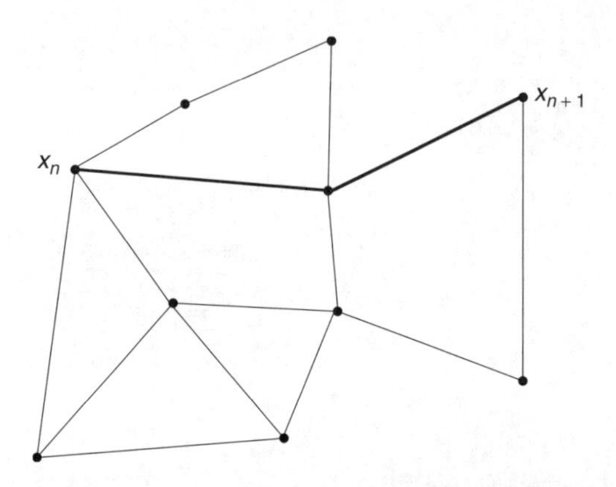

Figure 7.3 The space graph mobility model: waypoints are nodes in a spatial graph; trajectories correspond to shortest paths in the spatial graph.

mobility sub-domain. This model is very similar to the vehicular mobility model introduced in Tian et al. (2002).

- **Random walk on torus**: this is the classical time- and space-continuous random walk model on a bounded domain (say, a rectangle), where a wrap-around border rule is used. The trip selection rule dictates that a speed vector $\mathbf{V}_n$ and a trip duration S_n are chosen independently, according to some time-invariant probability distributions.

References

Blazevic L, Le Boudec JY and Giordano S 2004 A location-based routing method for mobile ad hoc networks. *IEEE Transactions on Mobile Computing* **3**, 97–110.

Jardosh A, Belding-Royer E, Almeroth J and Suri S 2003 Towards realistic mobility models for mobile ad hoc networking. *Proceedings of the ACM International Conference on Mobile Computing and Networking (MOBICOM)*, pp. 217–229.

Le Boudec JY and Vojnovic M 2006 The random trip model: Stability, stationary regime, and perfect simulation. *IEEE/ACM Transactions on Networking* **14**, 1153–1166.

Tian J, Hahner J, Becker C, Stepanov I and Rothermel K 2002 Graph-based mobility model for mobile ad hoc network simulation. *Proceedings of the ACM/IEEE Annual Simulation Symposium*, pp. 337–344.

Part Three

Mobility Models for WLAN and Mesh Networks

In Part Two of this book, we described the most representative "general-purpose" mobility models which can be used to gain an understanding of the performance trend in a mobile network without targeting specific application scenarios. In the remainder of this book, starting with this part, we will present mobility models aimed at resembling mobility patterns encountered in specific application scenarios. The focus of this part is on WLANs, a mature wireless technology which is becoming increasingly popular, and on their evolution as mesh networks currently being actively researched.

8

WLAN and Mesh Networks

In this chapter, we will briefly describe the main features of the WLAN concept and related technology, whose use is becoming widespread. We will also describe an extension of the WLAN design based on the mesh networking paradigm, which is currently being considered as a very promising solution to reduce installation costs and improve coverage in WLAN environments (and not just those). For the wireless network paradigms considered, we will first present state of the art and representative use cases, and then proceed to describe technological evolutions envisioned in the near future.

8.1 WLAN and Mesh Networks: State of the Art

A WLAN (Wireless Local Area Network) is a network used to connect two or more devices (called *stations* in WLAN terminology) in a geographic vicinity. The main purpose of the WLAN technology is to replace cables in the setup and operation of a local area network.

The spatial domain spanned by a single WLAN varies from relatively small environments (a room, or a house), to medium-size environment (a building), and up to relatively large areas covering a few square kilometers (e.g., a university campus). It is important to note that, even in their largest deployments, WLANs cover only a small fraction of a cellular network cell, whence the name *local* in the WLAN acronym.

The WLAN concept was developed in the 1980s, following experiments performed by Norman Abramson at the University of Hawaii when testing for the first time wireless communication between computers. A milestone in WLAN development occurred in 1985, when the Federal Communication Commission in the USA announced experimental industrial, scientific and medical (ISM) frequency bands for commercial applications of the spread spectrum technology which was being investigated for realizing

Mobility Models for Next Generation Wireless Networks: Ad Hoc, Vehicular and Mesh Networks, First Edition. Paolo Santi.

WLANs. Since then, several efforts have been made to design an effective and affordable WLAN technology. These efforts were canalized in the establishment of the IEEE 802.11 working group (IEEE 2011) in 1990, whose goal was (and still is) to develop a standard for WLANs. Since then, the family of IEEE 802.11 protocols, and the WiFi commercial alliance offspring promoting and certifying the interoperability of IEEE 802.11-based WLAN solutions, have led to the development and deployment of WLAN networks. Nowadays, WLANs have become extremely popular, providing Internet connectivity in the home and corporate environments, in public spaces, etc. (see next section).

8.1.1 Network Architecture

Three different types of network architectures fall under the umbrella of WLANs: *Independent Basic Service Set*, *Infrastructure Basic Service Set*, and *Extended Service Set*.

In WLAN terminology, a Basic Service Set (BSS) is a set of stations that can communicate with each other through wireless links. The difference between an Independent BSS and an Infrastructure BSS lies in the network architecture: in the former case, all the stations comprising the BSS communicate through peer-to-peer wireless links, thus realizing an ad hoc network as defined in Section 1.2 – see Figure 8.1; in the latter case, one of the stations in the BSS has the role of access point (AP), whose purpose is to coordinate communications in the BSS, and possibly to provide connectivity to external networks such as the Internet. It is important to observe that in Infrastructure BSSs communication between two stations in the BSS is possible only via the AP (see Figure 8.2). Each Infrastructure BSS is uniquely identified through an identifier called the BSSID, corresponding to the MAC address of the AP.

The Extended Service Set (ESS) is composed of a number of connected Infrastructure BSSs, where the APs of each BSS are connected by a

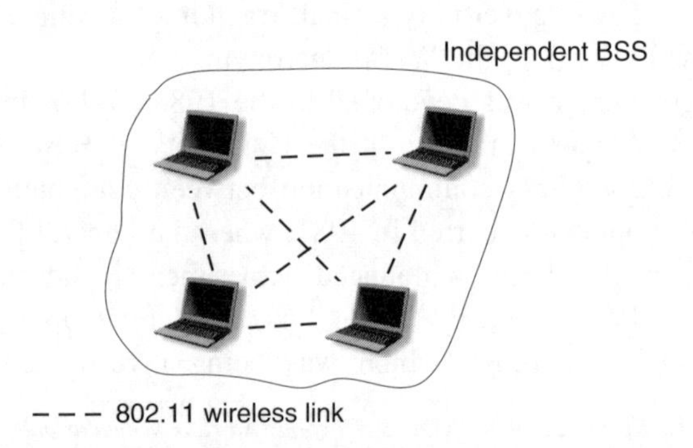

Figure 8.1 Architecture of an Independent BSS.

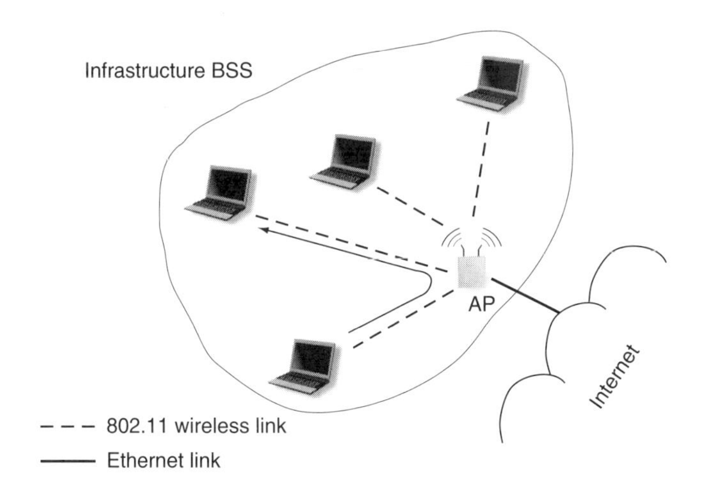

Figure 8.2 Architecture of an Infrastructure BSS.

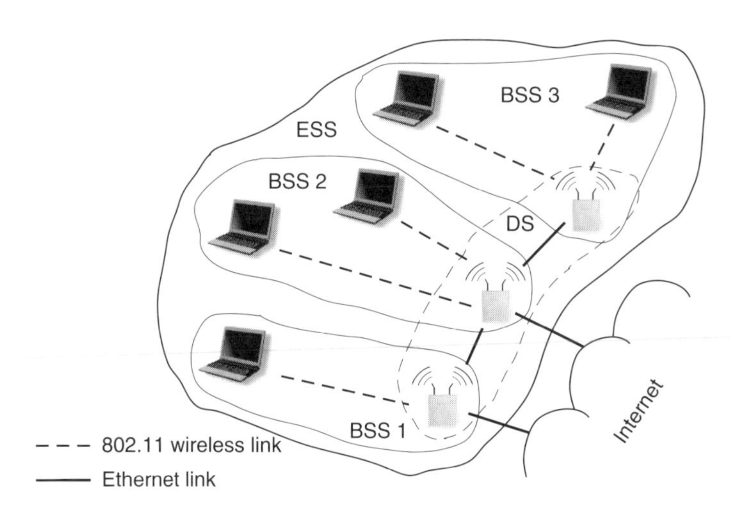

Figure 8.3 Architecture of an Extended Service Set.

Distribution System (DS), which is typically realized through wired connections such as Ethernet – see Figure 8.3. The main purpose of an ESS is to extend WLAN coverage, allowing a station to move between different BSSs without losing connectivity (roaming). Similar to Infrastructure BSSs, ESSs are associated with an ID, a 32-byte character string called the SSID.

A fourth, emerging network architecture that is being considered within the WLAN world is the mesh architecture, which is essentially obtained when the DS used to connect different Infrastructure BSSs is realized through wireless, instead of wired, links. The resulting network architecture is shown in

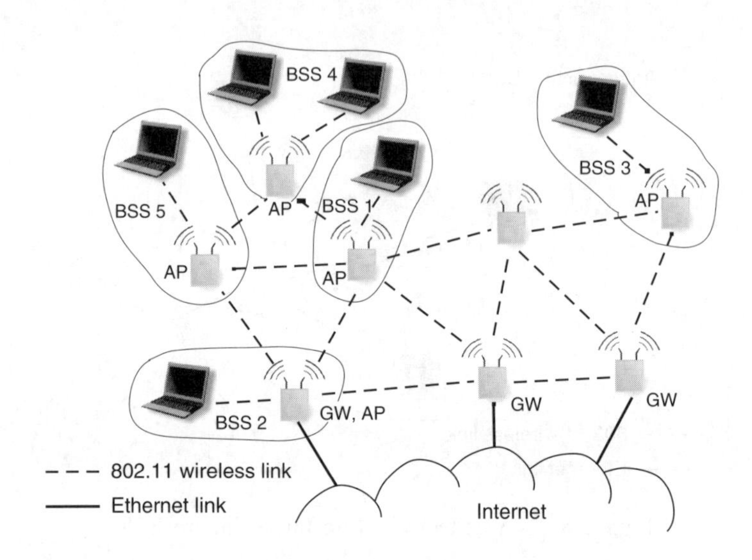

Figure 8.4 Architecture of WLAN mesh network.

Figure 8.4. A wireless mesh WLAN is formed from two types of stations: client stations and mesh stations. Mesh stations are connected to each other through wireless links in a mesh topology. Some of the mesh stations can provide additional functionalities, such as AP functionality to provide connectivity to a set of mesh clients, and the gateway functionality to provide access to external networks such as the Internet. Notice that both the AP and the gateway (GW) functionalities are optional for a mesh station (see Figure 8.4). Differently from previous architectures, WLAN mesh networks are not yet standardized; Working Group 802.11s is currently finalizing the standardization of the 802.11-based WLAN mesh network architecture and protocols (see below).

8.1.2 The IEEE 802.11 Standard

The family of IEEE 802.11 standards defines the procedures for operating WLANs, mostly covering the definition of the physical and Medium Access Control (MAC) layer in the International Organization for Standardization's Open System Interconnection (ISO/OSI) model.

The first version of the standard, now dubbed the "802.11 legacy," was released in 1997, and considered two possible ways of setting up a WLAN wireless link, namely, through infrared or radio technology. The possible data transfer rates were 1 and 2 Mbps. Infrared transmission was dropped in the succeeding versions of the standard, since most of the manufacturers preferred radio technology for implementing the first version of 802.11.

A major milestone in the evolution of the family of IEEE 802.11 standards was the release of IEEE 802.11b and 802.11a in 1999, which were major evolutions of the first version released in 1997.

IEEE 802.11b operates in the ISM frequency bands centered around 2.4 GHz, which can be used by other appliances such as microwave ovens, cordless phones, Bluetooth devices, and so on. Thus, reduced transmission quality might be experienced if an 802.11b link is operated in an environment where other appliances operating in the ISM bands exist. The 802.11b physical layer defines four possible raw data rates ranging from 1 to 11 Mbps, and uses DSSS (Directed-Sequence Spread Spectrum) modulation; 11 channels 22 MHz wide can be used for transmission. Since the center frequencies of channels are only 5 MHz apart, 802.11b channels partially overlap. As can be seen from Figure 8.5, interference-free transmissions in an 802.11b network occur only if concurrent transmissions use channels at least 25 MHz apart from each other.

The IEEE 802.11a standard operates in the 5 GHz frequency band, and uses orthogonal frequency-division multiplexing (OFDM) modulation. Eight different raw data rates ranging from 6 to 54 Mbps can be used for transmission. The standard defines 12 non-overlapping channels 20 MHz wide. The price to pay for the higher data rate is a lower transmission range, which is much shorter than that typically achieved by 802.11b devices. This drawback of IEEE 802.11a has limited its popularity, especially since the release of the 802.11g standard which increased 802.11b data rates up to 54 Mbps.

IEEE 802.11g was released in 2003 and is essentially an evolution of 802.11b aimed at increasing data rates. With the adoption of the more effective OFDM modulation, the achieved data rates are comparable to those of 802.11a, up to 54 Mbps.

In 2009, an amendment to the standard, named IEEE 802.11n, was released, which introduced use of multi-antenna systems. IEEE 802.11n operates on both the 2.4 GHz and the less crowded 5 GHz bands. When operated with four antennas at both ends of a link, and using two contiguous 20 MHz channels as a single 40 MHz channel, IEEE 802.11n achieves raw data rates as high as 600 Mbps.

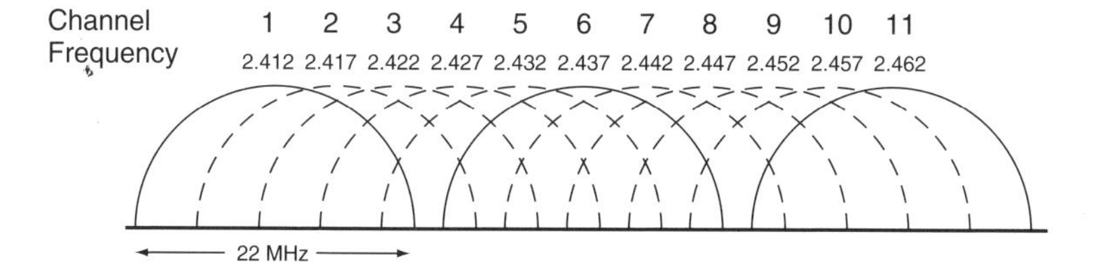

Figure 8.5 IEEE 802.11b/g channel allocation.

Table 8.1 Main features of the various IEEE 802.11 protocols

802.11 protocol	Freq. (GHz)	Bandwidth (MHz)	Data rate (Mbps)	Modulation	Range indoor–outdoor (m)
a	5	20	6, 9, 12, 18 24, 36, 48, 54	OFDM	30–100
b	2.4	22	1, 2, 5.5, 11	DSSS	45–150
g	2.4	22	6, 9, 12, 18 24, 36, 48, 54	OFDM, DSSS	45–150
n	2.4, 5	20, 40	up to 600 Mbps	OFDM	70–250

The differences between the various versions of the standard lie essentially in the definition of the physical layer, while the MAC layer procedures are the same. In particular, 802.11 MAC is based on CSMA/CA (Carrier Sense Multiple Access – Collision Avoidance), a fully distributed medium access technique designed to avoid collisions in accessing the wireless channel.

A summary of the main features of the various versions of the 802.11 standard is given in Table 8.1. A comment on the data rates reported in the table is in order: those reported in Table 8.1 are the *nominal* data rates potentially achieved by the standard as measured at the physical layer. Indeed, application-layer data rates are much lower, due to several factors such as MAC layer overhead, sub-optimal transmission quality, contention in accessing the wireless channel, and so on. Typically, the actual data rates measured at the application layer can be considered to be approximately half the nominal values reported in Table 8.1.

In addition to the ones mentioned above, other amendments and extensions to the standard have been released or are in advanced status of definition. A list of the most significant amendments and extensions to the standard is given in Table 8.2. It is worth mentioning 802.11s, currently in draft status

Table 8.2 List of the most significant amendments to the IEEE 802.11 standard

Amendment	Description
e	Enhancements for QoS support
h	Redefine spectrum management at 5 GHz for Europe
i	Enhanced security
j	Extensions for Japan
p	Redefine physical layer for vehicular environments
s	(Draft) extension for mesh networking

and expected to be released by the end of 2011, defining protocols and procedures for WLAN mesh networking.

8.2 WLAN and Mesh Networks: User Scenarios

In this section, we describe representative user scenarios of WLANs and their forthcoming extension into WLAN mesh networks.

8.2.1 Home WLAN

A user scenario which is becoming increasingly popular is that of a WLAN used to connect appliances in the home environment and giving access to the Internet. De facto, WiFi has replaced dial-up modems as the prominent technology for Internet access in the home. The benefits of WiFi over its wired counterpart are evident: a single wireless router with Internet access – a very cheap device, often provided for free by the ISP as part of the subscription – can be used to provide Internet connectivity in the entire home. So, a user with a laptop can freely move within the house or in the immediate surroundings and still have access to the Internet. More importantly, Internet connection can easily be shared among a few users (e.g., members of a family) with negligible performance degradation. In fact, given that asymmetric digital subscriber line (ADSL) connection to the ISP has a typical (maximum) data rate of 20 Mbps, and considering that IEEE 802.11a/g provide a data rate of 54 Mbps (IEEE 802.11n indeed provides much higher data rates), the home WLAN is not going to be the bottleneck in the Internet connection as long as the number of users sharing the connection is in the order of 3–4.

The home WLAN can be used not only for Internet connectivity, but also to connect other appliances in the home environment. A first example is printers, scanners, and other computer peripherals, which are nowadays increasingly equipped with a WiFi interface for wireless operation. A second example is smart phones, also increasingly equipped with a WiFi interface. Thus, the home WLAN can be used to transfer multimedia content (audio files, pictures, etc.) between laptops, PCs, and smart phones. Finally, other home appliances such as TVs equipped with WiFi interfaces are starting to appear on the market, turning the vision of a home WLAN where a multitude of devices interact into reality.

8.2.2 Campus/Corporate WLAN

Another typical user scenario is that of a WLAN deployed on a university campus or in a corporate environment. From an architectural point of view, these WLANs are much larger and more complex to manage than the home WLAN described above, which is typically composed of a single AP. In campus/corporate environments, several *hundreds* of APs can be used to provide seamless Internet and intranet connectivity over the entire campus/corporate area.

A typical user of this class of WLANs (e.g., a university student) uses the WLAN as the wireless counterpart of the campus wired network to access the Internet, connect with other computers and peripherals on campus, access intranet services, and so on. Since the security level provided by the wireless connection is perceived to be (and typically is) much lower than that provided by a wired connection, the range of services accessible through the WiFi connection might be a subset of the full range of campus/corporate services.

The main purpose of a campus/corporate WLAN is to allow an authorized user to access campus/corporate services in a seamless manner while freely moving in the coverage area. Seamless connectivity is typically realized both horizontally within the WLAN (e.g., implementing soft handoff procedures between adjacent APs to support user mobility) and vertically between the different types of networks available within the campus/corporate environment – for example, allowing the user to switch between wireless and wired connection without losing connectivity.

Clearly, the setup and management of a campus/corporate WLAN is much more complex than that of a home WLAN. For instance, radio resource allocation strategies should be defined to optimally allocate 802.11 channels to the APs so as to minimize interference between adjacent APs. However, nowadays these issues can be successfully tackled, and the WLAN in recent years has become the prominent wireless technology in campus/corporate environments.

8.2.3 Public Area Hotspots

The term *hotspot* refers to an AP (or a few APs) that is used to provide Internet connectivity in a public area such as an airport, a train station, public park, shopping center, and so on. Typically, restricted Internet access is available for free, while access to the full Internet is subject to payment of a flat or hourly rate to the ISP running the hotspot.

The most typical example (at least for the author of this book, a frequent traveler) is that of someone waiting for a departing flight at an airport. If the reader has ever tried to turn on the WiFi card of her/his laptop in an airport, she/he will know very well that one or more hotspots are typically available. In most airports, a public hotspot provides free access to the airport webpage for information on departing flights, shopping facilities, restaurants, ground transportation, and so on. Furthermore, one or more ISPs typically provide paying, full Internet connectivity. In very few airports (such as the tiny and tidy Luleå airport in Sweden), full Internet access is given for free through the public hotspot!

8.2.4 Community Mesh Network

Different from the ones described so far, the user scenario described in this subsection is not yet mature and still relatively unpopular. As mesh networking technology matures, the one described here will very likely become as popular as the scenarios described above.

A community mesh network is a WLAN mesh network used to connect members of a community, such as people living in the same neighborhood, citizens of a municipality, etc. In principle, the community mesh network can be used to share Internet connectivity between community members; however, its main purpose is to give access to *community services* such as local news, content sharing, messaging services, local VoIP, etc.

A prominent example of community mesh network is Seattle Wireless (Team 2011b), started in 2000 with the purpose of providing free WiFi access and sharing the costs of Internet connectivity in the city of Seattle, USA. Currently (April 2011), the network is composed of about 80 free APs spread all over Seattle, and is growing continuously.

Another important example of community mesh networks is WLAN mesh networks used to extend Internet coverage to rural villages in developing countries such as India. For instance, the RuralNet research project (Team 2011a) demonstrated the feasibility of using IEEE 802.11 technology to realize a wireless mesh network connecting rural villages on the Gangetic Plains. An important outcome of the project was the successful testing of IEEE 802.11 technology for setting up reliable links over very long distances (several kilometers), which was achieved through the use of high-gain directional antennas.

8.3 WLAN and Mesh Networks: Perspectives

What is the future for WLANs and mesh networks? What challenges are these related technologies facing, and how can they be addressed?

The evolutionary trends of WLANs and mesh networks in the near future can be outlined quite clearly. Two major trends can be identified: (i) increasing data rates and communication range; and (ii) transition to a meshed network architecture.

For (i), increasing data rates are the prominent evolution of any communication technology, and it blends naturally with similar increasing trends in storage capacity and computational power of devices: as more data can be processed and stored on devices, communication technologies are expected to increase data transmission rates accordingly, so as to avoid bottlenecks in a network.

Recent inclusion of multi-antenna technology in the IEEE 802.11 standard is the right answer to the need of increased data rate, because, in principle, the data rate on a link can be made to increase almost linearly with the number of antennas on the link. Another major advantage of multi-antenna systems is that they can also be used to extend the communication range, even though this typically requires partially reducing the benefits in terms of improved data rates.

For (ii), we expect that once the mesh architectural extension of the WLAN concept has been standardized (expected by the end of 2011), meshed WLAN will gradually replace traditional WLAN architectures. In fact, a meshed architecture displays several advantages over traditional WLAN architecture, such as reduced installation costs, increased coverage, and better reliability.

As for the challenges, we believe that the major ones are related to the proliferation and widespread use of WLAN technology itself, and the consequent coexistence issues. So far, WLAN technology has been mostly conceived "in isolation," implicitly assuming that a WLAN is operating in a relatively free and isolated wireless environment. For instance, when procedures for addressing congestion on the wireless channel are defined, the implicit assumption is that congestion is caused by the stations comprising the WLAN itself – we call this *internal congestion*. However, with an increased pervasiveness of WiFi deployments in the environment, it is expected that *external congestion*, that is, congestion due to devices operating in the same frequency bands which are not part of the WLAN, will become predominant. External congestion is much more difficult to address than internal congestion, since it cannot be directly controlled and managed by the APs forming a single WLAN. Sophisticated WLAN coexistence procedures will be identified, defining a sort of "coexistence etiquette" that all WLAN operators will follow in order to ensure peaceful coexistence of a number of heterogeneous WLANs in the vicinity.

Another major challenge in WLAN mesh networks is related to the design of algorithms and protocols for maximizing the data transport capacity in the (wireless) distribution system, which constitutes the wireless backbone of the network. In fact, this part of the network is in charge of transporting data from the gateways to the mesh clients and vice versa, and thus should be able to transport a very large amount of data if the size of the network is even moderate. Since the mesh stations forming the wireless network backbone share the same wireless medium, efficient radio resource allocation algorithms should be defined to effectively use the available spectrum, which will likely be able to manage multiple radios per station operating on several channels.

8.4 Further Reading

Only a very short description of WLAN and mesh networking technology and related issues has been presented in this chapter. The reader interested in gaining a better understanding of these topics is referred to the several books and survey papers available, such as Gast (2005) and Perahia and Stacey (2008) for 802.11 WLANs, and Akyildiz et al. (2002) and (Zhang et al. 2007) for mesh networking.

References

Akyildiz I, Su W, Sankarasubramanian Y and Cayirci E 2002 A survey on sensor networks. *IEEE Communications Magazine* **40**(8), 104–112.

Gast M 2005 *802.11 Wireless Networks: The Definitive Guide*. O'Reilly Media, Sebastopol, CA.

IEEE 2011 *http://www.ieee802.org/11/*.

Perahia E and Stacey R 2008 *Next Generation Wireless LANs: Throughput, Robustness, and Reliability in 802.11n*. Cambridge University Press, Cambridge.

Team R 2011a *http://www.cse.iitb.ac.in/br/iitk-webpage/dgp.html*.

Team S 2011b *http://seattlewireless.net*.

Zhang Y, Luo J and Hu H eds. 2007 *Wireless Mesh Networking: Architectures, Protocols and Standards*. Auerbach Publications, Boca Raton, US.

9

Real-World WLAN Mobility

Given its importance within the realm of short-range wireless technologies, WLAN performance has been carefully investigated in the literature. In particular, since it is relatively easy to trace AP activity in terms of number of registered users, traffic load, etc., collection of real-world data traces and their analysis have become a common methodology in WLAN performance evaluation.

In this chapter, after briefly describing the main features of the WLAN traces available in the literature (most typically through the CRAWDAD website), and the typical methodologies followed by researchers to collect, post-process, and analyze these traces, we will present the main features of a WLAN that concern the mobility of users. These features have been used in the literature to guide the design of trace-based mobility models, whose purpose is to faithfully mimic user mobility behavior observed in WLAN traces.

9.1 Real-World WLAN Traces

As commented above, tracing user behavior in a WLAN environment is a relatively simple task: since APs typically have a direct connection to a wired network (e.g., the campus or corporate network), and users need to be registered with an AP in order to use WLAN services, instructing APs to periodically report information such as number of registered users, their IDs, traffic load, etc., is relatively simple. As we will see, much more cumbersome is the task of post-processing and analyzing the possibly massive amount of logging data generated.

Three types of WLAN environments have prominently been traced in the research literature:

1. Public WLANs deployed at conferences to give Internet access to attendees. An example of such trace is the one analyzed in Balachandran

Mobility Models for Next Generation Wireless Networks: Ad Hoc, Vehicular and Mesh Networks, First Edition. Paolo Santi.

et al. (2002), which was collected during the ACM SIGCOMM Conference of 2001. Conference WLAN traces are characterized by a relatively small size in terms of both number of APs composing the network and number of registered users, and by a very short time duration – typically limited to a few days. Furthermore, the type of mobility and user behavior displayed in a conference environment is peculiar to that specific environment, hence conclusions on user mobility patterns cannot be easily generalized to larger WLAN deployments such as campus/corporate WLANs.
2. Private WLANs deployed in corporate environments to give access to corporate networks and services. An example of such a trace is the one analyzed in Balazinska and Castro (2003), which was collected from a WLAN covering three different corporate buildings. Corporate WLANs typically have a relatively large size, comparable to that of campus WLANs for the number of APs comprising the network and the number of registered users. As for the time duration of the traces, this can vary from a few weeks to several months.
3. Private WLANs deployed in campus environments to give access to campus networks and services. As the reader might guess, this is the most common type of real-world WLAN trace available in the literature. Examples of such traces are the ones analyzed in Kim et al. (2006), Tang and Baker (2000), and Tuduce and Gross (2005), which were collected at ETH Zurich, at Dartmouth College, and at the Computer Science Building of Stanford University, respectively. These traces are characterized by a very large network size in both number of APs and registered users, and by a long time duration up to several months.

WLAN traces are collected by instructing the APs to report information to a central server connected through the wired network interface. Typically, SNMP, TCPDump, or syslog traces, or combinations of them, are collected by the central server. These traces contain, among other things, information about the MAC address of registered users, thus allowing univocal identification of a user – indeed, a device – within the network. Other collected information is the time when a user associated and de-associated (in case explicit de-association procedures are part of the WLAN setting at hand) with an AP.

The features of the real-world WLAN traces mentioned above are summarized in Table 9.1. In the table, collected traces vary from very small WLANs with relatively few users to very large WLANs with a few thousand users. Also, the timespan of the collected traces varies from three days to one year.

A significant amount of post-processing is needed to make the traces suitable for further analysis. Typically, post-processing is needed to identify all users registered in the traces, and to determine user activity patterns for each of them. For instance, user activity and inactivity periods are identified,

Table 9.1 Main features of some real-world WLAN traces

Reference	Type	APs	Users
Balachandran et al. (2002)	Conference	4	195
Balazinska and Castro (2003)	Corporate	177	1366
Tang and Baker (2000)	Campus	12	74
Kim et al. (2006)	Campus	560	198 (only VoIP users)
Tuduce and Gross (2005)	Campus	166	4762

Reference	Trace type	Duration
Balachandran et al. (2002)	SNMP, TCPDump	3 days
Balazinska and Castro (2003)	SNMP	4 weeks
Tang and Baker (2000)	SNMP, TCPDump	12 weeks
Kim et al. (2006)	SNMP, TCPDump, syslog	12 months
Tuduce and Gross (2005)	SNMP, syslog	18 weeks

where a user is assumed to be active if the user is registered with an AP and sending/receiving traffic in a suitably defined time period. General statistics about user activities are obtained, possibly categorizing users based on the type of application-layer traffic sent/received. This allows the possible inclusion/exclusion of a fraction of the users from the data trace based on predefined criteria. For instance, Kim et al. (2006) focused their analysis only on data referring to VoIP users. Similarly, users displaying only a minimal amount of activity (i.e., users active for at least a fixed percentage of the overall trace duration) can be retained in the analysis, and so on. Finally, periods of time during which very low user activity is registered (e.g., nighttime) might be filtered out from the traces in order to focus the analysis on periods of intense network usage.

Another important step in the trace post-processing phase is clearly identifying the transitions of a specific user between different APs. In fact, due to many factors such as AP congestion level, varying radio signal quality, etc., a user's association with an AP can repeatedly change even if the user is not moving. These kinds of repeated association and de-association with nearby APs are called *ping-pong transitions* and should be filtered out from the trace since they are not related to the mobility of users. Techniques for detecting and removing ping-pong transitions from WLAN traces will be described in the next chapter.

Finally, an important post-processing step which is necessary prior to making collected traces publicly available is anonymization, which essentially consists of decoupling user identities from the MAC addresses of WLAN devices. This step is fundamental to preserve user privacy, which must be maintained even if the users being traced are informed about the tracing experiment.

9.2 Features of WLAN Mobility

Due to the network architecture, and to the features of the collected traces, mobility in WLAN environments is typically characterized in terms of user association patterns with the different APs in the network. Thus, mobility is not *explicitly* characterized in terms of location and trajectory of a mobile user (a notable exception to this is the mobility characterization of VoIP users performed by Kim et al. (2006)), rather it is *implicitly* characterized through the analysis of migration patterns between APs in the network. The implicit assumption is that, if a user changes her/his association from an AP A to another AP B, that user has changed her/his position. It is important to observe that this assumption is not necessarily true: as we commented at the end of the previous section, due to ping-pong transitions a change in AP association is not necessarily caused by a user's physical movement. However, after ping-pong transitions have been filtered out, the assumption "change in AP association = user physical movement" is reasonably accurate in practice.

Another important observation about WLAN user mobility is that it is often *discontinuous* in the time domain. In other words, a user can be associated with an AP A at time t_1, then de-associate from A, and later on (at time $t_2 \gg t_1$) "reappear" in the network associated with a different AP, say B. This mobility pattern is very common, for example, in a campus network, where students move from, say, dormitory to classrooms, and turn off their laptops while moving from one location to the other.

To avoid possible problems or difficulties in the trace analysis phase due to time-discontinuous mobility, an artificial AP modeling the OFF state – corresponding to a user's inactivity status – is introduced in the model. The time-discontinuous mobility pattern described above can then be turned into a time-continuous mobility pattern by adding a transition between AP A and the OFF state at time t_1, and a transition from the OFF state to AP B at time t_2.

The most relevant WLAN mobility features analyzed in the literature are the following:

1. *Prevalence*, defined as the overall fraction of (active) time a user spends with an AP, for each AP in the network. The purpose of the prevalence metric is to measure a user's attitude to register with the various APs: if a user visits an AP frequently or for a relatively long time, the prevalence of this AP in the user's trace will be high. When computed across all users,

the prevalence metric gives rise to the prevalence matrix $P(i, j)$, with element p_{ij} corresponding to the prevalence metric of user i on AP j.

2. *Prevalence distribution*, defined as the histogram of the prevalence values of the various users. The prevalence distribution can be easily computed starting from the prevalence matrix, and allows characterization of the aggregate attitude of users on registering with relatively few (or many) APs.
3. *Persistence*, also called *session duration*, defined as the amount of time a user stays associated with an AP before moving to another AP, or transitioning to the OFF state. In contrast to prevalence, which accounts for the *total* time a user spends with an AP, persistence measures the length of a single session with a specific AP.
4. *Number of visited APs*, defined as the total number of APs that a user visits during the entire trace.

The three metrics above, if suitably combined, allow a thorough characterization of user mobility patterns, for instance, categorizing users as highly mobile (users with low prevalence values, short session duration, and a large number of visited APs), moderately mobile, and almost stationary (users with a few, relatively high prevalence values, long session duration, and small number of visited APs).

User prevalence and prevalence distribution have been shown to have relatively stable patterns in both campus and corporate WLANs. Figure 9.1 shows the user prevalence distributions obtained from the traces in Balazinska and Castro (2003) and Tuduce and Gross (2005), which are indeed very similar. The shape of the distribution clearly indicates a bimodal distribution,

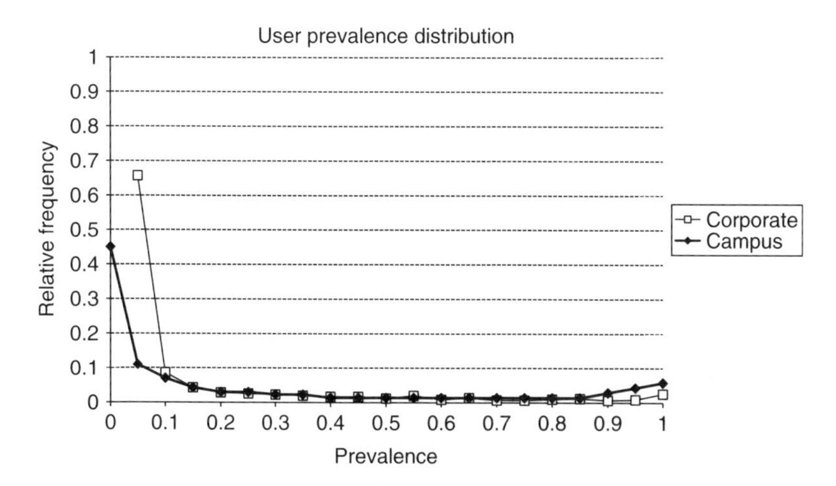

Figure 9.1 User prevalence distribution of traces collected by Balazinska and Castro (2003) (corporate) and Tuduce and Gross (2005) (campus).

with a higher concentration around very small prevalence values, and a lesser, yet noticeable, concentration around very large prevalence values. This shape of the distribution seems to indicate that users spend most of the time in one or a few APs (giving rise to the low peak of the distribution around prevalence values greater than or equal to 0.9), while they visit other APs only for a very short time (giving rise to the high peak of the distribution around prevalence values less than or equal to 0.2).

The session duration distributions of four different mobility traces in both corporate and campus WLANs are shown in Figure 9.2. More specifically, the figure reports two plots derived from the corporate trace collected by Balazinska and Castro (2003), and two portions of the trace collected on the ETH campus reported in Tuduce and Gross (2005). The two corporate traces distinguish between user behavior when a user is located in her/his home building – defined as the building in which the user spends most of the time – or is a guest in other buildings.

The plots clearly show that, independently of the environment and the specific trace, the session duration distribution obeys a power law (corresponding to a line in log–log scale). The only difference between the four plots is in the inclination of the line, corresponding to the exponent of the power law. The analysis of the session duration distribution indicates that the vast majority of sessions are very short (20 minutes or less), but there exists a relatively fat tail of long, or even very long, sessions (up to a few hours). It is also interesting to compare the session distribution in the corporate home and guest configuration. In the former case, sessions tend to be longer than in the guest configuration, hinting that users tend to be relatively

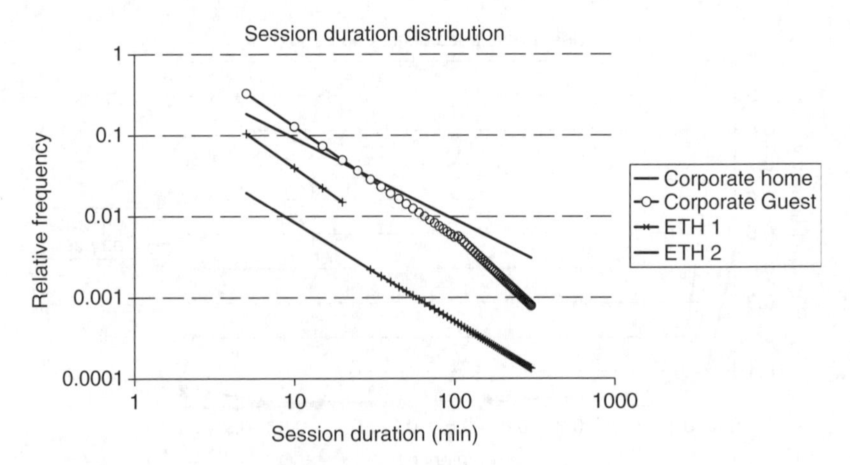

Figure 9.2 Session duration distribution of traces collected by Balazinska and Castro (2003) (corporate) and Tuduce and Gross (2005) (ETH). The x- and y-axes are in logarithmic scale.

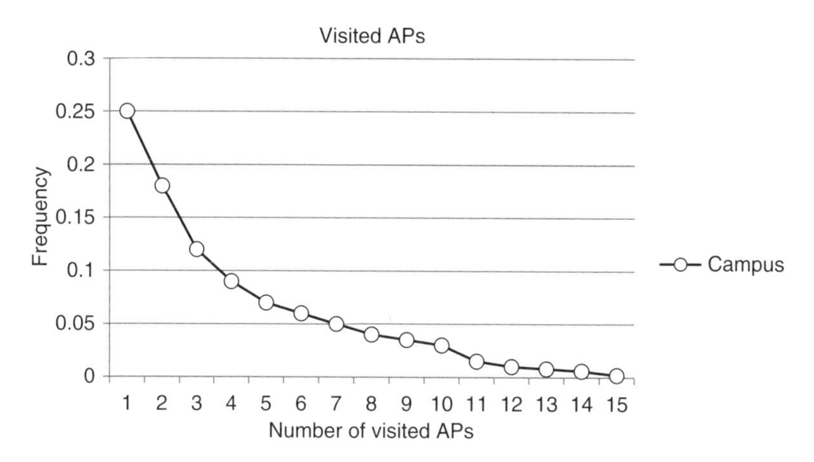

Figure 9.3 Distribution of the number of visited APs in a portion of the trace collected by Tuduce and Gross (2005) (campus).

stationary when they are in the home location and relatively more mobile when roaming outside.

The distribution of the number of visited APs in a portion of the ETH trace collected by Tuduce and Gross (2005) is shown in Figure 9.3. Careful observation of the distribution reveals that the majority of users visit only up to three APs, but only a few users visit more than ten APs. Similar conclusions can be drawn from an analysis of the corporate trace reported in Balazinska and Castro (2003). Thus, we can conclude that the degree of mobility in a WLAN is very limited, considering that traces are collected over a relatively long period of time. The reader, however, has to bear in mind that WLAN traces do not allow continuous tracking of a user's position over a period of time, but just a user AP association pattern. Hence, the above conclusion on user mobility does not imply that WLAN users have a limited degree of physical mobility, rather that *when they access the WLAN*, they tend to have a relatively repetitive association pattern exploiting only a minimal part of the APs available in the network.

References

Balachandran A, Voelker G, Bahl P and Rangan PV 2002 Characterizing user behavior and network performance in a public wireless LAN. *Proceedings of ACM Sigmetrics*, pp. 195–205.

Balazinska M and Castro P 2003 Characterizing mobility and network usage in a corporate wireless local-area network. *Proceedings of ACM MobiSys*.

Kim M, Kotz D and Kim S 2006 Extracting a mobility model from real user traces. *Proceedings of IEEE Infocom*.

Tang D and Baker M 2000 Analysis of a local-area wireless network. *Proceedings of ACM Mobicom*, pp. 1–10.

Tuduce C and Gross T 2005 A mobility model based on WLAN traces and its validation. *Proceedings of IEEE Infocom*, pp. 664–674.

10

WLAN Mobility Models

Mobility models for WLAN environments can be broadly classified into two categories. The first category comprises models whose goal is to closely resemble user/AP registration patterns observed in WLAN traces. Models in this category, such as those presented in Jain et al. (2005), Lee and Hou (2006), Lelescu et al. (2006), Chinchilla et al. (2004), and Tuduce and Gross (2005), can be used, for instance, to predict the next user AP association given the current association. Physical user mobility in these models is then not explicitly modeled, although it can be deduced from a user's AP association pattern. The second category of models is instead aimed at modeling user physical mobility. In this case, the history of a user's AP associations is used to generate a realistic trace of the user's physical mobility in the environment. By aggregating traces of several users, a statistical model of user physical mobility in the WLAN environment can be derived and used to faithfully reproduce the physical movement of users. Examples of models in this category are given in Hsu et al. (2004) and Kim et al. (2006).

In this chapter, we will present two representative mobility models for each of these two categories of WLAN models. More specifically, we will present:

1. The model introduced in Lee and Hou (2006), which we name LH, whose purpose is to generate synthetic user/AP association patterns resembling features observed in real-world traces, in particular concerning temporal correlation of user/AP association patterns;
2. The model introduced in Kim et al. (2006), which we name KKK, whose purpose instead is to extract and faithfully reproduce traces of *physical* user mobility based on observed user/AP association patterns.

Before proceeding further, a comment is in order about the motivating reasons for designing WLAN mobility models. In contrast to other application

Mobility Models for Next Generation Wireless Networks: Ad Hoc, Vehicular and Mesh Networks, First Edition. Paolo Santi.

scenarios, a relatively large number of real-world WLAN traces nowadays are publicly available and can be used to test the performance of WLAN algorithms/protocols. Despite this, synthetic models able to generate realistic WLAN traces are very useful, since they allow changes in parameters and network configuration (e.g., growing the number of users beyond the maximum number registered in a WLAN trace, or increasing the number of APs in the network), which are only partially possible if an existing real-world trace is used instead.

10.1 The LH Mobility Model

The LH model was introduced by Lee and Hou (2006) to faithfully reproduce WLAN traces. Different from other existing models, the LH model attempts to predict not only the most probable AP association given the current user association, but also the *time* at which the transition will take place. Bringing the temporal dimension into the model was a major advance with respect to the state of the art.

As in other approaches, the LH model uses Markov chains to model transitions between the APs in the network. However, the model introduces a temporal dimension in the Markov chain by including a parameter that models the time at which transitions between APs take place.

More formally, given a network composed of m APs, which we assume to be numbered from 1 to m, the behavior of a user in the network is modeled by the following semi-Markov model:

$$(X_n, T_n),$$

where $X_n \in \{1, \ldots, m\} \bigcup \{0\}$ is the nth user association, called *state* in the following, and T_n is a continuous random variable taking values in $[0, +\infty)$ representing the time at which the transition between state X_{n-1} and state X_n takes place ($T_0 = 0$). Note that the set of possible states is augmented with state 0, modeling a user who is currently disconnected from the WLAN (recall the discussion in Section 9.2).

The model described is a semi-Markov chain, since the distribution of random variable T_n is *not* exponential; instead, Lee and Hou (2006) make no specific assumption about the distribution of T_n, which is empirically derived from the analysis of an existing WLAN trace. Their choice of not assuming an exponential distribution for variable T_n is motivated by the observation, first made by Balazinska and Castro (2003) and reported in the previous chapter (see Figure 9.2), that the session duration distribution is a power law. Lee and Hou, however, make the assumption that the semi-Markov process is *time homogeneous*: that is, the distribution of random variable T_n does not change during the period in which the mobility model is built.

In order to fully characterize the previously defined semi-Markov chain, the following transition probabilities need to be evaluated:

$$Q_{ij}(t) = P(X_{n+1} = j, T_{n+1} - T_n \leq t | X_n = i) = p_{ij} H_{ij}(t),$$

where $P(X_{n+1} = j, T_{n+1} - T_n \leq t | X_n = i)$ is the probability of making a transition to state j within time t starting from state i, $p_{ij} = \lim_{t\to\infty} Q_{ij} = P(X_{n+1} = j | X_n = i)$ is the state transition probability between states i and j (independent of transition time), and $H_{ij}(t) = P(T_{n+1} - T_n \leq t | X_{n+1} = j, X_n = i)$ is the sojourn time in state i when the next state is j.

Note that the semi-Markov chain model allows great flexibility in modeling sojourn time at APs, since different sojourn time distributions can be used depending on the specific transition the user will perform. In other words, the sojourn time distribution when the user is associated with AP i and the next state is known to be AP j is in general different from that when the user is known to make a transition to another AP $z \neq j$ starting from i.

Let $D_i(t)$ denote the probability distribution of the sojourn time in state i, independently of the next state. It can be seen that

$$D_i(t) = P(T_{n+1} - T_n \leq t | X_n = i) = \sum_{j=0}^{m} Q_{ij}(t).$$

Given $D_i(t)$, the homogeneous semi-Markov chain modeling WLAN user behavior can be formally defined as $Z = (Z_t, t \in \mathbb{R}_0^+)$, with transition distributions defined as

$$\phi_{ij}(t) = P(Z_t = j | Z_0 = i) = (1 - D_i(t))\delta_{ij} + \sum_{l=0}^{m} \int_0^t \dot{Q}_{il}(\tau)\phi_{lj}(t-\tau)\, d\tau.$$

In this formula, $\phi_{ij}(t)$ must be interpreted as the probability of finding a user in state j after time t, starting from state i, and δ_{ij} is the Kronecker delta function defined as

$$\delta_{ij} = \begin{cases} 1 & \text{if } i = j \\ 0 & \text{otherwise} \end{cases}.$$

10.1.1 *Estimating the Transition and Steady-State Probabilities*

In order to define the semi-Markov chain modeling user behavior, the following quantities need to be derived:

1. The probabilities p_{ij} that a user in state i makes a transition into state j, independently of the transition time, for each possible pair (i, j).

2. The aggregate distribution $D_i(t)$ of the sojourn time in state i, independently of the next state, for each possible state i.

The above values have been estimated in Lee and Hou (2006) by analyzing wireless network traces collected at Dartmouth College between November 1, 2003, and June 30, 2004, and available through the CRAWDAD website (Team 2011). The traces comprise data collected from 586 APs all over the 161 buildings of the Dartmouth College campus, and refer to 6202 tracked users in total.

Based on the estimated transition probabilities and sojourn time distributions, it is possible to estimate the steady-state user distribution, that is, the probability π_i of finding a user in state i at a random (and very large) instant of time t, as follows:

$$\pi_i = \frac{\bar{d}_i \tilde{\pi}_i}{\sum_{j=0}^{m} \bar{d}_j \tilde{\pi}_j},$$

where $\bar{d}_x$ is the average sojourn time in state x, and $\tilde{\pi}_x$ is the steady-state probability of making a transition to state x, which can be found by solving the following set of equations:

$$\tilde{\pi} = \tilde{\pi} P, \quad \sum_{j=0}^{m} \tilde{\pi}_j = 1,$$

where $\tilde{\pi} = (\tilde{\pi}_0, \ldots, \tilde{\pi}_m)$ is the vector of steady-state transition probabilities, and $P = [p_{ij}]$ is the matrix of transition probabilities p_{ij}.

As we will see next, characterization of the steady-state distribution of users across APs allows evaluation of the temporal correlation between user and AP association patterns.

10.1.2 Finding Temporal Correlation in User/AP Association Patterns

The steady-state user distribution across APs can be used to investigate whether, and to what extent, mobility patterns are correlated in time. The basic idea is to compute the steady-state distribution at different time intervals (e.g., a day, a week, or a month), and to use a similarity metric to estimate the degree of similarity between the distributions computed at the different time intervals. Since the steady-state distribution of users across APs is represented by an $(m + 1)$-dimensional vector $\tilde{\pi} = (\tilde{\pi}_0, \ldots, \tilde{\pi}_m)$, Lee and Hou (2006) suggest using a vector similarity metric, the *cosine distance*, to express the similarity between distributions.

The cosine distance metric is defined as follows. Let **A** and **B** be two n-dimensional vectors. The cosine distance between **A** and **B**, denoted

$sim(\mathbf{A}, \mathbf{B})$, is computed as

$$sim(\mathbf{A}, \mathbf{B}) = \frac{\mathbf{A} \cdot \mathbf{B}}{|\mathbf{A}||\mathbf{B}|} = \frac{\sum_{i=1}^{n} a_i b_i}{\left(\sqrt{\sum_{i=1}^{n} a_i^2}\right)\left(\sqrt{\sum_{i=1}^{n} b_i^2}\right)}.$$

If both **A** and **B** lie in the first quadrant of Euclidean space (as is the case with probability distributions, which can take only positive values), the cosine similarity metric takes values in [0, 1], with $sim(\mathbf{A}, \mathbf{B}) = 0$ expressing minimal similarity between the vectors (orthogonal vectors), and $sim(\mathbf{A}, \mathbf{B}) = 1$ corresponding to maximal possible similarity (identical vectors).

Before proceeding to estimate temporal correlations, Lee and Hou (2006) observed that, in order to obtain estimates of sojourn time distribution that accurately reflect user mobility behavior, a certain number of AP transitions not related to mobility must be removed from the traces. These transitions, called *ping-pong transitions*, might occur if a user is located in a position covered by multiple APs. In this situation, due to phenomena such as changes in network traffic load, changes in the radio environment, etc., a user's association can change even if the user is not physically moving. However, identifying ping-pong transitions in the trace is relatively easy, since they generate repetitive association patterns.

Lee and Hou (2006) classify ping-pong transitions as either $i \to j \to i \to j$ or $i \to j \to k \to i$ form (corresponding to a user lying within the coverage areas of two or three different APs, respectively – see Figure 10.1). The ping-pong transitions are then removed from the traces by associating a user with the prevalent AP for the entire duration of the ping-pong transition, where the prevalent AP is defined as the AP with which the user spends most of the time during the ping-pong phase.

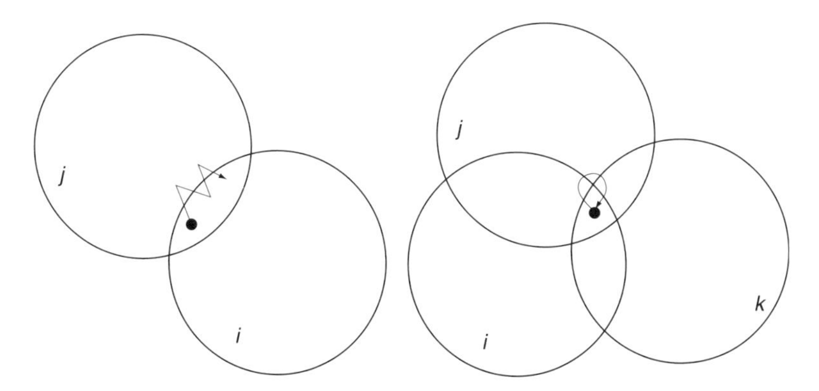

Figure 10.1 Examples of ping-pong transitions when the user is in the coverage area of two (left) or three (right) APs.

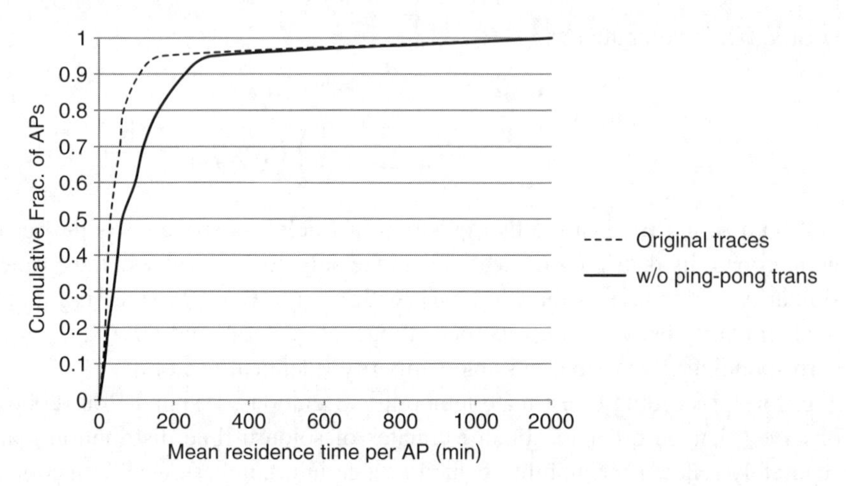

Figure 10.2 The distribution of mean residence time at APs before and after removal of ping-pong transitions.

An important observation made by Lee and Hou (2006) is that ping-pong transitions do occur frequently in WLAN traces (at least the ones analyzed by the authors): on average, about one-third of a user's transitions are indeed ping-pong transitions, and should be removed from the trace. If ping-pong transitions are not removed, the sojourn time distribution at APs can be erroneously underestimated. The effect of ping-pong transitions on sojourn time distribution is evident in Figure 10.2, showing the distribution of the mean residence time at APs derived from the original traces, and after removal of ping-pong transitions. The mean and median AP residence times are 51 minutes and 33 minutes, respectively, in the original data; after removal of ping-pong transitions, these times increase to 108 minutes and 73 minutes, respectively.

Let us now evaluate the similarity in user mobility at different time scales. In particular, Lee and Hou (2006) considered three time intervals: day, week, and month. For each time interval, these authors divided the original trace into a number of sub-traces of the corresponding interval. For instance, when considering monthly correlation in user mobility, the authors divided the trace into eight sub-traces, each corresponding to a different month. The steady-state distribution of users across APs is then computed for each sub-trace, giving a number k of different steady-state distributions $\tilde{\pi}^1, \ldots, \tilde{\pi}^k$, respectively. For each pair $(\tilde{\pi}^i, \tilde{\pi}^j)$, the cosine distance between $\tilde{\pi}^i$ and $\tilde{\pi}^j$ is then computed. Intuitively, if the cosine distance turns out to be relatively close to 1, this implies that user mobility (more specifically, the distribution of users across APs) in time interval i is very similar to user mobility in

Table 10.1 Monthly correlation in WLAN user mobility reporting, for each pair of monthly sub-traces, the cosine distance of the corresponding steady-state user/AP distribution

	m1	m2	m3	m4	m5	m6	m7	m8
m1	1	0.90	0.83	0.79	0.79	0.47	0.77	0.75
m2	–	1	0.84	0.78	0.80	0.47	0.75	0.73
m3	–	–	1	0.89	0.85	0.61	0.71	0.73
m4	–	–	–	1	0.92	0.72	0.72	0.73
m5	–	–	–	–	1	0.73	0.74	0.74
m6	–	–	–	–	–	1	0.48	0.49
m7	–	–	–	–	–	–	1	0.93
m8	–	–	–	–	–	–	–	1

time interval j, that is, user mobility patterns tend to repeat themselves in the time scale of interest (day, week, or month).

The cosine distance values for all possible pairs of monthly sub-traces derived from the eight month-long original data traces are given in Table 10.1. It is interesting to observe that the degree of similarity between monthly sub-traces is in general very high, especially for consecutive months. A relatively lower correlation value is displayed between sub-trace m6 and the other traces, which can be explained by the fact that this sub-trace contains the week corresponding to the spring break. Based on the results in Table 10.1, Lee and Hou (2006) conclude that user mobility displays a high degree of monthly correlation. Similar conclusions are drawn for daily and weekly correlation in user mobility patterns – see Lee and Hou (2006) for details.

10.1.3 Timed Location Prediction with the LH Model

While the characterization of the steady-state distribution of users across APs can be used to reveal interesting time correlation properties of user mobility patterns, it cannot be directly used to predict the most likely next association (also called *location*) of a user, given her/his current association. Location prediction is an important feature of a WLAN mobility model that can be used, for example, for network resource provisioning, network traffic load balancing, etc.

Lee and Hou (2006) present a methodology for location prediction based on the derived semi-Markov chain model. Indeed, the methodology allows estimation of not only the most probable next location given the current AP association, but also the most likely session duration at the current AP. The methodology used to derive location and session duration predictions is quite involved, so in the following we only summarize the intuition behind it. The interested reader can find details in Lee and Hou (2006).

In order to predict the next location and session duration, the transition probabilities $\phi_{ij}(t)$ of the semi-Markov chain must be estimated. We recall that $\phi_{ij}(t)$ represents the probability of finding the user in state j at time t, given that the user was in state i at time 0.

Estimating probabilities $\phi_{ij}(t)$ is a quite difficult task, which Lee and Hou (2006) tackle by considering a time-discrete approximation of the original transition probabilities. More specifically, these authors assume that time is discretized into relatively small time intervals of size h, and estimate transition probabilities $\phi_{ij}(k)$, corresponding to the probability of finding a user in state j at time $t = k \cdot h$, given that the user was in state i at time 0. Time discretization is used also to estimate the distributions required to compute $\phi_{ij}(k)$, namely, $Q_{ij}(kh)$ (transition probability between states i and j, independently of transition time) and $D_i(kh)$ (distribution of residence time in state i, independently of next location). Also, the distribution $H_{ij}(kh)$ (distribution of residence time in state i, given that next state is j), which is needed to estimate sojourn time at the current AP, is estimated through time discretization.

In order to estimate the transition probabilities, the following procedure is performed whenever a transition between state i and j occurs:

1. Update the state transition probabilities $P = [p_{ij}]$ and the time-discrete distribution $H_{ij}(kh)$, for $k = 1, \ldots, K$.
2. Compute $Q_{ij}(kh) = p_{ij} H_{ij}(kh)$, for $k = 1, \ldots, K$.
3. Compute $D_i(kh) = \sum_j Q_{ij}(kh)$, for $k = 1, \ldots, K$.
4. Compute $\phi_{ij}(k)$, for $k = 1, \ldots, K$.

Here K is the number of bins used to approximate the probability distributions.

The accuracy of the timed location prediction methodology has been evaluated in Lee and Hou (2006) using the Dartmouth WLAN trace. Every T_p seconds, where T_p is a tunable parameter, the semi-Markov mobility model is used to make predictions about the next possible state. When performing predictions, those made in the OFF state are not considered when collecting accuracy statistics. In fact, users tend to remain in the OFF state for a relatively long period of time, and estimating the next most likely state when a user is in the OFF state is a trivial task that succeeds most of the time, thus leading to an overestimate of the actual prediction accuracy.

Two methods are suggested to evaluate prediction accuracy: the 1-location estimator, according to which the next AP corresponds to the AP with the highest transition probability; and the 2-locations estimator, which considers also the AP corresponding to the next highest transition probability as a

Table 10.2 Prediction accuracy of the 1-location (1) and 2-locations (2) estimators for different values of the time parameters. Values of T_p, h, and K are expressed in seconds

T_p	$K = 6$, $h = 300$	$K = 12$, $h = 300$	$K = 6$, $h = 600$	$K = 12$, $h = 600$
600	(1) 0.84	(1) 0.84	(1) 0.80	(1) 0.80
	(2) 0.90	(2) 0.91	(2) 0.90	(2) 0.91
1800	(1) 0.69	(1) 0.71	(1) 0.70	(1) 0.69
	(2) 0.85	(2) 0.85	(2) 0.85	(2) 0.86

second possible next state. Note that also the current AP can be included in the outcome of the location estimators, corresponding to the situation in which no transition actually takes place.

The accuracy of the two location estimators described above for different values of the time discretization parameters h, K and estimation period T_p are given in Table 10.2. As can be seen from the table, the accuracy is quite high, especially for the 2-locations estimator.

10.2 The KKK Mobility Model

In contrast to the LH model described in the previous section, the mobility model introduced in Kim et al. (2006) has the goal of modeling the physical movement of users within the environment. More specifically, the KKK model is designed to generate synthetic trajectories in a physical space resembling those observed in a large WLAN covering a university campus. Since the physical movement of users is not directly traced in collected WLAN data, but only the history of their AP associations, a first difficulty faced by these authors was extracting a reasonably accurate trajectory of a user's movement starting from the AP associations trace. The analysis of the salient features of these user trajectories allowed Kim et al. (2006) to derive a relatively simple synthetic model generating trajectories with similar features.

10.2.1 Extracting Physical Movement Trajectories from WLAN Traces

Extracting a physical movement trajectory from an AP association trace is a very challenging task. First of all, as we commented in Chapter 9, most WLAN users are indeed *nomadic*, rather than *mobile*: different from a mobile user, a nomadic user uses the WLAN device while residing in a certain place for a relatively long time; this user can then move to a different location (where she/he will also likely spend a relatively long time), but typically the WLAN device is turned off during the traveling phase. Thus, extracting accurate physical movement trajectories from traces of nomadic users is very difficult, essentially due to lack of data.

To get around this difficulty, Kim et al. (2006) decided to base their model only on the observation of a specific class of users, namely, VoIP users: in fact, unlike other classes of WLAN users, VoIP users are likely to behave as mobile (i.e., having the VoIP device turned on also during the traveling phase), instead of nomadic, entities. More specifically, the Kim et al. (2006) considered a trace referring to 198 VoIP users collected at Dartmouth College in a 13-month period from the beginning of June 2003 to the end of June 2004.

Even if only mobile users are traced, extracting a physical movement trajectory from an AP association trace remains a difficult task. The most intuitive methodology for transforming AP traces into a physical movement trajectory would be to locate the user in the vicinity of the AP she/he is registered with. Unfortunately, directly applying this methodology would work poorly, since the AP a user is registered with is not necessarily the closest one to the user. In fact, APs might use different power levels to transmit; even if the same transmit power level were set in all APs, radio signal propagation in the environment would be highly heterogeneous, possibly leading a user to get better signal quality from a relatively far-away AP. Finally, different network interfaces might have different aggressiveness in changing associations from a relatively weak to a relatively strong AP: if the network card is not very aggressive, the current AP association can be retained for a relatively long time, even if an AP with much better signal quality is available.

Kim et al. (2006) suggest three methods for extracting a physical movement trajectory from an AP trace, all of which build upon the fact that the exact (x, y) coordinate of each AP on the Dartmouth College campus is known and reported in the traces:

1. *Triangle centroid*: this algorithm uses the locations of the past three AP associations of a user as the vertices of a triangle; then, user position is estimated as the centroid of the triangle (see Figure 10.3, left).
2. *Time-based centroid*: this algorithm considers a fixed time window of length q seconds; the user's location is updated every p seconds, considering the positions of the AP associations that happened during the past q seconds. More specifically, denoting (x_i, y_i) as the position of the ith AP observed in the q seconds time window, we have

$$x_u(t) = \frac{1}{h}\sum_{i=1}^{h} x_i, \qquad y_u(t) = \frac{1}{h}\sum_{i=1}^{h} y_i,$$

   where $x_u(t)$ and $y_u(t)$ represent the x and y coordinates of user u at time t, and h is the total number of AP associations observed in the time

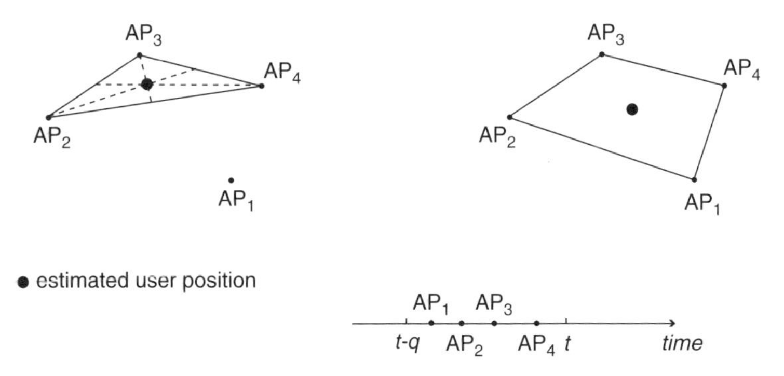

Figure 10.3 Estimated user position using the triangle centroid (left) and time-based centroid (right) algorithms.

window $(t - q, t]$. If no association occurred during the past q seconds (i.e., $h = 0$), the user's location is kept unchanged, that is,

$$x_u(t) = x_u(t - p), \qquad y_u(t) = y_u(t - p).$$

See Figure 10.3 (right) for an example of location estimation with the time-based centroid algorithm.

3. *Kalman filter*: a suitably defined Kalman filter is used to estimate the current user position based on a set of measurements perturbed by some random error, where a measurement is given by the AP position when a new association occurs. Details on how the Kalman filter is formally defined can be found in Kim et al. (2006).

By comparing the physical trajectories derived from the above algorithms to a few GPS traces obtained from volunteers moving on the campus, Kim et al. (2006) conclude that the Kalman filter is the most accurate method, and it is then used to extract physical movement trajectories from the WLAN data traces.

Before extracting the physical trajectories, Kim et al. (2006) pre-processed the traces in order to distinguish between relatively mobile and relatively stationary sub-traces. First of all, for each user, only data collected within working hours (defined as the interval between 8 AM and 6 PM) is considered. Then, the authors classified each workday sub-trace as either *mobile* or *stationary* according to the following criterion: for each workday, the maximum Euclidean distance (*diameter*) between the locations of any two APs occurring in the sub-trace is computed. A workday sub-trace is then classified as stationary if its diameter is less than 100 m, otherwise it is classified as mobile. The analysis of the whole trace reveals that only 46% of the workday sub-traces are classified as mobile.

10.2.2 Extracting Pause Time

Another task to be accomplished in the process of extracting physical movement trajectories from WLAN traces is accurately estimating the pause time at the various locations. In other words, the time elapsing between a user's association with access point AP_1 and her/his next association with AP_2 must be broken down into a pause time at AP_1, followed by a travel time from AP_1 to AP_2 (see Figure 10.4).

The algorithm proposed in Kim et al. (2006) to estimate pause times is as follows. The "raw" user speed when moving from access point AP_1 to the next access point AP_2 is computed as

$$v_r = \frac{|l_2 - l_1|}{t_2 - t_1},$$

where $|l_2 - l_1|$ is the Euclidean distance between AP_2 and AP_1, and t_i is the time of association with access point AP_i.

The raw user velocity is then compared to a "normal" speed range estimated from the assumption that the user is a pedestrian. More specifically, the normal speed range is defined as the interval $[min, 10]$ m/s, where two possible values of min (0.1 and 0.5 m/s) are considered in Kim et al. (2006). If v_r is within normal range, then it is assumed that the user did not stop at AP_1, but was simply traveling between AP_1 and AP_2 during time interval $t_2 - t_1$ (i.e., pause time at AP_1 is assumed to be 0). If the raw speed v_r is *above* the normal speed, it is assumed that the change in AP association is not reflecting a real user movement, and that portion of the trace is discarded. Finally, if the raw speed is *below* the normal speed, a pause time at AP_1 is computed as follows:

$$p_1 = t_2 - t_1 - \frac{|l_2 - l_1|}{\bar{s}},$$

where $\bar{s}$ is the average speed of the user, which is computed as an exponentially weighted moving average; more specifically,

$$\bar{s} = 0.25 v_r + 0.75 \bar{s}',$$

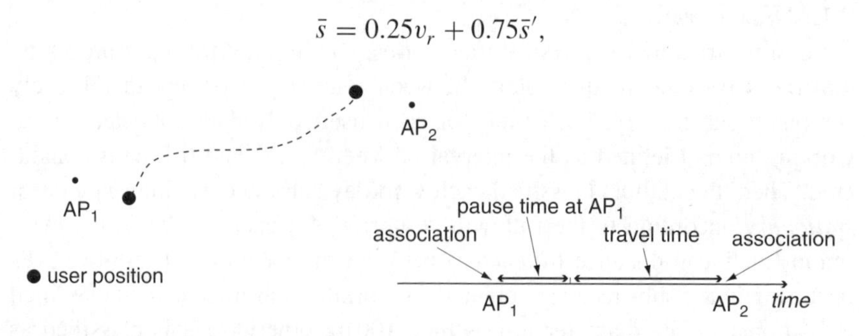

Figure 10.4 Breakdown of a time interval into pause time at an AP and travel time between two consecutive APs.

where $\bar{s}'$ is the user average speed as computed in the previous segment of the trajectory.

Kim et al. (2006) then proposed a heuristic to aggregate consecutive pause times in case the location of a user remains confined within a relatively small range (called *clustering range*) – see Kim et al. (2006) for details.

Similar to the physical trajectory extraction method, the pause time extraction algorithm is evaluated in Kim et al. (2006) by comparing the estimated pause times to ground truth data obtained from a few volunteers performing a pause time+travel trip within the Dartmouth campus. The comparison reveals the efficacy of the proposed algorithm extended with the clustering range heuristic in estimating pause times.

10.2.3 Dealing with Stationary Sub-Traces

The methods described to extract physical movement trajectories and pause times are relevant only for mobile sub-traces. For stationary sub-traces, the degree of user movement is so low (recall that a workday sub-trace is classified as stationary if its diameter is less than 100 m) that the user recorded in the trace is assumed to be stationary for the entire duration of the trace. Thus, determining user location and pause time is a relatively easy task in this case: the user location is determined according to the triangle centroid method and remains fixed for the entire duration of the sub-trace; accordingly, the pause time at the estimated location is assumed to be equal to the duration of the sub-trace.

10.2.4 Finding Hotspot Locations

A final ingredient of the KKK mobility model is the definition and positioning of hotspot locations, namely, popular locations. For mobile sub-traces, the authors identify hotspot locations within the Dartmouth College campus using a simple technique that builds upon the definition of pause time: whenever a user pauses at an AP, a 2-D Gaussian distribution of "popularity weight" is applied at the current user location, thus creating a small "popularity mountain" centered at the user's location. The "height" of the mountain (peak of the Gaussian distribution) is directly proportional to the pause time duration. Popularity mountains for each visit are then added up across all users, and hotspot regions are identified as those region of the space where the popularity weight is consistently higher than a certain threshold (see Kim et al. (2006) for details). In the case of stationary sub-traces, the 2-D Gaussian distribution is not weighted with the duration of the stay at the current (fixed) location.

Based on the analysis of data reported in both mobile and stationary sub-traces, Kim et al. (2006) identify five hotspot regions on the Dartmouth

College campus, corresponding to the School of Engineering, the main Dartmouth Library, the Computer Science Department, the office building of campus network administrators, and a hotel containing a restaurant (see Figure 10.5). The distribution of the initial user location across the five hotspots is displayed in Figure 10.6, with region 0 – called the *cold*

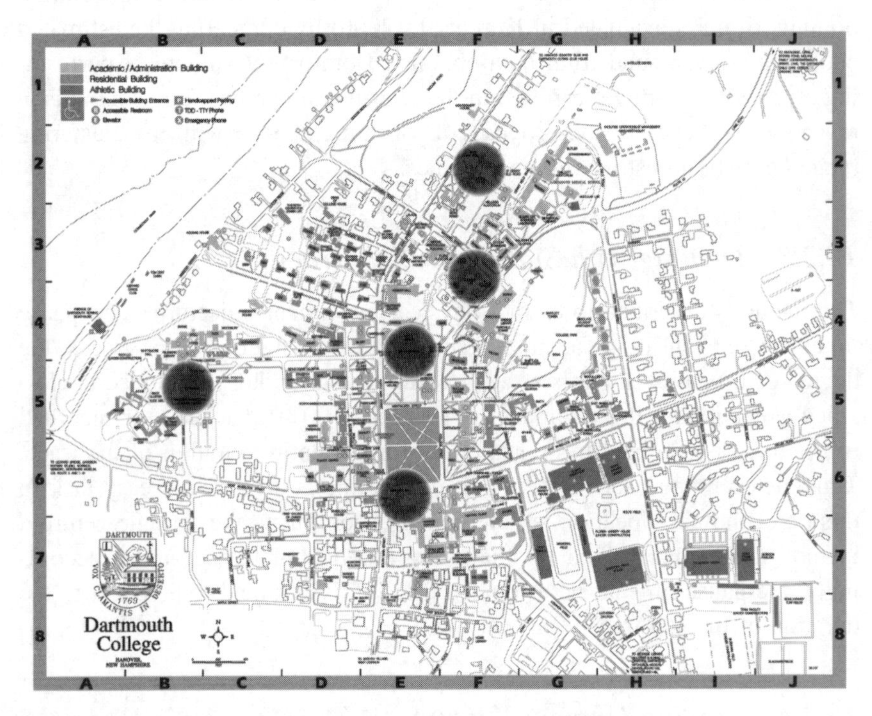

Figure 10.5 Dartmouth College campus map with the locations of the five hotspot regions.

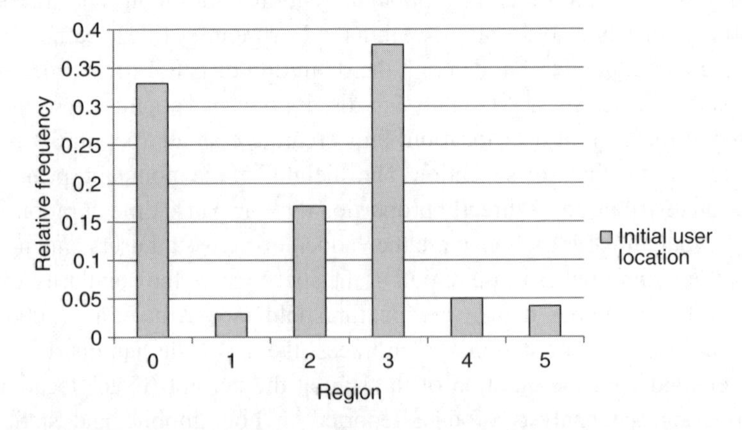

Figure 10.6 Distribution of initial user location across hotspots.

region – modeling an initial location anywhere outside the five hotspot regions. As can be seen from the figure, approximately 67% of the users start their walk from a hotspot region.

The definition of hotspot regions is also useful for characterizing movement between the hotspots. More specifically, based on WLAN data traces, an $n \times n$ transition matrix M can be computed, with element m_{ij} corresponding to the probability of moving to location j starting from location i. Note that the transition matrix includes not only the five hotspots and the cold region, but also a "virtual" location corresponding to the OFF state.

10.2.5 Mobility Modeling

Based on the above-described algorithms for physical movement trajectory extraction, pause time characterization, and hotspot localization, the following model for synthetically generating user trajectories is defined in Kim et al. (2006):

1. Assign each user whose mobility has to be reproduced to either the stationary or mobile class based on the ratio derived from WLAN traces; that is, a user is classified as mobile with probability $p = 0.46$ and stationary with probability $1 - p$.
2. If the user is stationary:
   (a) assign an initial location according to the initial user location distribution, as reported in Figure 10.6;
   (b) determine a start time for the trace according to the distribution defined in Table 10.3 – see the following page;
   (c) determine the duration of stay according to the distribution defined in Table 10.3.
3. If the user is mobile:
   (a) assign an initial location according to the initial user location distribution, as reported in Figure 10.6;
   (b) determine a start time for the mobility trace according to the distribution defined in Table 10.3;
   (c) given the current location i, determine the next location j based on probabilities in the corresponding row of transition matrix M;
   (d) determine the number of waypoints on the way from i to j using a Gaussian distribution centered on the average number of waypoints observed on trajectories between i and j (obtained from WLAN data trace analysis), and let k be the selected number of waypoints;
   (e) select uniformly at random k points in the rectangle with locations i and j at opposite corners (see Figure 10.7), order the k selected points based on distance from i and let $a_1, \ldots, a_k$ denote these points, and then generate a trajectory connecting i to a_1, a_1 to $a_2, \ldots,$ and a_k to j;

(f) select the speed of movement based on the speed distribution defined in Table 10.3;
(g) when the user reaches destination j, select a pause time at j according to the pause time distribution defined in Table 10.3;
(h) repeat until the duration of the trace elapses, where the trace ending time is generated according to the distribution reported in Table 10.3.

The distributions used to generate the start and end times of the trace, duration of the stay and pause time, and movement speed, are derived from an analysis of the WLAN data trace, and are summarized in Table 10.3.

We recall that the exponential distribution is defined by the following *pdf*:

$$f(x) = e^{ax+b},$$

Table 10.3 Probability distributions used in the KKK mobility model. The pause time and speed distributions are derived using $min = 0.1$ m/s

Class	Characteristic	Unit	Mean	Distribution	Parameters
Stationary	Start	Hour	3.3	Exponential	$a = -0.175$, $b = -1.599$
Stationary	Duration (after 30 min)	Hour	5.1	Uniform	[0.36, 9.84]
Mobile	Start	Hour	1.9	Exponential	$a = -0.438$, $b = -0.872$
Mobile	End	Hour	8.5	Exponential	$a = 0.523$, $b = -6.637$
Mobile	Pause	Hour	0.718	Log-normal	$\mu = -1.880$, $\sigma^2 = 5.085$
Mobile	Speed	m/s	0.76	Log-normal	$\mu = -0.741$, $\sigma^2 = 0.788$

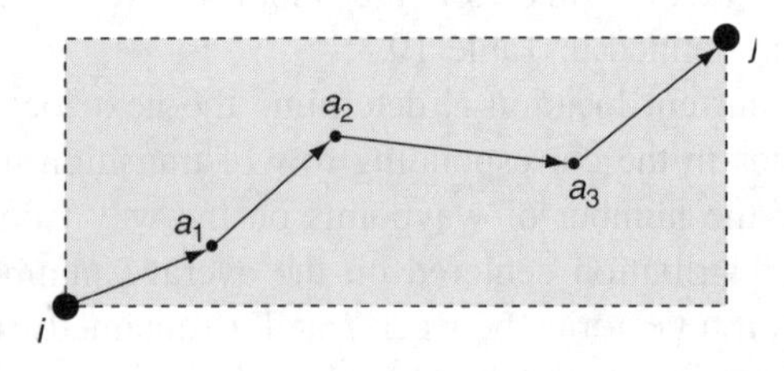

Figure 10.7 Generation of the mobility trace starting from an initial location i and the final location j.

and the log-normal distribution is defined by the following pdf:

$$f(x) = \frac{\frac{e^{-[\ln(x-\mu)^2]}}{2\sigma^2}}{x\sigma\sqrt{2\pi}}.$$

Kim et al. (2006) validate the KKK model against the WLAN data trace. Clearly, to validate the model a mobility parameter different from those used to define the model itself should be used. The authors use the number of visitors within a certain region in each hour of the workday as a relevant parameter, and validate the KKK model by comparing visitor numbers as computed from a synthetic KKK trace to those derived from the WLAN data trace. They find that the KKK model generates reasonably accurate movement traces, with the median relative error in estimating visitor numbers around 17%.

10.3 Final Considerations and Further Reading

In this chapter, we have presented two relevant mobility models for WLAN environments. Besides acquainting the reader with two specific mobility models, the aim of this chapter was to give also an idea of the probabilistic/statistical tools and methodologies used to extract a mobility model from WLAN data traces.

As the reader might have noticed, extracting a mobility model from WLAN traces is not a trivial task, because it requires a significant amount of pre- and post-processing of the traces, as well as considerable modeling efforts. Yet, it is possible to define synthetic trace generators able to faithfully reproduce traces observed in a certain WLAN environment. On the downside, the models described in this chapter – as well as the other WLAN mobility models introduced in the literature – are representative only of the specific WLAN environment from which they are derived, namely, different traces collected on the Dartmouth College campus. If the goal of a network designer were to model mobility in a different WLAN environment – say, a different campus, or a corporate instead of university WLAN – a different tuning of the model parameters, if not a complete redefinition of the models, would be required.

As commented briefly at the beginning of this chapter, several other mobility models derived from WLAN trace analysis have been proposed in the literature. Among the most representative ones, we cite the Model T and Model T++ models introduced in Jain et al. (2005) and Lelescu et al. (2006), respectively, and the model introduced in Tuduce and Gross (2005). The reader interested in gaining a better understanding of WLAN mobility modeling can start from these models.

References

Balazinska M and Castro P 2003 Characterizing mobility and network usage in a corporate wireless local-area network. *Proceedings of ACM MobiSys*.

Chinchilla F, Lindsey M and Papadopouli M 2004 Analysis of wireless information locality and association patterns in a campus. *Proceedings of IEEE Infocom*.

Hsu WJ, Merchant K, Shu HW, Hsu CH and Helmy A 2004 Preference-based mobility model and the case for congestion relief in WLANs using ad hoc networks. *Proceedings of the IEEE Vehicular Technology Conference (VTC), Fall*.

Jain R, Lelescu D and Balakrishnan M 2005 Model T: An empirical model for user registration patterns in a campus wireless LAN. *Proceedings of the ACM International Conference on Mobile Computing and Networking (MOBICOM)*, pp. 170–184.

Kim M, Kotz D and Kim S 2006 Extracting a mobility model from real user traces. *Proceedings of IEEE Infocom*.

Lee JK and Hou J 2006 Modeling steady-state and transient behaviors of user mobility: Formulation, analysis, and application. *Proceedings of the ACM International Symposium on Mobile Ad Hoc Networking and Computing (MobiHoc)*, pp. 85–96.

Lelescu D, Kozat U, Jain R and Balakrishnan M 2006 Model T++: An empirical joint space-time registration model. *Proceedings of the ACM International Conference on Mobile Ad Hoc Networking and Computing (MobiHoc)*, pp. 61–72.

Team C 2011 *http://crawdad.cs.dartmouth.edu/index.php*.

Tuduce C and Gross T 2005 A mobility model based on WLAN traces and its validation. *Proceedings of IEEE Infocom*, pp. 664–674.

Part Four

Mobility Models for Vehicular Networks

In this part of the book, after having briefly introduced the field of vehicular networks and related technology, we will describe the two main classes of models for vehicular mobility, namely macro- and micro-mobility models. We will then present representative examples of micro-mobility models, which are more relevant to vehicular networks, and close this part by describing existing tools and approaches for integrating vehicular mobility and wireless communication in the simulation of a vehicular network.

11

Vehicular Networks

In this chapter, we will describe in brief the main features of an emerging class of next generation wireless networks, namely, vehicular networks. After describing the reasons for the growing interest in vehicular networks from the research, as well as the governmental and industrial, communities, we will describe the state of the art. We will then present some representative use cases and conclude the chapter by commenting on envisioned technological evolutions in the near future, and the related challenges.

11.1 Vehicular Networks: State of the Art

Vehicular networks are short-range wireless networks formed by vehicles traveling on the road. Besides communicating with nearby vehicles, a vehicle could also communicate with radio devices installed on road side equipment such as traffic lights, lamps, etc. An important observation regarding terminology is that by the term "vehicular network" we mean a network where short-range wireless links are established either between *different* vehicles – V2V communications – or between a vehicle and a road side unit (RSU) – V2I communications. That is, we are not concerned with in-vehicle wireless networks, such as those formed by various Bluetooth devices co-located in a vehicle, which are also sometimes called "vehicular networks."

A typical vehicular network architecture in an urban scenario is depicted in Figure 11.1: V2V and V2I links coexist and form a relatively dense network of vehicles, especially close to road traffic hotspots such as junctions, intersections, and so on. On the other hand, in highway scenarios the vehicular network is typically relatively less dense and mostly formed of V2V links, except for occasional connections with RSUs – see Figure 11.2.

Mobility Models for Next Generation Wireless Networks: Ad Hoc, Vehicular and Mesh Networks, First Edition. Paolo Santi.

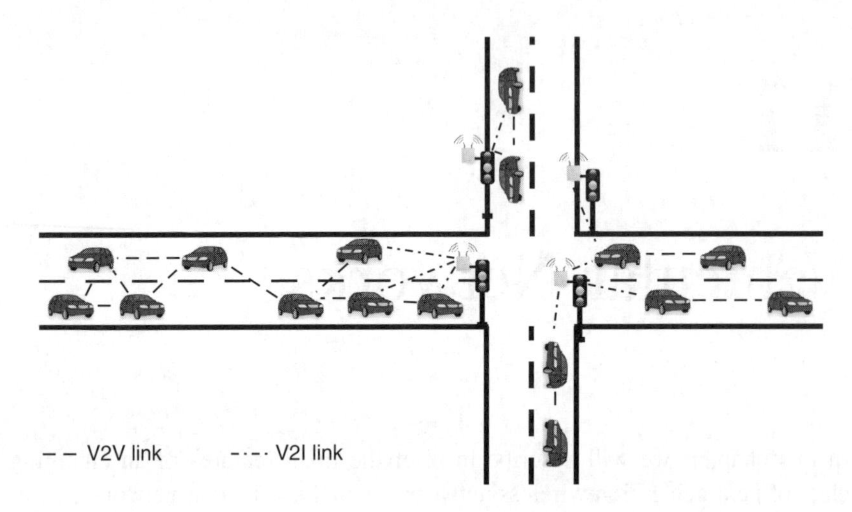

Figure 11.1 A typical vehicular network architecture in an urban scenario.

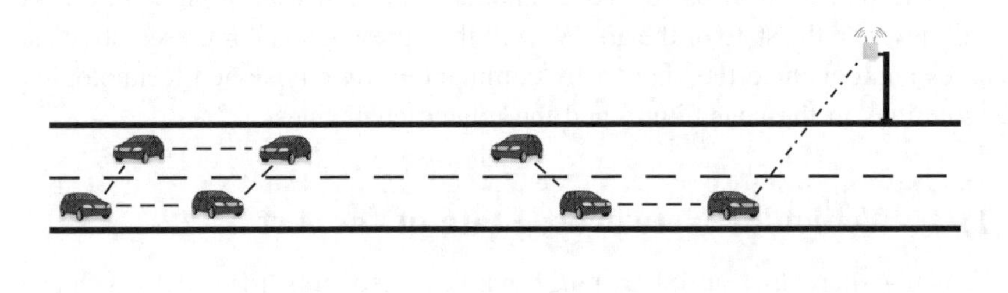

Figure 11.2 A typical vehicular network architecture in a highway scenario.

From an architectural point of view, vehicular networks are hybrid networks sharing features with MANETs – the part of the network formed by V2V links – and opportunistic networks – the occasional occurrence of V2I links. Also, network density can vary from highly dense, mostly connected networks – the urban scenarios – to sparse, mostly disconnected networks – the suburban and highway scenarios. In the urban setting, RSUs can be thought of also as a light infrastructure that can be exploited to optimize network operation.

A distinguishing feature of vehicular networks, independently of the specific scenario, is a high degree of mobility, which results in a highly dynamic network topology. As we will see in the following, dealing with highly dynamic topologies challenges the wireless communication and networking solutions that are being developed for vehicular networks.

11.1.1 Motivation

What are the reasons behind the increasing attention and the considerable research and development efforts being devoted to vehicular networks? Some statistical data may help the reader understand the main driving forces behind the development and consolidation of vehicular networking technology.

Despite alternative means of transportation such as trains, metros, aircraft, etc., becoming relatively more commonly used in recent years, road transport still dominates, by far, other means of transport. For instance, in Europe 45% of goods and 87% of passengers are transported along roads (Commission 2004). The number of passengers and goods traveling on roads keeps increasing, while the road network is destined to increase at a much lower rate. Thus, the density of vehicles traveling on roads is expected to increase in the near future, with a potentially negative impact in terms of traffic efficiency (fuel wastage, air pollution, etc.) and societal costs (number of accidents, road maintenance costs, etc.).

Although the density of vehicles is expected to increase in forthcoming years, most governments (e.g., in the USA, Europe, and Japan) have launched plans to *reduce* the enormous costs related to road traffic. To give the reader an idea of these costs, the World Health Organization estimates that 1.2 million people are killed every year in road accidents, the ninth leading cause of death in the world and the leading cause of death among adolescents and young adults (World Health Organisation 2008). The World Health Organization also estimates that, if no corrective actions are undertaken, road accidents will become the fifth leading cause of death in the world by 2030 (Organization 2009). In economic terms, the World Health Organization estimates the direct cost of road crash injuries to be between 1 and 2% of the gross domestic product of each country, amounting to about US$ 518 billion worldwide.

How is it possible to *reduce* the costs related to road transportation while the density of vehicles on the roads is expected to *increase* in forthcoming years? The consensus is that large-scale deployment of intelligent transport systems will be key to achieving such an ambitious goal.

Intelligent transport systems are a class of systems that integrate information and communication technologies with pre-existing transport infrastructure, vehicles, and users. Examples of these systems are induction loops, infrastructure variable message signs, traffic management centers, etc. Among intelligent transport systems, vehicular networks – also called *cooperative vehicular systems* – are considered a prominent player, due to their enormous potential of improving safety conditions on the roads and increasing traffic efficiency in general. This explains why government agencies have launched funding programs specifically aimed at the development and consolidation of vehicular communication and networking technology. Similarly, the automotive and electronics industries are devoting considerable resources to the development of vehicular networking solutions aimed at improving road safety conditions.

11.1.2 Standardization Activities

Although in principle different wireless technologies might be used to realize V2V and V2I links, a derivation of the well-known WiFi standard is currently being considered as the prominent radio technology for vehicular networks. The IEEE 802.11 working group recently (July 2010) released an amendment to the standard, called IEEE 802.11p, aimed at extending the standard for vehicular communications (both V2V and V2I).

IEEE 802.11p has been designed mainly to address the challenges related to a highly dynamic network topology at the MAC layer, and to challenging radio propagation conditions at the PHY layer.

At the MAC layer, procedures for setting up and operating a wireless link between two nodes have been significantly simplified with respect to IEEE 802.11a/b/g. In particular, the concept of Basic Service Set (recall Section 8.1), which introduces considerable complexity during the setup of a wireless link, has been dropped: in IEEE 802.11p, nodes can establish a wireless link on-the-fly, without the need for setting up an Independent BSS or joining a pre-existing BSS first. With this choice, the delay in accessing the wireless channel has been considerably reduced in IEEE 802.11p with respect to WiFi solutions, which is especially important considering the highly dynamic topology typically characterizing a vehicular network, with several links appearing/disappearing in the network in just a few seconds.

The IEEE 802.11p PHY layer is a modification of the 802.11a PHY layer, operating in a set of frequencies – slightly above those used in IEEE 802.11a – exclusively devoted to vehicular communications. In particular, 75 MHz of spectrum between 5.850 and 5.925 GHz are reserved in the USA for vehicular communications, while in Europe only 30 MHz of spectrum between 5.875 and 5.905 GHz have been reserved for the same purpose. Exclusively reserving a set of frequencies for vehicular communications is, on the one hand, tangible proof of government interest in vehicular networking technology and, on the other hand, an important action for ruling out the negative effect on communication performance of interference originated by other types of networks operating in the same range of frequencies – as would have been the case if the populated 2.4 GHz frequency band were used instead.

The main difference between the IEEE 802.11p and 802.11a PHY layers is channel width, which has been reduced from 20 to 10 MHz. The reason for using narrower channels is to combat the multi-path delay spread and Doppler effects, caused by high mobility and road environments. However, the standard allows use of 20 MHz wide channels by transmitting on two adjacent

Table 11.1 Main features of the IEEE 802.11a and 802.11p PHY layer specifications

Protocol	Freq. (GHz)	Bandwidth (MHz)	Data rate (Mbps)	Modulation	Range indoor – outdoor (m)
a	5.2 – 5.825	20	6, 9, 12, 18 24, 36, 48, 54	OFDM	30 – 100
p	5.850 – 5.925	10	3, 4.5, 6, 9 12, 18, 24, 27	OFDM	up to 1 km (outdoor)
p	5.850–5.925	20	6, 9, 12, 18 24, 36, 48, 54	OFDM	up to 1 km (outdoor)

10 MHz channels. The main features of the IEEE 802.11p PHY layer are given in Table 11.1, together with those of IEEE 802.11a. As can be seen from the table, except for a narrower channel width and slightly different range of transmission frequencies, the IEEE 802.11p PHY layer shares many features with those of 802.11a. However, a major difference between the two specifications lies in the much longer transmission range required for IEEE 802.11p, which can be achieved due to the higher transmission powers allowed in the 802.11p bands: up to 44.8 dBm (around 30 W) in the USA, compared to 20 dBm (around 100 mW) allowed in the 802.11a frequency bands.

Another important difference between IEEE 802.11a/b/g (WiFi) and 802.11p is multi-channel operation; in WiFi, no specific rules for channel assignment are defined in the standard, so a WiFi device willing to discover other devices within transmission range must scan all available channels. This implies that discovery of other devices in the surroundings is a relatively slow process in WiFi. In vehicular networks, on the other hand, fast and reliable establishment of wireless links between vehicles is mandated by the very stringent application requirements (see next section). To improve the reliability and responsiveness of the vehicular wireless link, two actions have been undertaken: first, as explained above, a set of channels has been exclusively reserved for vehicular networks, to reduce interference from other types of networks; second, use of the available channels has been at least partially regulated, imposing specific requirements on the different channels in the IEEE 802.11p frequency band. In particular, one channel has been defined as the *control channel*, which is used both to quickly set up a wireless link between nearby vehicles and to exchange status information required by active safety applications (see the next section). So, different from WiFi networks, in IEEE 802.11p all vehicles must periodically tune their radios to the control channel to discover new vehicles and exchange status information, while the other available channels – called *service channels* – are left

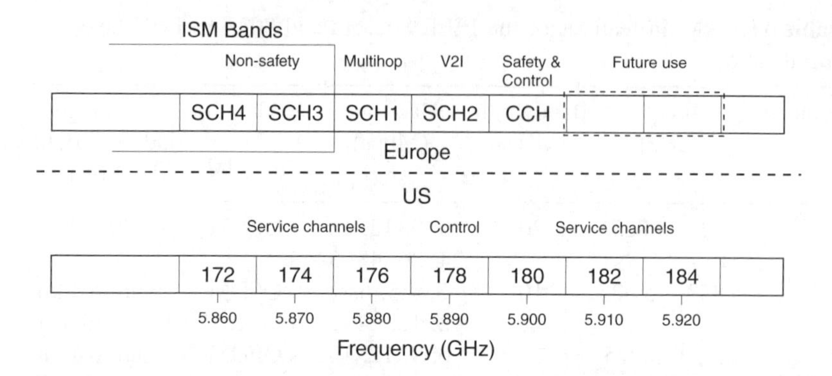

Figure 11.3 IEEE 802.11p channel allocation in Europe and the USA.

for communications related to other types of services. Channel allocation is slightly different in Europe and the USA (channel allocation for vehicular networks in Japan has not yet been decided), and is reported in Figure 11.3.

The IEEE 802.11p standard is part of a wider class of currently ongoing standardization activities in vehicular networks, encompassing not only the PHY and MAC layers, but also the upper levels of the network protocol stack. In particular, we mention the CALM (Communications Access for Land Mobiles) architecture promoted by ISO Technical Committee 204 Working Group 16 (204IT 2011) and the WAVE (Wireless Access in Vehicular Environment) initiative promoted by the IEEE in the USA (1609WG 2007), and the ITS (Intelligent Transportation Systems) communications architecture which is being developed by ETSI in Europe (ETSI 2009). In all these wider standardization efforts, IEEE 802.11p is being considered as a possible (if not the only) wireless communication technology for realizing V2V and V2I links.

11.2 Vehicular Networks: User Scenarios

In contrast to other emerging short-range radio technologies for which user scenarios are often not well defined, or sometimes even emerge after the technology has been realized, in vehicular networks user scenarios and the resulting application requirements have been clearly defined in advance. In fact, one of the first tasks of the above-mentioned standardization working groups (CALM, WAVE, and ITS) was the definition of representative use cases for vehicular networks and the characterization of the resulting application requirements. These use cases have been selected considering strategic, economic, performance, legal, and organizational requirements, as well as the needs of both users and stakeholders.

For instance, the ETSI ITS working group defines four classes of applications for vehicular networks: *active road safety*, *cooperative traffic efficiency*, *cooperative local services*, and *global Internet services*. In turn, active road safety applications can be divided into two classes: *cooperative awareness applications*, which are based on the period exchange of status information between vehicles, and *road hazard warning applications*, which are instead event-driven applications triggered by the occurrence of certain, potentially dangerous, events.

In the following, we briefly describe a representative use case with corresponding requirements for each of the above-mentioned groups. Application requirements are expressed in terms of:

1. *Minimum packet frequency*: the minimum frequency at which vehicles should locally broadcast their status, namely, position, speed, acceleration, etc.
2. *Maximum latency time*: the maximum latency allowed between the packet generation at the application layer and the actual transmission through the wireless channel.
3. *Absolute/relative position accuracy*: the required absolute or relative accuracy of the positioning system.
4. *Authentication/security requirements*: needed especially for emergency vehicles and commercial applications.
5. *Availability of digital map information*: accurate knowledge of the specific surrounding road environment in the form of a digital map might be requested.

11.2.1 Active Safety Applications

An instance of the class of cooperative awareness active safety applications is *intersection collision warning* (ICW). The goal of this application is to warn drivers about a potential collision when approaching an intersection which can be regulated by traffic lights or not. The warning is delivered to the driver through some user-friendly interface, for example, a danger sign blinking on the instrument panel.

It is important to observe that in this application, and in active safety applications in general, the goal is to give the driver early warning of a possible dangerous situation; however, full responsibility for possible actions taken to avoid the dangerous situation (e.g., braking) is left to the driver. In other words, active safety applications are not designed to replace the driver, but to assist her/him in improving road safety conditions. This is achieved by extending a driver's situation-awareness beyond the human eye, which is enabled by the exchange of short-range radio messages between nearby vehicles.

Intersection collision warning can be based on either V2V or V2I communications. When based on V2V communications – typically, when the intersection is not regulated by traffic lights – vehicles periodically broadcast status packets to detect each other's position, speed, direction, etc. In this case, the processing of broadcast packets from surrounding vehicles is distributively performed onboard vehicles, as well as detection of a potentially dangerous situation. When V2I communications are used instead – typically, when the intersection is regulated by traffic lights – RSUs installed at the intersection receive status messages from approaching vehicles, and elaborate these messages in order to detect dangerous conditions. If such a condition is detected, a warning message is locally broadcast in the vicinity of the intersection.

Independently of whether the application is realized through V2V or V2I communications, it is important that warning messages are delivered to the drivers early enough, so that safe deceleration of vehicles can take place, hopefully avoiding collisions at the intersection. It is also important that *all* approaching vehicles receive the warning in time, otherwise an even more dangerous traffic situation might actually occur. Thus, very stringent requirements on the lower layers of the network architecture are imposed by an intersection collision warning application, especially in terms of latency to access the wireless channel. These requirements, together with those of the other application described in this section, are summarized in Table 11.2.

Among other cooperative awareness active safety applications, we mention emergency vehicle warning, slow-vehicle indication, and motorcycle approaching indication.

An instance of the class of road hazard warning active safety applications is *emergency electronic brake lights* (EEBL). The goal of this application is to locally broadcast warning messages to following vehicles whenever a vehicle brakes hard. Thus, this application can be thought of as the radio-enabled counterpart of a feature already realized in some existing vehicles, which automatically turns on the emergency lights during hard braking. Due to the

Table 11.2 Requirements of ICW and EEBL applications

Application	Requirement	Value
ICW	Min. freq. periodic msg	10 Hz
	Max. latency	100 ms
	Max. Tx range	300 m
	Digital map	Required
EEBL	Min. freq. periodic msg	10 Hz
	Max. latency	100 ms
	Max. Tx range	300 m
	Digital map	Not required

exchange of radio messages, EEBL can extend the scope of the warning notification beyond human eyes, with potentially considerable benefits in terms of improved safety conditions. Similar to the ICW application, the warning is delivered to the driver through some user-friendly interface, for example, by a blinking danger sign on the instrument panel.

In contrast to ICW, EEBL typically exploits only V2V communications. However, requirements imposed by the application to the lower layers of the network architecture are similar to those imposed by ICW, especially for the low latency and high communication reliability requirements – see Table 11.2.

Among other applications, we mention wrong-way driving warning, stationary vehicle indication, traffic condition warning, road works warning, and collision risk warning.

11.2.2 Cooperative Traffic Efficiency Applications

An example of cooperative traffic efficiency applications is an *enhanced route guidance and navigation application* which is able to provide the driver with real-time information about traffic conditions, availability of nearby parking spots, etc. Different from existing navigation applications, which typically compute the route at the beginning of the trip and do not modify the suggested route while traveling, this "next generation" navigation application should be able to dynamically recompute the best route based on the acquired real-time traffic information.

While in principle both V2V and V2I communications can be used to realize this application, the presence of a light infrastructure – RSUs – is recommended to effectively implement real-time broadcasting of up-to-date traffic information. Also, real-time traffic information can be provided through access to external networks, for example, through a cellular network connection.

It is important to note that vehicles in this type of navigation application can be thought of as both *providers* and *consumers* of information: for instance, a vehicle might estimate travel times between selected checkpoints when traveling, and deliver these measurements to a traffic monitoring system through, for example, V2I communication with an RSU, acting in this case as information provider. On the other hand, (a driver onboard) a vehicle acts as information consumer when, for example, exploiting real-time traffic information provided by the traffic monitoring system to dynamically recompute the best route to the destination.

As other examples of cooperative traffic efficiency applications, we mention traffic light optimal speed advice, contextual speed limit notification, and limited access warning and detour notification.

11.2.3 Cooperative Local Services

An example of cooperative local service is *ITS local electronic commerce*: when traveling in a neighborhood, a driver might automatically receive advertisements for, say, restaurants and shops in the vicinity, electronic coupons and discounts for local services, and so on. Also, the driver might electronically purchase some goods/services in the neighborhood.

It is important to note that this type of application is perceived – at least by governments and the automotive industry – as a relatively less important application, since its is not related to road safety and traffic efficiency. Indeed, if not adequately implemented, ITS local electronic commerce applications might actually distract the driver from driving.

Other examples of applications in this class are media downloading, point of interest notification, and automatic access control and parking management.

11.2.4 Global Internet Services

This class of applications encompasses services that can be implemented only through access to the Internet. Such access can be realized either through vehicular networking technology – especially in urban scenarios where an RSU-based infrastructure is present – or through other types of networks such as cellular networks. Applications in this class include insurance and financial services, fleet management, loading zone management, vehicle software provisioning and updates, etc.

As with cooperative local services, global Internet services are also perceived as a relatively less important class of applications with respect to safety- and traffic-related applications. Further, all services in this class must be implemented in such a way that the driver is not distracted from the main task of driving.

11.3 Vehicular Networks: Perspectives

Unlike WLANs, which are nowadays a well-established and very popular wireless technology, vehicular networks are still not fully mature from a technological viewpoint: the IEEE 802.11p standard has been released only very recently, and only a few off-the-shelf 802.11p products are currently available. Most of them are still in prototypal form and relatively expensive.

Consequently, we believe the main evolution trend for vehicular networks in the near future will be a maturing of the communication and networking technology, which will hopefully require only a few years. Concurrently with technological evolution, vehicular applications will be developed and tested in the field, starting from the most compelling classes of applications such as active safety and cooperative traffic efficiency.

As for the challenges, we believe two classes will be addressed in forthcoming years. A first class is related to the very stringent requirements imposed by active safety applications on vehicular communications, which, as commented in the previous section, are very strict, especially in terms of latency and reliability of communication. Considering that vehicular networks are characterized by a high degree of mobility, a challenging radio environment, and the fact that the radio channel might saturate in the presence of high vehicle density, fulfilling these requirements will likely require a leap forward with respect to the current state of the art in communication and networking technology.

A second class of challenges is related to diffusion of the vehicular networking technology itself. While it appears credible that in the near future governments will impose the adoption of short-range vehicular technology for new vehicles through regulation (similar to what was done for seat belts, airbags, antilock braking system (ABS), etc.), there will nevertheless be a very long period of time during which the penetration rate of onboard vehicular networking technology will be very low. Just to give the reader an idea, it has been estimated that, even if all new vehicles were to be equipped with vehicular networking technology starting now, at the current market rate it would require more than 10 years to achieve a 100% penetration rate of the new technology in the USA. Thus, the applications mentioned in the previous section, and especially those related to active safety, should be implemented in an environment where short-range-radio-enabled vehicles coexist with non-equipped vehicles. The implications on safety conditions of this coexistence are not fully clear to date, and must be carefully addressed in order not to waste the enormous potential of vehicular networking technology to improve safety conditions also in this long transition phase.

11.4 Further Reading

This chapter presented only a very short description of vehicular networking technology and related issues. The reader interested in gaining a better view of this technology and related research activities is referred to the books available on this topic, among which we mention Hartenstein and Laberteaux (2010) and Emmelmann et al. (2010). A well-written and relatively short introduction to vehicular networks can be found in the first two chapters of Ribes (2010).

References

1609WG IS 2007 IEEE trial-use standard for wireless access in vehicular environments (WAVE). *IEEE Std 1609.1/.1/.3/.4*.

204IT 2011 CALM. *http://www.iso.org/iso/iso_technical_committee?commid=54706*.

Commission E 2004 Energy and transport. Report 2000-2004. *Official Publications of the European Communities KO-59-04-629-EN-C*.

Emmelmann M, Bochow B and Kellum C 2010 *Vehicular Networking: Automotive Applications and Beyond*. John Wiley & Sons, Ltd, Chichester.

ETSI 2009 Intelligent Transportation Systems (ITS); Communications; Architecture. *ETSI TS 102 665 V0.0.9*.

Hartenstein H and Laberteaux K 2010 *VANET Vehicular Applications and Inter-Networking Technologies*. John Wiley & Sons, Ltd, Chichester.

Organization WH 2009 *Global Status Report of Road Safety: Time for Action*. WHO, Geneva.

Ribes M 2010 Adaptive communication protocols for cooperative vehicular systems. *PhD Thesis*, *University Miguel Hernández of Elche*.

World Health Organization 2008 World health statistics 2008. WHO Library Cataloguing-in-Publication Data.

12

Vehicular Networks: Macroscopic and Microscopic Mobility Models

Modeling mobility in a vehicular network is a very challenging task. In fact, vehicular mobility displays several features that, on the one hand, can be difficult to model and, on the other hand, have a deep impact on a vehicle's mobility behavior. Examples of distinguishing features of vehicular mobility are:

1. *Geographically constrained movements*: vehicles are (luckily!) not allowed to move arbitrarily in the space, but are forced to move along pre-existing paths – the roads. Hence, geographically constrained movement is a very important feature of vehicular mobility, which cannot be overlooked in the definition of the mobility model if an acceptable level of realism is sought.
2. *Obedience to traffic rules*: very strict rules govern movement of vehicles along roads, such as speed limits, traffic junction rules, lane changing rules, etc. Thus, at least the most representative such rules should be incorporated into the mobility model in order to improve accuracy.
3. *Driver behavior*: different drivers might display very different driving styles concerning, for example, willingness to respect traffic rules, aggressive/non-aggressive driving behavior, etc. Including driver behavior in the mobility model is perhaps the most difficult task in vehicular mobility modeling, which explains why a driver behavioral model is often not included in the vehicular model definition. However, at least for certain applications (e.g., when the mobility model is used to assess the effectiveness of active safety applications in reducing the number of road

Mobility Models for Next Generation Wireless Networks: Ad Hoc, Vehicular and Mesh Networks,
First Edition. Paolo Santi.

accidents), driver behavior should not be disregarded in the definition of the model.

Given these features, it is evident why the "model accuracy" vs. "simulation running time" tradeoff, which is present in next generation wireless network mobility modeling in general, becomes especially critical in vehicular network models. On the one hand, using simplistic (but fast!) mobility models typically generates highly inaccurate simulation results, due to the fact that some of features 1, 2, or 3 above are not included in the model. On the other hand, a full-fledged vehicular mobility model including accurate road maps, traffic rules, and driver behavior models is typically cumbersome and computationally intensive, and difficult to use in vehicular network simulation. Thus, an adequate tradeoff between model accuracy and simulation running time should be carefully evaluated, depending on the network designer's needs.

One possible way of addressing this tradeoff is to take into consideration the geographical scope of interest. According to geographical scope, vehicular mobility models can be classified as *macro-* or *micro-*models. In this chapter, we will briefly describe the main features of models in each of these classes.

12.1 Vehicular Mobility Models: The Macroscopic View

Macroscopic vehicular mobility models are used to model traffic flow in relatively large geographical regions – in the order of hundreds or even thousands of square kilometers. Macroscopic mobility models are mostly used by transportation engineers to estimate the amount of traffic flowing along the main roads in a region of interest. In turn, accurate traffic flow estimates can be used, for example, to optimize road traffic management, to study the effects of introducing new arterial roads, etc.

Given the relatively large geographical scope of interest, vehicular mobility is modeled at a coarse granularity. In particular, individual vehicles are not modeled; instead, relationships between traffic flow characteristics of interest in the road network. such as vehicle density, mean speed, etc., are modeled through a suitably defined system of partial differential equations. While driver behavior is typically not modeled in macroscopic mobility models, some traffic rules such as speed limits can be used as constraints on the system of equations modeling traffic flow in the road network.

The process of simulating vehicular mobility using a macroscopic mobility model is typically composed of four steps, as depicted in Figure 12.1:

1. *Map creation*: in this step, the region of interest is subdivided into a number of subregions, corresponding, say, to neighborhoods of a city,

suburbs, villages, etc.; typically, a center of gravity is defined for each subregion, called the *point of interest* in the following. Also in this step, a map of the main roads connecting the various points of interest is created. An example of a macroscopic mobility map is shown in Figure 12.2.

2. *Definition of the traffic demand matrix*: in this step, a traffic demand matrix, T, is defined. T is an $m \times m$ matrix, where m is the number of points of interest in the region, and element (t_{ij}) represents the amount of traffic originating at the ith point of interest which is destined for the jth point of interest. Typically, the traffic demand matrix is enriched with temporal information, that is, the amount of traffic generated at a certain point i and destined for point j at different times of the day is defined. Since the traffic demand matrix constitutes the input to the macroscopic vehicular traffic model, it is important that the matrix is generated in an accurate way. This is typically done by combining usage of statistical data such as density of population, density of workplaces, etc., with samples of traffic measurements.
3. *Route assignment*: after the traffic demand matrix is defined and traffic demands at the various points of interest are generated, a routing algorithm must be defined to route traffic from origin to destination. This is typically done using the well-known Dijkstra shortest path algorithm.
4. *Trip phase*: when the three steps above have been performed, the actual simulation of traffic flow can start, and statistics of interests to the designer (e.g., average trip time, average and maximum flow on the roads, average travel speed, etc.) can be computed.

If the goal is, for instance, to estimate the impact on traffic of introducing a new road in the road system, the four-step process above can be repeated several times using different maps, until the "optimal" location of the new road is identified.

Given the relatively large geographical scope and lack of modeling individual vehicles, macroscopic mobility models are not very useful in vehicular network simulation, other than generating realistic traffic input values for microscopic mobility models (see next section). For this reason, in the following we will restrict our attention to the class of microscopic mobility

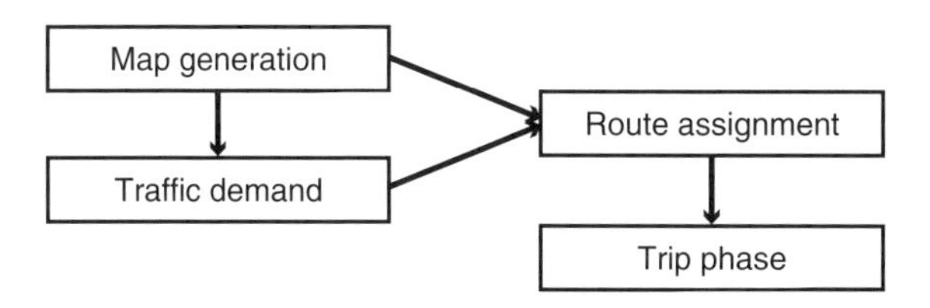

Figure 12.1 The four steps in macroscopic vehicular mobility simulation.

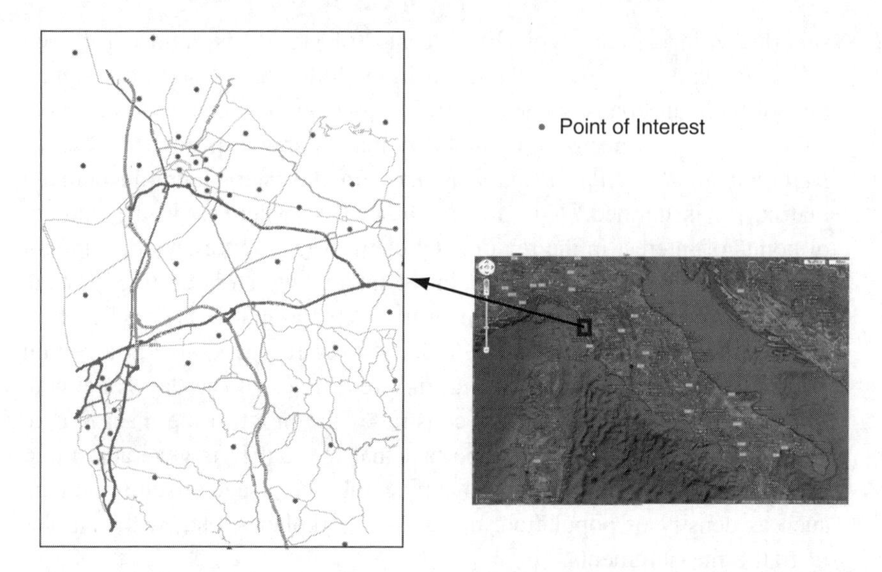

Figure 12.2 Example of a macroscopic mobility map of a region surrounding the city of Pisa, Italy (courtesy of Aleph SrL).

models, which are briefly introduced in the next section and treated in detail in the next chapter.

12.2 Vehicular Mobility Models: The Microscopic View

Unlike macroscopic models, microscopic mobility models aim to give a detailed view of vehicular mobility, where the behavior of a single vehicle is modeled. Clearly, given that each vehicle traveling on the road is modeled, with current computational capabilities the simulation of a moderate number of vehicles (in the order of several thousand at most) can be undertaken. So, unless a scenario with very low vehicle density is modeled, this upper bound on the number of modeled vehicles imposes an upper bound on the size of the geographical region of interest. With current technology, microscopic mobility models can be used to simulate regions of the size of a large metropolitan area (tens of square kilometers).

The necessary steps for simulating vehicular mobility with a microscopic mobility model are similar to those for macroscopic simulation:

1. *Map generation*: the first step consists of acquiring or generating a digitized road map of the region of interest. In contrast to macroscopic models, the digital map is typically quite accurate, including not only main roads

but also secondary ones. The level of detail of a digital map used in microscopic vehicular mobility models is similar to the one provided by Web-based tools such as Google Maps (Google 2011) or MapQuest (MapQuest 2011) at a level close to the maximum level of detail – see Figure 12.3 for an example.

2. *Definition of the traffic demand*: similar to the macroscopic case, traffic demands should be defined. However, the relatively small geographical scope of interest dictates some changes on how traffic demand values are generated. First, a typical assumption in microscopic modeling is that traffic can originate and be destined for outside the simulated region. In the usual approach, ingress/egress points at the borders of the region of interest are defined. Then, traffic demands are generated at ingress points and/or points of interest within the region, with destinations defined to be either egress points or another point of interest within the map. In order to generate realistic estimates of traffic originating and destined for outside the simulated region, the output of the macroscopic mobility model can be used to integrate the relatively small road system of interest with regional-scale traffic.
3. *Route assignment*: this step is also similar to the corresponding step in macroscopic models, the only relevant difference being that now routes are defined at the granularity of a single vehicle. That is, for each vehicle, given the trip origin, destination, and starting time as defined in the traffic demand matrix, a *best route* to a destination is computed using a Dijkstra-like shortest (or fastest) path algorithm.
4. *Trip phase*: when the map has been generated, the traffic demand defined, and the routes computed, the actual simulation of vehicular mobility can start, with relevant statistics such as average travel time (at single-vehicle granularity), average travel speed, and so on. It is important to observe that, different from macroscopic models, in microscopic mobility models traffic rules (lane changing, intersection rules, overtaking, etc.), and sometimes even driver behavior models, are included in the model definition. This implies that the movement of a vehicle is *directly* influenced by the movement of other vehicles. Hence, a supposedly best route as computed by the routing algorithm in the third step might turn out to be sub-optimal once the trip takes place due to interactions with other vehicles.

12.3 Further Reading

As mentioned at the end of Section 12.1, in the next chapter we will describe in detail some representative microscopic mobility models. The reader

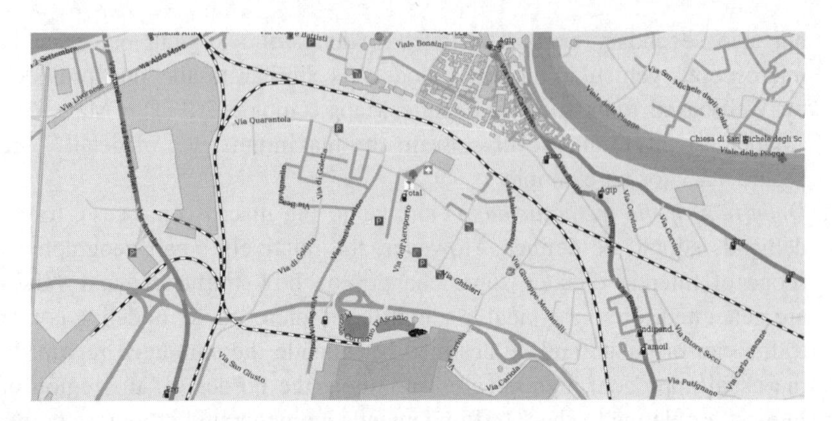

Figure 12.3 Example of a microscopic mobility map, reporting a portion of the city of Pisa, Italy (courtesy of Francesca Martelli).

interested in gaining a better understanding of macroscopic vehicular mobility models (also called traffic flow models) is referred to the several surveys (Bellomo et al. 2002; Coscia et al. 2007) and books (Kerner 2009) on the topic.

References

Bellomo N, Delitala M and Coscia V 2002 On the mathematical theory of vehicular traffic flow I: Fluid dynamic and kinetic modeling. *Mathematical Models and Methods in Applied Sciences* **12**, 1801–1843.

Coscia V, Delitala M and Frasca P 2007 On the mathematical theory of vehicular traffic flow II: Discrete velocity kinetic models. *International Journal of Non-Linear Mechanics* **42**, 411–421.

Google 2011 *http://maps.google.com*.

Kerner B 2009 *Introduction to Modern Traffic Flow Theory and Control*. Springer, Berlin.

MapQuest 2011 *http://www.mapquest.com*.

13

Microscopic Vehicular Mobility Models

In the previous chapter we saw how the problem of modeling vehicular mobility can be addressed at different geographical scopes, namely, at a macroscopic and microscopic level. In this chapter, we will focus our attention on the class of microscopic mobility models, which are by far the most relevant in the simulation of vehicular network performance.

We start by presenting three simple microscopic models, which can be thought of as relatively simple extensions of the well-known random walk and random waypoint mobility models. More specifically, we will present the graph-based mobility model introduced in Tian et al. (2002), and the Freeway and Manhattan mobility models introduced in Bai et al. (2003). Then, we will describe in detail a more complete model, the SUMO model, developed and maintained by the Institute of Transportation Systems at the German Aerospace Center. Unlike the models mentioned earlier, SUMO comprises tools for the automatic acquisition of digital road maps from different sources, and very accurate traffic rule and driver behavior modeling. Finally, in the last part of the chapter we will discuss the challenges related to accurately simulating both vehicular mobility and wireless communication, and present a representative tool aimed at effectively integrating vehicular mobility and wireless communication simulation.

13.1 Simple Microscopic Mobility Models

13.1.1 The Graph-Based Mobility Model

The graph-based mobility model was introduced in Tian et al. (2002) to model the mobility of humans in a urban environment. Although the model

Mobility Models for Next Generation Wireless Networks: Ad Hoc, Vehicular and Mesh Networks, First Edition. Paolo Santi.

was originally defined for pedestrian mobility, it can be extended to model vehicular mobility in a straightforward way.

The first step in the graph-based mobility model is the definition of a graph representing all possible paths in the simulated region. More specifically, the vertices of the graph represent locations that mobile nodes can visit – corresponding to points of interest in vehicular mobility modeling terminology – while the edges represent connections between these locations. Connections are defined in a quite abstract way in this model, and could be intended for instance to represent pathways between buildings on a campus in case pedestrian mobility is modeled, or roads if the goal is to model vehicular mobility.

Tian et al. (2002) assume that the path graph is given as input to the model, and that it is a connected graph. The path graph can be either randomly generated or extracted from a map of the region of interest. An example of a randomly generated path graph is shown in Figure 13.1.

The second step in the mobility simulation process consists of generating the initial positions of mobile nodes, which are assumed to be located at a randomly chosen vertex in the path graph. Once initial positions are chosen, nodes start moving according to a random waypoint-like pattern – see Figure 13.1:

1. A destination vertex is chosen uniformly at random in the path graph.
2. Then, the node starts moving toward the destination along the shortest path (in terms of Euclidean distance) connecting the current location with the destination vertex; the speed of movement is chosen uniformly at random in the $[v_{min}, v_{max}]$ interval, and it is kept constant during the whole trip to the destination.
3. On arrival at the destination, the node pauses for a short random time chosen uniformly in the interval $[t_{min}, t_{max}]$, then it starts a new trip.

While the graph-based mobility model accounts for geographically constrained movements and, in this respect, can be considered an improvement over other synthetic mobility models such as random walk and random waypoint, the model does not account for the other two distinguishing features of vehicular mobility described at the beginning of Chapter 12, namely, traffic rules and driver behavior. The models presented next take a step forward in accurate vehicular mobility modeling, since they also include simple traffic rules in the model definition.

13.1.2 The Freeway Mobility Model

The freeway mobility model was introduced in Bai et al. (2003) to model the movement of vehicles along a freeway. Similar to the graph-based mobility model, the model takes as input a map of freeways in the region

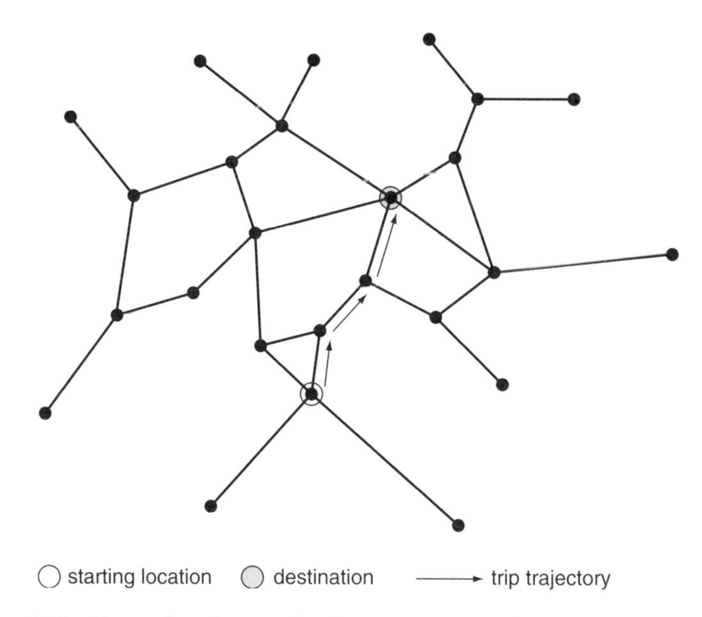

Figure 13.1 Example of a randomly generated path graph modeling a city center, and of a trip between randomly chosen source and destination.

of interest. Multi-lane freeways are modeled using parallel edges, as shown in Figure 13.2. The freeway map can be either extracted from a real map, or randomly generated.

The freeway map input implicitly defines ingress and egress points, that is, points where vehicles enter and leave the simulated region, as in the figure. Vehicles enter the simulated region from a randomly chosen ingress point (corresponding to a freeway lane), then they move along the selected lane until they exit the simulated region from the corresponding egress point. It is important to observe that, while freeway lanes are allowed to cross each other in the map (see Figure 13.2), these crossings do not represent junctions that can be used to enter in a different lane.

Since no rules for lane changing are defined in the model, the velocity of a vehicle is strongly related to the velocity of the preceding vehicles in the same lane – that is, a car-following rule is defined. Furthermore, in order to improve the realism of the model, the current velocity of a vehicle is made dependent also on its velocity at the previous time step (note that the freeway mobility model is a discrete-time mobility model).

More formally, the rules used for determining a vehicle's velocity are as follows:

1. *Dependence on past velocity*:

$$V_i(t+1) = V_i(t) + random(-1, 1) \cdot a,$$

where $V_i()$ represents the vehicle's velocity in the direction of the corresponding freeway lane, $random(-1, 1)$ represents a real number chosen uniformly at random in the $(-1, 1)$ interval, and a represents the maximum possible acceleration/deceleration.

2. *Car following*: let j be any vehicle preceding vehicle i in its lane; if $D_{i,j} \leq SD$, then $V_i(t) \leq V_j(t)$, where $D_{i,j}$ is the distance between vehicles i and j, and SD is a suitably defined parameter representing the safety distance between vehicles.

The car-following rule must be interpreted as imposing an upper bound on the velocity chosen using the first rule: if the chosen velocity of a vehicle according to this rule for the next time step is v_i and the vehicle is within the safety distance of a preceding vehicle j traveling at speed $v_j < v_i$, then the selected velocity of vehicle i at the next time step is v_j instead of v_i.

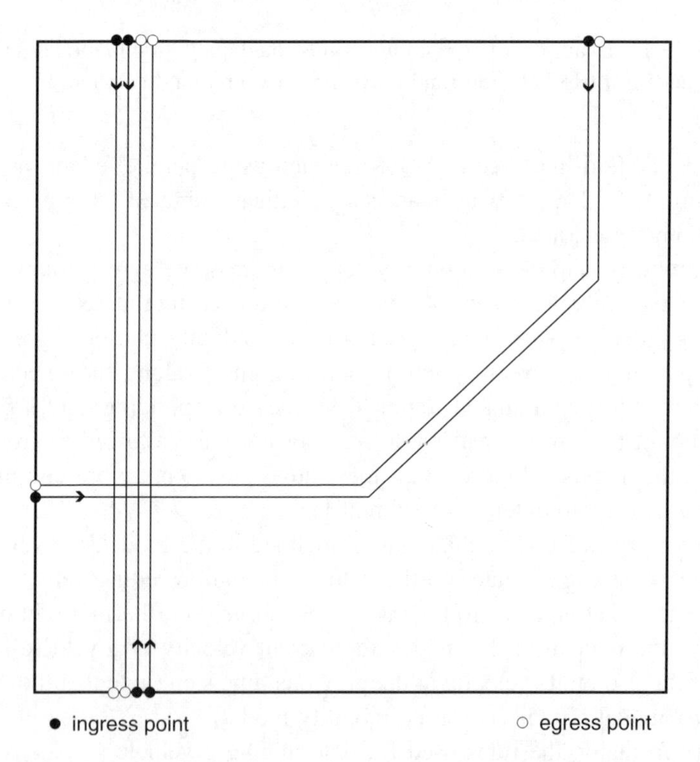

Figure 13.2 Example of freeway map and corresponding ingress/egress points in the freeway mobility model.

13.1.3 The Manhattan Mobility Model

As in the previous models, the Manhattan mobility model introduced in Bai et al. (2003) also takes as input a map representing possible paths within the simulated region. As the name suggests, the map in the Manhattan mobility model is assumed to have the square block regular pattern typical of Manhattan-like models – see Figure 13.3. Each road in the Manhattan map has a single lane in each direction. Similar to the freeway mobility model, the path map implicitly defines ingress/egress points, and vehicles enter the simulated region from a randomly selected ingress point. However, once a vehicle has entered the simulated region, its trajectory is not deterministic as in the freeway mobility model; instead, its trajectory is essentially a Manhattan-like random walk: on reaching an intersection, a vehicle can proceed in the same direction with probability 0.5, or make a left or right turn with the same probability of 0.25. If a vehicle reaches the border of the simulated region, it can exit the region from one of the egress points.

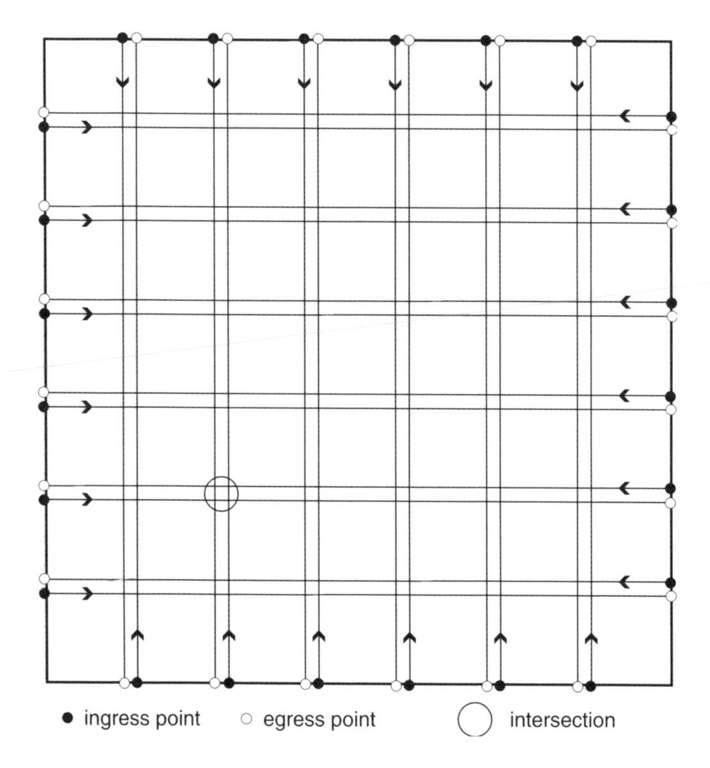

Figure 13.3 Example map in the Manhattan mobility model.

The rules for setting a vehicle's speed are the same as those in the freeway mobility model: the velocity of a node at the next time step is dependent on its current velocity (the first rule in the freeway mobility model), provided it does not exceed the velocity of the preceding vehicle in the same lane if within the safety distance (the second rule in the freeway mobility model).

13.2 The SUMO Mobility Model

The models presented in the previous section only partially account for the distinguishing features of vehicular mobility described at the beginning of Chapter 12. In particular, while geographically constrained mobility is considered in all models, traffic rules and driver behavior are disregarded in the graph-based mobility model, while only a simple car-following rule is included in the freeway and Manhattan mobility models. Furthermore, other aspects of the mobility models described in the previous section can be considered as not accurately reflecting reality, such as the process of generating an individual trip, which is completely random and does not reflect traffic patterns observed in the real world. It is therefore clear that the models described above cannot be used to model vehicular mobility if the designer's goal is to gain an understanding of, say, vehicular networking protocol performance under realistic traffic conditions. To this end, a more accurate microscopic mobility model is needed, which is indeed able to reproduce vehicular traffic patterns observed in the real world. In turn, this requires on the one hand that the mobility model accounts for relatively complex traffic rules governing traffic lights and intersections, lane changing, lane merging, vehicle overtaking, etc., and, on the other hand, sophisticated tools should be provided to import and/or generate realistic road networks and traffic demand matrices.

A representative example of an advanced microscopic vehicular mobility model with these characteristics is the SUMO (Simulation of Urban MObility) mobility model (Team 2011a). SUMO was originally designed and is being maintained by members of the Institute of Transportation Research at the German Aerospace Center. SUMO is aimed at enabling relatively fast and accurate simulation of road networks comprising tens of thousands of edges – approximately corresponding to the size of a relatively large metropolitan area.

A notable feature of SUMO as compared to commercial microscopic models is that it is an open source model, thus it can be downloaded and installed for free (under a general public license) from the official SUMO homepage. This feature makes SUMO a very appealing choice for accurately modeling vehicular mobility, without the need to buy expensive commercial vehicular traffic simulators.

The main features of SUMO are summarized below:

1. Discrete-time, continuous-space vehicle movement.
2. Multi-modal mobility: a trip can be composed of different portions, with a different transportation means (e.g., bus, train, and on foot) associated with each portion.
3. Models multi-lane roads with rules for lane changing and lane merging.
4. Rules for unregulated (e.g., stop signs) and regulated (traffic light) intersections.
5. Driver behavior modeling.
6. Personalized output generation.

The steps necessary for setting up a SUMO simulation are pictorially represented in Figure 13.4 and briefly described below:

1. *Build the road network*: in this step, the road network is built and given as input to the mobility model.
2. *Build the traffic demand*: in this step, individual trips for each considered vehicle are generated.
3. *Route computation and (optional) dynamic user assignment (DUA)*: in this step, routes for each individual trip are computed. In case a traffic optimization procedure is considered, dynamic user assignment is computed in this step – see below for details.
4. *Perform simulation and get desired output*: after the three steps above have been performed, the actual simulation can start, and the relevant output statistics are generated.

In the remainder of this section, we will describe in detail each of the SUMO building blocks shown in Figure 13.4.

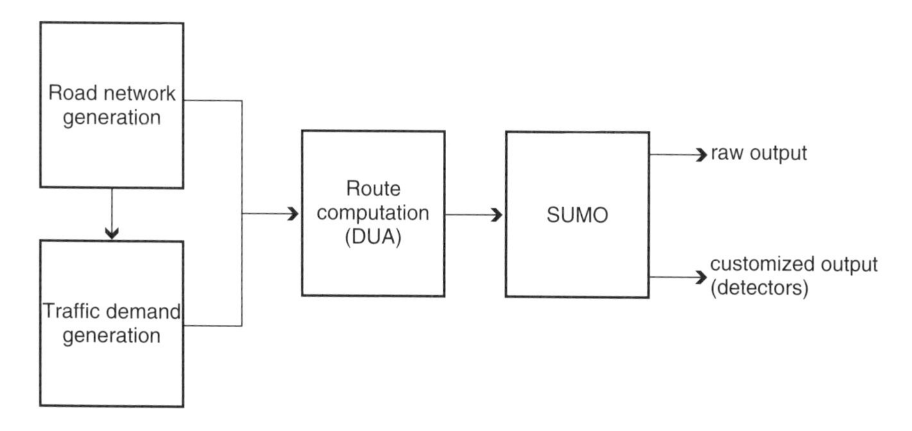

Figure 13.4 Setting up a simulation with SUMO.

13.2.1 Building the Road Network

SUMO provides two tools for building the road network: NETGEN and NETCONVERT. NETGEN is a road network *generator*, and allows three types of synthetic networks to be built: grid networks, spider networks, and random networks. Grid networks are Manhattan-like networks, where the user can decide on the number of intersections in the x- and y-axes, and the distance between intersections. Spider networks are defined by the number of axes dividing them, the number of concentric layers they are made of, and the distance between layers – see Figure 13.5 for an example of a spider network. Finally, random networks are randomly generated networks where certain parameters such as maximum and minimum road length, minimum angle between roads, etc., can be specified.

While synthetic road networks generated by NETGEN can be useful for an initial performance assessment, they are inadequate for accurately mimicking traffic patterns in a real-world situation. For this purpose, the NETCONVERT tool should be used instead. NETCONVERT allows both the definition of user-specified road networks – a tedious but sometimes unavoidable task – and importing existing road networks from several external sources such as VISUM, shape files, and digital map databases like OpenStreetMap. In the former case, an XML file called a *SUMO network file* can be defined to describe a digital road map. In the latter cases, NETCONVERT converts external digital maps into a native SUMO network file.

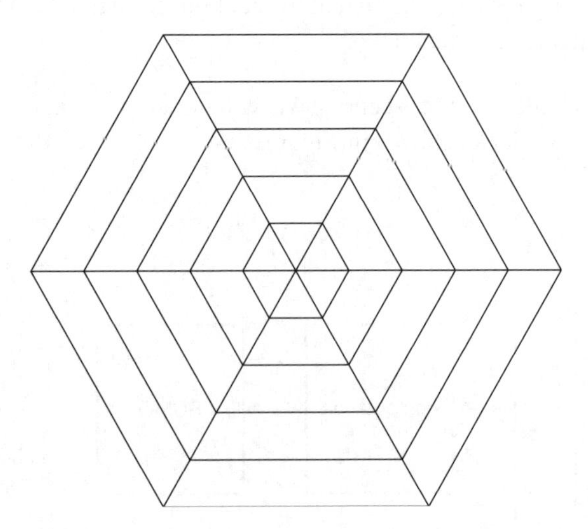

Figure 13.5 Example of spider road network with six axes and five concentric layers.

13.2.2 Building the Traffic Demand

There are different ways to build the traffic demand in SUMO. The most direct – but cumbersome – way is to define a route in the road network for each individual vehicle, including the trip starting time. Alternatively, traffic demands can be derived from user-defined flows and turning percentages at intersections. In turn, these values can be obtained from statistical data, traffic measurements, and so on. A third way of building the traffic demand is to generate routes between randomly selected origination and destination points, similar to what was done in the graph-based, freeway, and Manhattan mobility models. Finally, traffic demand values can be imported from existing data or from data generated by a macroscopic mobility model. These data can be expressed either in matrix form or as a trace of routes.

13.2.3 Route Computation

Once traffic demand is defined, routes can be computed. The basic approach for building routes is to use the Dijkstra shortest path algorithm to compute the shortest – or fastest, if speed limits are considered – route between each origin/destination pair. It should be noted, though, that due to traffic congestion, the best route as computed by the Dijkstra algorithm is not necessarily the one actually leading to the shortest possible travel time. In order to compute the best route considering also the effect of traffic congestion, which, in turn, depends on other vehicles' routing decisions, the Dijkstra algorithm must be iteratively executed several times, changing the weights on the edges of the road network reflecting the current vehicle's routing decisions at each iteration. This iterative process is terminated when equilibrium conditions are reached, corresponding to the best possible routes accounting also for traffic congestion. This more sophisticated approach to route computation is called *dynamic user assignment* and is an optional feature available in SUMO.

Static route computation is much faster than dynamic user assignment, and it is therefore suitable for simulating traffic in very large road networks where dynamic user assignment would be computationally too demanding. Static route computation is also the right choice for modeling situations in which a driver does not have a full view of the traffic situation in the area, and takes routing decisions based on stable information such as minimum distance to destination. On the other hand, dynamic user assignment is the right choice if the goal is to estimate, for example, the effects on traffic of introducing a new road in the network; in this case, it is useful to gain an understanding of the average travel time under the best possible conditions, namely, equilibrium conditions.

13.2.4 Running the Simulation and Generating Output

The final step is to run the simulation and generate output. Several models for traffic rules and driver behaviors are implemented in SUMO. In particular, SUMO defines rules for lane changing and lane merging, vehicle overtaking, car following, and sophisticated intersections for multi-lane roads. A driver behavioral model called the *intelligent driver model* (Treiber et al. 2000) is also implemented.

While the SUMO driving rules are designed to avoid accidents, these can be artificially created by requiring a vehicle to stop in the road for a sufficiently long time. As a consequence of accidents or other unexpected events, vehicle rerouting might be needed during the simulation. This can be done in SUMO by invoking the rerouter module, which changes a vehicle's route at run time.

During the simulation phase, several statistics of interest are collected by SUMO and returned as output of the simulation. These include vehicle-related data (e.g., a raw dump of a selected vehicle's position over time), road-related data (e.g., average flow traversing a certain road segment), trip-related data (e.g., average trip duration), and traffic light-based data (e.g., number of changes of each traffic light). Besides this standard output, personalized output can be obtained through the definition of *detectors*, which can be thought of as user-defined checkpoints in the road network where a certain quantity of interest can be tracked over time.

13.3 Integrating Vehicular Mobility and Wireless Network Simulation

We have so far described microscopic vehicular mobility models, highlighting their complexity if traffic rules and driver behavior are included in the model (cf. the SUMO model). However, what was described above represents only a part of the tasks to be accomplished if the goal of a network designer is to evaluate the performance of a vehicular network. In fact, a major difference between the simulation of other types of short-range wireless networks and the simulation of vehicular networks is that in most cases applications based on vehicular communications running onboard vehicles are expected, or even *designed*, to influence driver behavior.

Consider for instance the case of an active safety application such as the ones mentioned in Chapter 11 – say, intersection collision warning (ICW). The purpose of such an application is to influence driver behavior, and in particular to allow the driver to reduce speed when approaching the intersection and come to a complete stop before it. In turn, changes in the mobility pattern of the warned vehicle will trigger similar changes in the mobility patterns of following vehicles. Hopefully, combinations of these modified mobility patterns will result in preventing a possible accident at the approaching intersection. Thus, the mobility pattern of vehicles should be modified on-the-fly in order to account for the effects of an ICW application running on the vehicular network. Another example of possible effects of vehicular network-based applications on mobility pattern is that of a smart navigation system, which exploits real-time information about traffic status gathered through the vehicular network to dynamically change the route to the destination in order to minimize travel time.

Note that the above situation – that is, applications that influence mobility patterns of the wireless nodes – is peculiar to vehicular networks: in all other application scenarios of short-range wireless networks considered in this book, with the partial exception of wireless sensor networks with intentional mobility, user applications running on the wireless nodes have no direct effect on a node's mobility pattern. Consider, for instance, a typical WLAN scenario. In this case, the user displays her/his own mobility patterns and is simply interested in staying connected to the network while moving.

What are the implications for the simulation procedure of the above observation about the intertwining between vehicular network applications and vehicle mobility patterns? The most important implication is that if accurate performance evaluation of vehicular network applications is sought, a close *run-time* interaction between the mobility and networking components of the simulator should be realized. In particular, it is not possible to feed the networking simulator with pre-computed mobility traces (as done extensively, for instance, in the simulation of opportunistic networks – see Part Six of this book).

Different architectural choices can be undertaken when designing a vehicular network simulator, which can be roughly divided into two categories. The first category comprises solutions in which the mobility and networking component of the simulator are closely integrated, essentially forming a single simulator. This is the case, for instance, for the vehicular network simulators presented in Harri et al. (2006) and Mangharam et al. (2005). Simulators in this class are typically realized by extending open source network simulators such as Ns2 with mobility models resembling vehicular mobility. However, due to the enormous challenges related to accurately modeling vehicular mobility as described earlier in this chapter, the mobility models integrated into the network simulator are typically relatively simple and inaccurate, thus impairing simulation accuracy. The alternative approach that has been pursued in the literature is to couple existing vehicular mobility and networking simulators by developing suitable tools/interfaces allowing the two simulators to interact with each other at run time. This is the approach undertaken, for instance, by Eichler et al. (2005), Lochert et al. (2005), and Wegener et al. (2008). The potential advantage of this approach is that both the mobility and networking component of the coupled simulator can be very accurate, if adequately chosen. On the other hand, simulation running time could become an issue.

In the remainder of this section, we will describe in detail a tool, called TraCI (Wegener et al. 2008), for coupling vehicular mobility and networking simulators. While in principle the tool could work in combination with arbitrary mobility and networking simulators, in the following we present its use with the SUMO vehicular mobility and Ns2 networking simulators, which are both relatively accurate and open source simulators.

13.3.1 The TraCI Interface for Coupled Vehicular Network Simulation

TraCI (*Tra*ffic *C*ontrol *I*nterface) is a tool for interlinking vehicular traffic and wireless networking simulators presented in Wegener et al. (2008). The main idea of TraCI is to run vehicular mobility and networking simulation in a stepwise fashion, alternating the simulation of a mobility step with that of a networking step, and having the outcome of each step given as input to the next step.

The stepwise approach to simulation realized through TraCI is pictorially represented in Figure 13.6. At each round, a mobility step is performed taking as input the outcome of the mobility step and the networking step in the previous round; the generated outcome is fed both to the networking step in the same round and to the mobility step in the next round. Similar to the mobility step, the networking step takes as input the outcome of the networking step in the previous round, and feeds the output to the mobility and networking step in the next round.

In order to realize the stepwise simulation approach described above, the authors extend SUMO and Ns2 so that they can interact according to the client–server paradigm through a Transmission Control Protocol (TCP) connection. More specifically, a *TraCI-Server* component is added to SUMO,

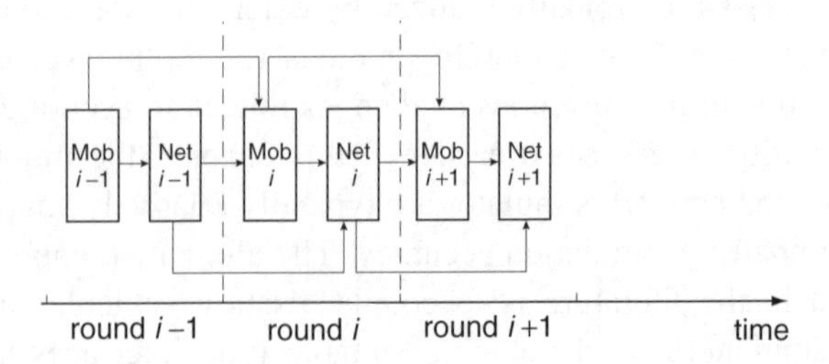

Figure 13.6 The TraCI stepwise simulation approach for coupling mobility and networking simulation.

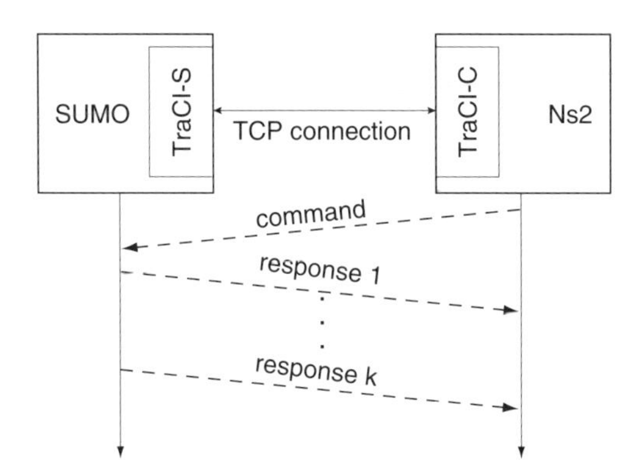

Figure 13.7 The TraCI client–server architecture.

and a *TraCI-Client* component is added to Ns2 (see Figure 13.7). The rationale for this choice is that the vehicular network application – which can potentially influence mobility patterns, as described above – runs in Ns2; hence, it is Ns2 that must be able to influence the SUMO simulation, which occurs through invocation of specific commands generated by the client application extending Ns2 and that are served by the server application extending SUMO.

More specifically, the TraCI-Client can send the TraCI-Server a number of commands, which are grouped in three functional categories:

1. **Simulation management**: this group comprises a command to invoke the execution of a simulation step on the mobility simulator.
2. **Atomic mobility primitives**: this group comprises commands used to modify mobility patterns of individual vehicles. Examples of mobility primitives are *change speed*, *change lange*, *change route*, etc. A minimal set of such primitives allowing the modeling of a wide range of vehicular networking applications is identified in Wegener et al. (2008) by considering the VANET application taxonomy proposed by the Car-to-Car Communication Consortium (Baldessari et al. 2007).
3. **Environmental commands**: this group comprises commands used to get scenario parameters, mostly at simulation setup time. For instance, there is a command to acquire the x and y dimensions of the simulated region – which is maintained in SUMO. Other commands in this group are also used at run time, such as the *driving distance* command to obtain the path and estimated distance for traveling between two points on the map. This command can be used, for instance, to feed a smart navigation application running in the networking simulator.

A number of response messages are sent back from the TraCI-Server to the TraCI-client in response to specific commands. These include:

- *move node* messages, reporting the status (position) of the nodes in the map;
- *status* messages, reporting to the client whether a certain command (e.g., *change speed* at a certain node) has been correctly executed.

For more details on TraCI and examples of its use, the reader is referred to Team (2011b).

References

Bai F, Sadagopan N and Helmy A 2003 IMPORTANT: A framework to systematically analyze the impact of mobility on performance of routing protocols for ad hoc networks. *Proceedings of IEEE Infocom*, pp. 825–835.

Baldessari R, Bodekker B, Deegener M, Festag A, Franz W, Kellum C, Kosch T, Kovacs A, Lenardi M, Menig C, Peichl T, Rockl M, Seeberger C, Strassberger M, Stratil H, Vogel H, Weyl B and Zhang W 2007 Car-2-Car Communication Consortium – manifesto (version 1.1). *http://www.car-2-car.org/index.php?id-570*.

Eichler S, Ostermaier B, Schroth C and Kosch T 2005 Simulation of car-to-car messaging: Analyzing the impact on road traffic. *Proceedings of MASCOTS*, pp. 507–510.

Harri J, Filali F, Bonnet C and Fiore M 2006 Vanet-MobiSim: Generating realistic mobility patterns for VANETs. *Proceedings of ACM VANET (poster proceedings)*, pp. 96–97.

Lochert C, Caliskan M, Scheuermann B, Barthels A, Cervantes A and Mauve M 2005 Multiple simulator interlinking environment for IVC. *Proceedings of ACM VANET (poster proceedings)*, pp. 87–88.

Mangharam R, Weller D, Stancil D, Rajkumar R and Parikh J 2005 GrooveSim: A topography-accurate simulator for geographic routing in vehicular networks. *Proceedings of ACM VANET*, pp 59–68.

Team S 2011a SUMO – Simulation Of Urban Mobility. *http://sumo.sourceforge.net/*.

Team T 2011b Traffic Control Interface. *http://sourceforge.net/apps/mediawiki/sumo /index.php?title = TraCI*.

Tian J, Hahner J, Becker C, Stepanov I and Rothermel K 2002 Graph-based mobility model for mobile ad hoc network simulation. *Proceedings of the ACM/IEEE Annual Simulation Symposium*, pp. 337–344.

Treiber M, Hennecke A and Helbing D 2000 Congested traffic states in empirical observations and microscopic simulations. *Physical Review E* **62**(2), 1805–1824.

Wegener A, Piòrkowski M, Raya M, Hellbruck H, Fischer S and Hubaux JP 2008 TraCI: An interface for coupling road traffic and network simulators. *Proceedings of the ACM Communications and Networking SimulationSymposium (CNS)*, pp. 155–163.

Part Five

Mobility Models for Wireless Sensor Networks

The focus in this part of the book is on wireless sensor networks, a wide class of emerging short-range wireless networks. After briefly introducing the state of the art, use cases, and prospects in wireless sensor networking, we will present representative examples of the two main classes of mobility models for wireless sensor networks: *passive* mobility models, aimed at modeling the movement of wireless sensors caused by external forces (e.g., wind, animal movement, ocean currents, etc.); and *active* mobility models, aimed instead at modeling the movement of wireless sensors endowed with self-propelled motion capabilities.

14

Wireless Sensor Networks

In this chapter, we will briefly describe the main features of a wide class of emerging short-range wireless networks called *wireless sensor networks*. After presenting the state of the art in wireless sensor networking and related technologies, we will describe a few representative use cases. Finally, we conclude the chapter with a short overview of the envisioned technological evolution and challenges in wireless sensor networking.

14.1 Wireless Sensor Networks: State of the Art

Wireless sensor networks are a class of short-range wireless networks composed of a number of *sensor nodes* and one or more *base stations* (also called *sinks*) communicating with each other through a wireless transceiver.

A sensor node is typically composed of several parts: a wireless transceiver for communication, one or more sensors (e.g., temperature, pressure, humidity, accelerometer, etc.), a microcontroller unit, and an energy source, typically a battery. In some cases, sensor nodes are also equipped with actuators allowing the node to undertake some action (e.g., reducing illumination) in response to specific environmental conditions (e.g., a too high illumination level). If a network is composed of nodes, some of which also include actuators, it is called a *wireless sensor and actuator network*. The size, complexity, and cost of a sensor node vary widely depending on the specific application scenario: there can be sensor nodes as small as a coin costing a few dollars, as well as sensor nodes as large as a shoebox costing hundreds of dollars.

Sink nodes are relatively more powerful nodes, which typically are not equipped with sensors/actuators, but they have a more powerful processing unit, a possibly more powerful wireless transceiver, and other network interfaces usually allowing connection to the Internet. Furthermore, sink

Mobility Models for Next Generation Wireless Networks: Ad Hoc, Vehicular and Mesh Networks, First Edition. Paolo Santi.

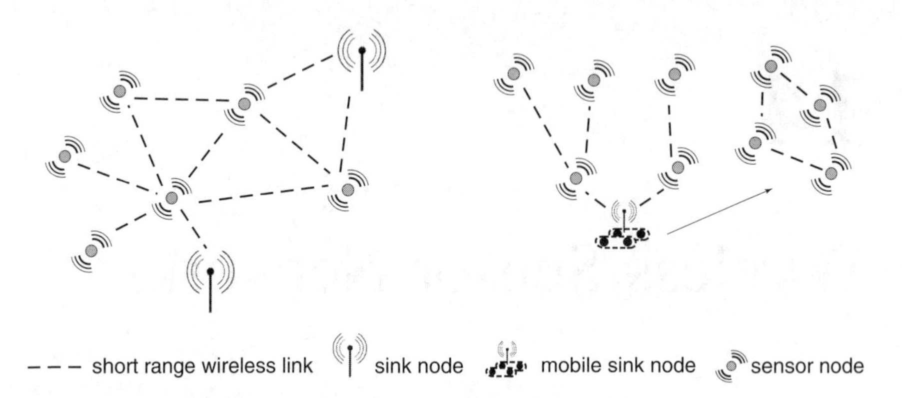

Figure 14.1 Typical wireless sensor network architectures.

nodes are typically power-plugged, so energy consumption, which is a primary constraint for sensor nodes, is not critical for sink nodes. A special class of sink nodes are the *mobile sinks* – typically realized by mounting the sink node on a mobile robot – whose purpose is to travel throughout the wireless sensor network deployment area and gather the data collected by the sensor nodes.

A typical wireless sensor network architecture is shown in Figure 14.1: several sensor nodes are located in a geographical region and use wireless transceivers to set up wireless links between themselves. One or more sink nodes are co-located in the monitored region; these nodes act as data collection points, and possibly also realize network coordination functionalities. Due to severe energy constraints on the battery-equipped sensor nodes, a wireless transceiver on the sensor nodes transmits at low power but, consequently, the transmission range is relatively short – in the order of few tens of meters typically. Then, multi-hop communication is commonly a prerequisite for transmitting data gathered at remote sensor nodes to the sink node(s). In most application scenarios, both the sensor nodes and the sink nodes are stationary – Figure 14.1, left. In other cases, though, sensor nodes (or, at least, a portion of them) are mobile, for instance, when sensor nodes are installed on animals to monitor their movement. Finally, in some cases the sensor nodes are stationary while the sink nodes are moving, with the purpose of visiting the sensor nodes and directly gathering their collected data – Figure 14.1, right.

The wireless sensor network concept was introduced in the late 1990s, and has become more and more popular in recent years as technology has turned this concept into reality. Currently, several wireless sensor network hardware and software platforms are available on the market, and significant standardization activities have been going on in the past few years (see below). Wireless sensor network prototypes have been extensively deployed for research purposes, and wireless sensor network-based solutions are becoming available on the market. However, as we will see in the next section, the application scenarios are so different that wireless sensor network-based solutions must be almost completely redesigned for each different scenario, negatively impacting the profitable use of this technology.

14.1.1 Hardware and Software Platforms

A number of hardware and software platforms for wireless sensor networking have become quite popular in recent years, at least among the research community. Indeed, due to the fact that sensor nodes are embedded devices, these platforms typically are strongly integrated. Thus, buying a certain type of sensor node usually "forces" the designer to use a specific embedded operating system and software platform.

The most popular hardware platform for sensor nodes is that of Motes, initially designed at the University of California, Berkeley and nowadays produced by a number of companies including Intel, Crossbow, etc. Mote nodes are composed of a relatively small motherboard (the size of two AA batteries or even less) including a microcontroller unit and wireless transceiver, and one or more daughter boards hosting different types of sensors.

The embedded operating system which was developed for Mote nodes at UC Berkeley is called TinyOS (Team 2011b), and is currently the most popular operating system for wireless sensor networks. TinyOS is an event-based operating system designed to account for specific features of wireless sensor networks, and includes support for energy-saving techniques, in-network data aggregation, procedures for gathering data to sink nodes, etc. Since the memory size on sensor nodes is very limited, special care has been taken in reducing the TinyOS memory footprint as much as possible. Currently, TinyOS can be used in combination with a number of hardware platforms including the family of Motes (Mica, Mica2, Mica2Dot, MicaZ, TMote Sky, iMote), the Eyes platform developed at Technische Universität erlin, and so on.

Another relatively popular operating system for wireless sensor networks is Contiki (Team 2011a), developed at the Swedish Institute for Computer Science (SICS). One of the main features of Contiki is that a relatively advanced programming model including "proto"-threads and dynamic memory allocation is made available to the user on top of the event-driven operating system kernel. Currently, Contiki supports a number of hardware platforms including the Mote family.

14.1.2 Standardization Activities

Several standards for wireless sensor networks are currently either ratified or under development. The most important standard, first approved in 2003, is IEEE 802.15.4 (IEEE 2011), which defines the PHY and MAC layer specifications for low-cost, energy-efficient wireless personal area networks. Since the emphasis in IEEE 802.15.4 is on reducing energy consumption by as much as possible, data rates are relatively low and the communication range relatively short (between 10 and 75 m). Furthermore, support for transmit power control is included in the specifications. At the PHY layer, the standard operates in the ISM radio bands: 868 MHz in Europe, 915 MHz in the USA and Australia, and 2.4 GHz in most regions worldwide. The data rates are 20/40 kbps in the 868 and 915 MHz bands (increased to 250 kbps in the 2006 release of the standard), and 250 kbps in the 2.4 GHz band. At the MAC layer, the standard implements CSMA/CA, in a manner similar to that defined in the IEEE 802.11 standard.

The IEEE 802.15.4 standard defines two types of network nodes: the *full-function device* (FFD), which can serve as the coordinator of a personal area network (PAN) or as a common node, and includes advanced communication capabilities such as relaying messages; and the *reduced-function device* (RFD), which is a simplified device with limited communication capabilities, and can communicate only with a FFD.

IEEE 802.15.4 allows the formation of two types of network topology (see Figure 14.2): the *star topology*, in which all devices communicate directly with a central device called a *PAN coordinator*, and the *peer-to-peer topology*, in which devices can set up an ad hoc network with peer-to-peer wireless links.

While IEEE 802.15.4 defines specifications for the lower layers of the network architecture, other standards have been defined for upper layer protocols. ZigBee is a specification (first released in 2005) for a suite of protocols encompassing the network layer and above, for networks based on 802.15.4 wireless links (Alliance 2011). As in IEEE 802.15.4, the main goal of ZigBee is to reduce the energy consumption of devices: just to give an idea of this emphasis on reduced power consumption, an individual device must have a battery lifetime of at least two years to pass ZigBee certification.

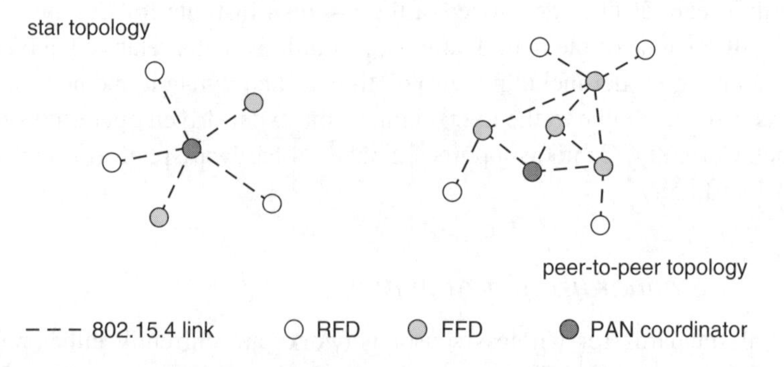

Figure 14.2 IEEE 802.15.4 network topologies.

Starting from the PHY and MAC layers defined in IEEE 802.15.4, ZigBee introduces four main components into the standard: the network layer, the application layer, ZigBee Device Objects (ZDOs), and manufacturer-defined application objects.

At the network layer, ZigBee uses ad hoc on-demand distance vector (AODV) routing (originally designed for ad hoc networks) to route messages within the network. The application layer is the highest level layer defined by the specification, and it comprises the majority of components added by the ZigBee specification: ZDOs and relative management procedures, together with application objects defined by manufacturers. The ZDO is responsible for defining the role of a device as a *coordinator*, *router*, or *end device*. The coordinator is the most capable ZigBee device: it forms the root of a network tree, and it might act as a bridge to other ZigBee networks. There is exactly one coordinator in each ZigBee network, which is the device that started the network originally. The router is a device which can act as an intermediate router, relaying messages from other devices. Finally, the end device is the simplest device, and it encompasses functionality to communicate with the parent node (either the coordinator or a router). Several individual ZigBee networks can be connected to form a larger wireless sensor network with mesh topology (see Figure 14.3).

Similar to what happens in WLANs with the WiFi Alliance promoting usage of 802.11-based WLANs, the ZigBee Alliance is a group of companies that maintain and promote usage of the ZigBee standard and underlying 802.15.4 technology.

Another standard for higher layer protocols which is based on 802.15.4 PHY and MAC layer specifications is WirelessHART (HARTCommunicationFoundation 2011),first released in 2007 and the wireless evolution of the pre-existing HART standard for process measurement, control, and asset management applications. Given the considered application scenario, emphasis in the standard is put on reliability, security, and effective power

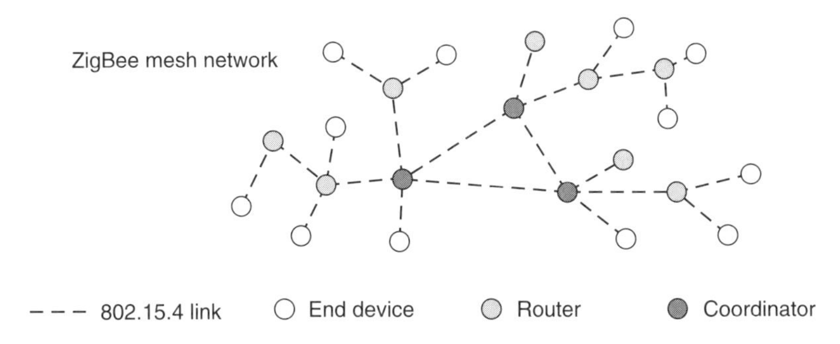

Figure 14.3 Example of ZigBee mesh network.

management. The standard operates in the 2.4 GHz band only, and ensures full compatibility with legacy HART devices.

14.2 Wireless Sensor Networks: User Scenarios

As mentioned previously, wireless sensor networks find application in many diverse user scenarios. Roughly, we can divide these scenarios into the following categories: environmental monitoring, industrial monitoring, health and well-being monitoring, precision agriculture, seismic and structural monitoring, intrusion detection, and tracking of objects, people, or animals. In the following, we will briefly describe representative user scenarios in each of these categories.

14.2.1 Environmental Monitoring

Wireless sensor networks (WSNs) can be used to remotely monitor relevant environmental parameters in relatively large geographical areas. For example, a wireless sensor network can be deployed in an urban area to monitor pollution levels. In this case, sensor nodes are equipped with sensors able to detect polluting gases and particles. The advantage of using a WSN instead of traditional systems to monitor pollution in a city is that, due to the use of wireless technology, sensor deployment is much cheaper, and finer grain monitoring can be obtained at affordable costs. In other cases, WSNs can be used to monitor environmental parameters in dangerous regions, such as volcanic regions, where the use of wiring is virtually impossible. Another example of possible WSN usage is to monitor land movements in order to prevent or quickly notify about landslide events. Similarly, WSNs can be used to monitor glaciers and the status of snow in order to quickly alert about possible avalanches. Finally, another user scenario worth mentioning is forest fire detection.

In environmental monitoring scenarios, data collected by the WSN (possibly suitably aggregated by the sensor nodes themselves) is conveyed to the sink node(s), and from there sent to the user for remote monitoring (continuous data monitoring model). In some cases, an automatic alert system could be implemented in order to quickly inform the human operator when environmental parameters exceed certain thresholds, corresponding to the occurrence of possibly dangerous events (event-driven model).

A special class of WSN that can be used for monitoring purposes is the class of underwater WSNs. In this class, communication between sensor nodes is realized through either acoustic or optical communications, since radio waves do not propagate underwater. The challenges faced by a network designer, especially concerning communication between sensor nodes, are quite different in an underwater environment compared to those faced in terrestrial applications.

14.2.2 Industrial Monitoring

WSNs can be used to monitor complex industrial processes, such as in oil refineries, water or wastewater monitoring, power plants, nuclear plants, etc. In these scenarios, use of wireless technology offers unique cost reduction opportunities: it is in fact well known that a major cost in industrial process monitoring is related to wiring costs. However, the security and reliability of the WSN comes to the fore in this class of applications, considerably challenging the WSN designer.

In many cases, WSNs employed in industrial monitoring are wireless sensor and actuator networks: for instance, if a dangerous pressure level is detected in a pipe in an oil refinery, a certain actuator can be activated to turn a valve and reduce pressure. In this way, immediate action can be undertaken by the wireless sensor and actuator network itself in response to a potentially dangerous situation, without requiring the intervention of a human operator.

14.2.3 Health and Well-Being Monitoring

Another important class of WSN user scenarios is related to monitoring the health and well-being of people. For instance, a small WSN can be attached to a patient suffering from heart disease to continuously monitor blood pressure, heart rate, etc. The data collected from the WSN can be transmitted wirelessly to a sink node located in the home. If some abnormal condition is detected, the sink node issues an alert message to the nearest hospital asking for medical assistance. This way, the patient is free to move around at home without any physical restriction, while her/his health status is continuously monitored. Another application of WSNs is in monitoring the status of participants in large sporting events, such as a marathon: by equipping each participant with a wireless device endowed with pressure, temperature, etc., sensors, the health status of each athlete can be remotely monitored. An application of WSNs related to well-being is when sensor nodes are attached to different parts of the body (wrist, elbow, shoulder, knee, etc.) in order to monitor whether a certain physical exercise is performed correctly.

14.2.4 Precision Agriculture

WSNs can be used to provide fine-grained monitoring of crops in agriculture. For instance, WSNs can be used to monitor the sugar content of grapes, so as to judge the right level of ripening needed for a certain wine production. WSNs can also be used to monitor the level of humidity in the ground surrounding crops, so as to optimize use of the irrigation system. Note that these applications would be extremely expensive and difficult to realize without using a wireless technology.

14.2.5 *Seismic, Structural, and Building Monitoring*

Another important class of WSN applications is concerned with monitoring large structures or buildings. For instance, wireless sensor nodes can be installed at specific points in a large structure such as a bridge, in order to monitor movement of the various parts comprising the structure, and to promptly detect possible critical conditions. This way, engineers could monitor the "health" status of the structure remotely, avoiding costly and time-consuming site visits. Another advantage of using WSNs for the purpose of structural monitoring is the availability of a large amount of continuously collected data, as compared to weekly or monthly data collected during physical site visits. The availability of such fine-grained and massive amounts of data allows a much more detailed study of the events experienced by the structure, as well as better monitoring of its structural health.

Wireless sensor and actuator networks can also be used to monitor and optimize the management of a large building: sensors can be used to monitor temperature, illumination, humidity, etc. within the building, and actuators can be used to undertake appropriate actions in response to some detected conditions. For instance, if the temperature in a room is too high and illumination too intense, window shutters in that room could be automatically closed. The ultimate goal of a wireless sensor and actuator network used in building monitoring and management is to reduce energy consumption in the building, while at the same time providing optimal environmental conditions inside it.

14.2.6 *Intrusion Detection*

WSNs can be used also to detect intruders in a certain area or building. For instance, movement sensors, typically complemented with some camera-equipped sensor node, can be installed on a fence surrounding a restricted access area, in order to detect possible intrusions. Typically, WSNs used for intrusion detection obey the event-driven model, that is, an alert message is sent to a remotely located human operator to warn of a possible intrusion. On receipt of the message, the human operator can activate and operate wirelessly controlled cameras, to check whether an intruder is actually trying to access the restricted area.

14.2.7 *Tracking of Objects, People, and Animals*

Another class of WSN applications is related to tracking objects, people, or animals. For instance, WSNs can be used to track the movement of objects in a large warehouse. Tracking objects or people finds application in military scenarios, for instance, to track the movement of soldiers in the field or the movement of tanks and other military vehicles along roads. Finally, WSNs can be used to track the movement of both domestic (e.g., herds) and wild animals. As a matter of fact, one of the first examples of a working WSN prototype is ZebraNet (Martonosi et al. 2004), a WSN composed of a small number of radio-equipped zebras living in the Sweetwaters Reserve in Kenya.

14.3 WSNs: Perspectives

While WSN technology is relatively mature, due especially to the introduction and maturing of the IEEE 802.15.4 and related standards, its applications are still mostly at the prototype stage. Thus, a major development that we envision in the coming years is the evolution of WSNs into a commercially profitable technology. As briefly mentioned earlier in this chapter, a factor currently slowing down the wide adoption of WSN-based solutions on the market is the numerous and diverse applications of WSNs (see Section 14.2) such that the WSN designer is forced to almost completely redesign the network for each different application. Thus, this market, which is potentially very large given the very wide range of possible applications, is partitioned into several different "niche" markets, almost in a one-to-one relationship with possible applications. Clearly, this market fragmentation is a factor negatively impacting the development of profitable WSN-based solutions.

In terms of technology, two major challenges are yet to be fully addressed. The first refers to the reliability and security of WSN technology: while reliability and security are an issue in wireless technologies in general, they are even more severe in WSNs, due to the use of low-cost, low-power network devices in a potentially harsh environment. It is important to observe that reliability and security are prerequisites for introducing WSN-based solutions in one of the potentially more profitable niche markets, namely, that of industrial monitoring applications. Thus, significant research efforts must be undertaken in order to improve the degree of reliability and security level provided by WSNs.

The second major technological challenge relates to the use of energy harvesting techniques to prolong WSN lifetime. In fact, WSN lifetime is perceived as another major factor currently limiting the widespread use of WSN-based solutions, and energy harvesting techniques – using, for example, micro solar panels, extracting energy from vibrations, etc. – are increasingly being considered as a possible way of integrating the energy provided by batteries and significantly extending WSN lifetime. How to optimize the WSN design in order to fully exploit the harvested energy is currently a very active research field, and it is likely to remain so for several years to come. In fact, WSNs have so far been designed to reduce energy consumption by as much as possible, but under the assumption that the (limited) energy is continuously available over time. Instead, the amount of energy that can be harvested from the environment changes significantly over time, often very quickly (think about the impact on solar energy production of a cloud obscuring the Sun). Thus, the WSN should be optimized to operate in different conditions for the amount of energy available, and methods should be devised to quickly detect changes in the amount of this energy.

14.4 Further Reading

This chapter presented only a very short description of wireless sensor networking technology and related issues. The reader interested in gaining a better understanding of this topic is referred to the several survey papers and books available in the literature, such as the survey by Akyildiz et al. (2002) and the more recent survey by Yick et al. (2008). Suggested books include those by Karl and Willig (2007), Akyildiz and Vuran (2010), and Zhao and Guibas (2004).

References

Akyildiz I and Vuran M 2010 *Wireless Sensor Networks (Advanced Texts in Communications and Networking)*. John Wiley & Sons, Chichester.

Akyildiz I, Su W, Sankarasubramanian Y and Cayirci E 2002 A survey on sensor networks. *IEEE Communications Magazine* **40**(8), 104–112.

Alliance Z 2011 *http://www.zigbee.org/*.

HARTCommunicationFoundation 2011 Wireless hart technology *http://www.hartcomm.org/protocol/wihart/wireless_technology.html*.

IEEE 2011 *http://www.ieee802.org/15/pub/TG4.html*.

Karl H and Willig A 2007 *Protocols and Architectures for Wireless Sensor Networks*. John Wiley & Sons, Chichester.

Martonosi M, Lyon S, Peh LS, Poor V, Rubenstein D, Sadler C, Juang P, Liu T, Wang Y and Zhang P 2004 *http://www.princeton.edu/mrm/zebranet.html*.

Team C 2011a Contiki operating system. *http://www.sics.se/contiki/about-contiki.html*.

Team T 2011b Tiny operating system. *http://www.tinyos.net*.

Yick J, Mukherjee B and Ghosal D 2008 Wireless sensor network survey. *Computer Networks* **52**(12), 2292–2330.

Zhao F and Guibas L 2004 *Wireless Sensor Networks: An Information Processing Approach*. Morgan Kaufmann, San Francisco.

15

Wireless Sensor Networks: Passive Mobility Models

As we saw in the previous chapter, WSNs find application in very diverse user scenarios. Although in many scenarios there is no mobility involved, applications of WSNs where mobility plays a role do exist. We can distinguish three different kinds of mobility in WSNs: (i) mobility of the sensor nodes; (ii) mobility of the sink node(s); and (iii) mobility of a monitored event/object. The first kind of mobility occurs when at least part of the sensor nodes is mobile. Examples of sensor node mobility are when sensor nodes are dispersed and free to move in the monitored region (e.g., sensor nodes floating on the surface of the ocean), or when they are installed on animals for wildlife tracking and monitoring. The second kind of mobility refers to a situation in which sink nodes are able to autonomously move in the monitored region, with the purpose of collecting data from the sensor network. Finally, the third kind of mobility occurs when a WSN is used for monitoring or tracking purposes, and works under the event-driven data model. In this case, modeling the occurrence and mobility of an event to be monitored (e.g., a gas leak and movement of the resulting gas plume) is useful for understanding the resulting data traffic pattern in the WSN. Similarly, when the WSN is used for target tracking, modeling target movement is useful for estimating the amount and pattern of data generated in the WSN during target tracking.

These types of WSN mobility typically occur in isolation: that is, if sensor nodes are mobile because, say, they are floating on the surface of the ocean, sink nodes are typically fixed – they are, for instance, installed on buoys. On the other hand, when mobile sink nodes are used to gather data collected by a WSN, the sensor nodes comprising the WSN typically are fixed, and so on.

Mobility Models for Next Generation Wireless Networks: Ad Hoc, Vehicular and Mesh Networks, First Edition. Paolo Santi.

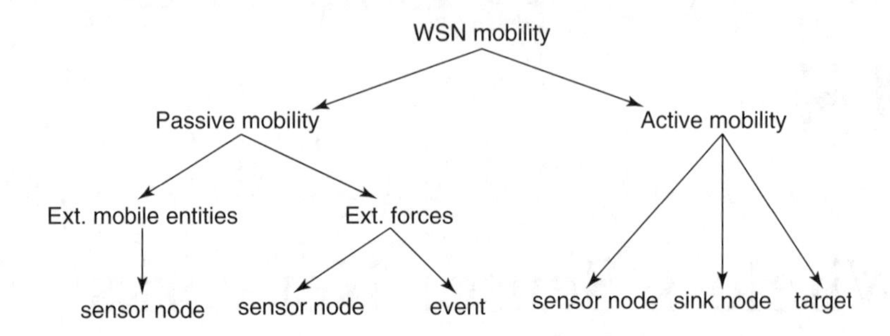

Figure 15.1 A taxonomy of mobility models for WSNs.

Cases in which at least two of the three types of WSN mobility coexist are relatively scarce, although possible.

Independently of whether mobility occurs in the sensor nodes, in the sink nodes, or in the target/monitored event, we can distinguish two radically different forms of mobility in WSNs: *passive* and *active* mobility. In the former case, mobile entities are not capable of performing autonomous movements, and their mobility is caused by some external force, such as animal movements, wind, ocean flows, and so on. In the latter case, mobile entities are instead capable of performing autonomous movements, that is, they can decide when to move and stop, and the destination/trajectory of their movements. In this chapter, we will present representative models of passive mobility in WSNs, while the next chapter will be devoted to active mobility models.

A taxonomy of mobility models for WSNs is displayed in Figure 15.1. Mobility can be either passive or active. Passive mobility can be caused either by mobility of the entity to which a sensor node is attached, or by external forces. Passive mobility caused by external forces can model movement both of sensor nodes and of a natural event to be monitored. Active mobility can refer to active mobility of sensor nodes, of sink nodes, or of a target being tracked by the WSN – see next chapter.

15.1 Passive Mobility in WSNs

As mentioned above, passive mobility in WSNs occurs when the mobile entities are not capable of performing autonomous movements, but move in response to external forces. Two main classes of passive mobility are present in WSNs, both of them referring to the case in which sensor nodes are mobile. The first class is when sensor nodes are installed on mobile entities (e.g., people, animals, moving vehicles, etc.), in which case the mobility of sensor nodes becomes equivalent to the mobility of the entity carrying the sensor node. The second class of passive mobility encompasses situations in which sensor nodes are not fixed, but are free to move in the environment in response to some external force such as wind, ocean flows, and so on. In this case, mobility of sensor nodes can be modeled through modeling the external force and the effects of this force on a sensor node. Models of natural forces can also be used to estimate the movement of an event (say, a gas plume) monitored by a WSN.

In the remainder of this chapter, we will present representative models for both classes of passive mobility: in the next section we will describe two mobility models for wildlife tracking applications, while in Section 15.3 we will describe a mobility model aimed at modeling the dispersion of sensor nodes floating in a large body of water.

15.2 Mobility Models for Wildlife Tracking Applications

In this section, we present two mobility models for applications of WSNs to wildlife tracking scenarios. In these scenarios, sensor nodes are installed on animals, so mobility models for this class of applications aim at closely resembling animal movements. The first is the model introduced in Juang et al. (2002) as an outcome of the ZebraNet project (Martonosi et al. 2004). The second model was introduced in Small and Haas (2003) to resemble movements of a population of whales.

Besides using more specific mobility models such as the ones described below, simpler general-purpose mobility models could be used as well in wildlife tracking applications, at least for the purpose of gaining a first understanding of WSN performance. In particular, the Lévy flight model described in Section 4.2.2, which, we recall, is a random walk-like model in which erratic behavior in a close neighborhood is occasionally complemented with long-distance travel, has been shown in the literature to resemble animal movement quite accurately.

15.2.1 The ZebraNet Mobility Model

The ZebraNet mobility model was derived in Juang et al. (2002) as an outcome of the ZebraNet project, whose purpose was the fine-grain monitoring of movements of a population of zebras in the Laikipia ecosystem of central Kenya. A first observation made in the ZebraNet project was that, due to the features of the zebra social system as observed by zoologists, installing sensor nodes on selected male individuals was sufficient to monitor movements of a much larger population of zebras. In fact, zebras are organized in so-called "harems," formed of a single adult male zebra, 4–5 females, and their young offspring. Although females typically initiate movements, the males often adjust the direction and speed of movement of the group.

Thus, installing a single sensor node on an adult male zebra is sufficient for fine-grained tracking of a group of about 10–12 zebras.

Zebra movement patterns can be characterized in terms of three main states: *grazing*, *graze–walking*, and *fast-moving*. Zebras spend most of their time in a grazing state, during which mobility is characterized by low movement rates and high turning angles. In the graze–walking state, zebras walk deliberately, with heads lowered, and clip vegetation as they move. Mobility in this state is characterized by higher movement rates and smaller turning angles than those displayed in the grazing state. Finally, zebras occasionally move much more quickly (fast-moving state), usually because of the presence of predators, or because an area's vegetation has been exhausted. Mobility in this state is characterized by high movement rates and small turning angles.

Besides the three-state behavior, another characterizing feature of zebra mobility is that zebras look for water about once per day. When they reach a water source, they remain there for a relatively short time, after which they start behaving normally (alternating between grazing, graze–walking, and fast-moving states) until the next day. Finally, zebras tend not to have long periods of motionless sleep, due to the need to keep watch and avoid predators. Hence, the mobility pattern described above applies 24 hours a day.

In order to monitor zebra movements, the following parameters were tracked in the ZebraNet project with a periodicity of three minutes, which is considered by zoologists as an optimal time interval for sampling animal movements (Altmann 1974): the *movement distance* and the *turning angle*. The movement distance is defined as the net distance between the position of the zebra at the beginning and end of a three-minute interval. Similarly, the turning angle is defined as the absolute value of the angle between the position of the zebra at the beginning and end of a three-minute interval.

The distribution of observed movement distances is shown in Figure 15.2. The three movement statuses can be identified quite well in the distance distribution, which appears to be bimodal. In the first mode – grazing – the mean net distance is 3.1 m, with a peak at 2 m; in the second mode – graze–walking – the mean net distance is 13 m with a peak at 14 m. Finally, the few outliers at the tail of the distribution correspond to the fast-moving status.

The distribution of observed turning angles is shown in Figure 15.3. Different from the movement distribution, the turning angle distribution does not present a shape allowing a clear distinction of the mobility states. In general, we can observe that most observed angles lie in the 30–90 degree interval, with a peak at 45 degrees.

Based on the above observations, Juang et al. (2002) define the following model to resemble the mobility of zebras. The mobility region is a square

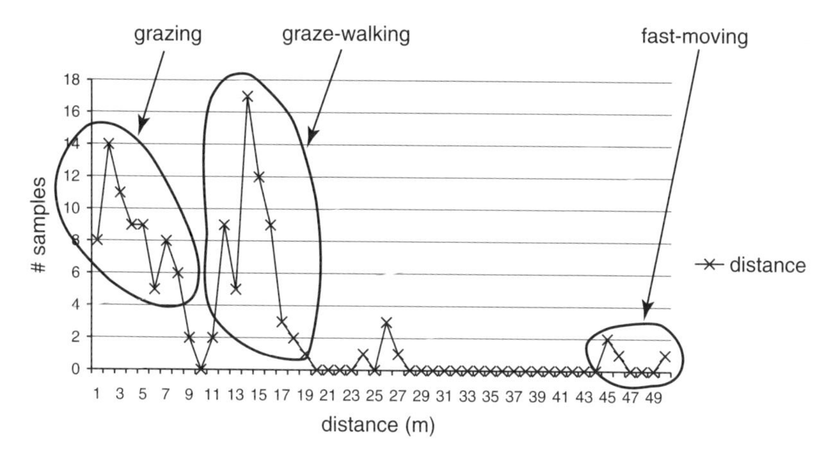

Figure 15.2 Distribution of movement distances covered by zebras.

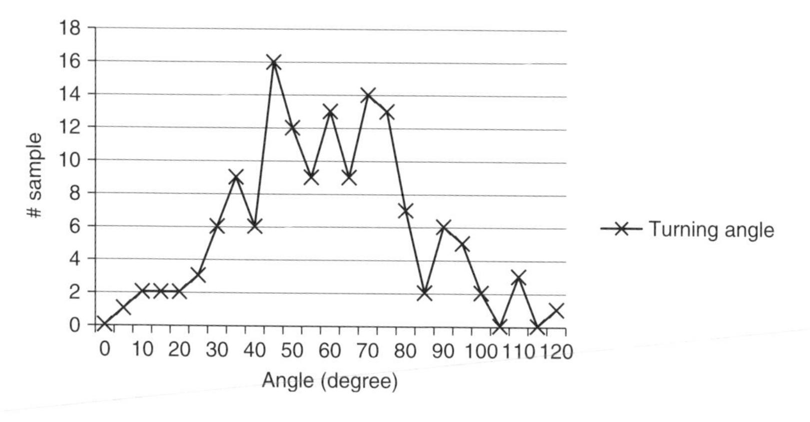

Figure 15.3 Distribution of zebra turning angles.

area of 20 × 20 km, approximately corresponding to the area of the Laikipia ecosystem. At the beginning of the simulation, 20 water sources are randomly located in the area, as well as 50 zebras. After initial deployment, each zebra (assumed to be the leading male of a group) starts moving independently of the others, according to the empirically derived distance and turning angle distributions shown in Figures 15.2 and 15.3. Zebras move at a base speed of 0.017 m/s while grazing, four times faster at 0.0723 m/s when graze–walking, and nine times faster at 0.155 m/s while fast-moving. Once per day, each simulated zebra becomes thirsty. When thirsty, the zebra moves into the graze–walking state and starts moving toward the closest water source.

15.2.2 The Whale Mobility Model

The whale mobility model was introduced in Small and Haas (2003) to model movement of radio-equipped whales. The idea of Small and Haas (2003) is to replace current whale tracking systems based on frequent offload of tracked data from single whales through satellite communications with a lower cost alternative based on WSNs. Their idea is to deploy a number of sink nodes in selected feeding grounds where whales are known to periodically return, and to gather data from whales through high data rate (compared to satellite communications) wireless links established between radio-equipped whales and sink nodes. Furthermore, collected data can be exchanged between whales when within each other's transmission range, thus allowing multi-hop data propagation to reach a sink node. Since whales are expected to come into contact with a sink node or another whale periodically, a relatively large amount of memory to store data needs to be installed on the sensor nodes deployed on whales. However, the higher costs induced by these higher memory needs are more than compensated by the reduced cost of the communication infrastructure.

In order to model mobility of whales in this application scenario, Small and Haas (2003) start from some observations made by biologists. In particular, it has been observed that whale movement is influenced by three main factors: *migration*, *grouping with other whales*, and *direction of the nearest feeding grounds*. Thus, mobility of whales in Small and Haas (2003) is modeled as the result of a sum of three suitably weighted vectors, representing the contribution to movement of migration, grouping, and feeding, respectively. More specifically, the weighted vector sum is used to determine the movement direction, while movement speed is selected according to the following rules: the speed of a whale is randomly chosen between $v_{min} = 2\,\text{m/s}$ and $v_{max} = 3\,\text{m/s}$ for whales outside the feeding grounds, and between $v_{min} = 0\,\text{m/s}$ and $v_{max} = 1\,\text{m/s}$ when within the feeding grounds. The movement status of each whale is updated every time step t, upon which new speed and direction values are generated for the next time step.

As mentioned above, direction of movement is computed as the weighted sum of three components. The first component is determined by the migration pattern. A migration direction is determined for each whale, which is defined as a preferred movement angle rather than an absolute direction. More specifically, the migration direction is expressed as the mean of a Gaussian distribution representing the migration angle. At each time step, a whale is assigned with a new migration direction randomly generated according to the Gaussian migration angle distribution, with a predetermined standard deviation. The role of the standard deviation in the migration pattern is quite evident: a larger standard deviation results in a more random migration pattern. The weight of the migration component of movement direction is fixed at 1.

The grouping component of movement direction is computed as follows. When a whale comes within a certain fixed range of other whales, it sets its grouping directional component in the direction of the closest female. When a whale becomes part of a group, its mobility is governed by the same migration and feeding behaviors of the other members of the group. The weight of the grouping direction is a Gaussian random variable, with the mean set to 0.8 and standard deviation set to 0.5. In order to allow whales to leave a group, a *maxGroupingTime* variable is introduced in the model, which is set to 3 in Small and Haas (2003). If the whales have grouped for longer than *maxGroupingTime*, then the weight of the grouping component of movement direction is set to 0 in the next time step.

The feeding component of the movement direction is generated as follows. First of all, a number of feeding grounds modeled as circular regions with a random radius between 20 and 30 km are generated in the movement region. Feeding grounds are assumed to be separated by a distance of about 200 km. The direction of the feeding component of the movement corresponds to the direction pointing to the closest feeding ground. The weight of the component is computed according to a variable *hungry*, modeling a whale's feeling of hunger. The variable is increased when the whale is swimming outside the feeding grounds, and decreased when it is swimming inside the feeding grounds. The weighting factor of the feeding component is computed according to a Gaussian distribution with mean $0.4 \cdot hungry$ and standard deviation $0.3 \cdot hungry$.

After the weighted sum of the three directional components has been computed, a motion consistency factor is introduced, in order to avoid motion patterns in which whales completely change direction at each time step. This factor takes values in the [0, 1] interval. Denoting by θ_i the direction at time step i, we have that the new direction θ_{i+1} at time step $i+1$ is computed as follows:

$$\theta_{i+1} = \theta_i \cdot cf + \theta'_{i+1} \cdot (1 - cf),$$

where $cf \in [0, 1]$ is the consistency factor, and θ'_{i+1} is the direction for time step $i+1$ as computed by the weighted directional sum described above. A pictorial view of the whale mobility model is shown in Figure 15.4.

15.3 Modeling Movement Caused by External Forces

In some application scenarios, movement of sensor nodes, or of a monitored event, can be caused by external natural forces, such as wind, ocean flows, and so on. In such cases, mobility models can be derived from models used

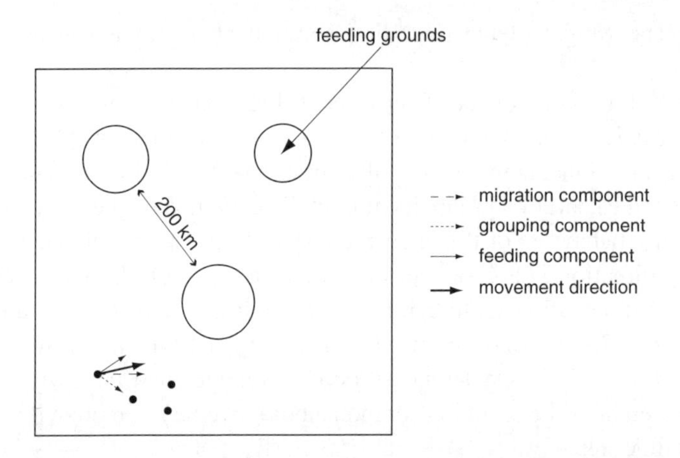

Figure 15.4 Pictorial view of the whale mobility model.

to describe the natural phenomena exerting the force. In this section, we present one of the few mobility models for WSNs aimed at modeling the effect of a natural force on sensor nodes. The model was introduced in Caruso et al. (2008) to model the effect of ocean flows on underwater mobile sensor networks, and is called the meandering current mobility model.

The application scenario considered in Caruso et al. (2008) is an underwater sensor network in which a number of floating or drifting sensor nodes are released underwater in a relatively small area, and are then left free to move with the underwater ocean flows. Sensor nodes are able to communicate with each other, if close enough, through acoustic wireless links. The purpose of the sensor network is to monitor the ocean currents and underwater habitat.

Caruso et al. (2008) exploit a physically inspired model of ocean flows to model mobility of floating/drifting sensor nodes. Definition of the model starts from the observation that oceans are a stratified, rotating fluid, hence vertical movements, although occasionally present, are negligible with respect to the horizontal ones. Thus, sensor nodes can be assumed to move over a two-dimensional horizontal surface, instead of in a three-dimensional volume. The goal of the model is to describe movement of sensors for a relatively long period of time (several days) in a relatively large region, spanning several kilometers.

An incompressible, two-dimensional flow can be described by a suitably defined *stream function* $\psi = \psi(x, y, t)$, from which the velocity components in x and y coordinates can be derived as follows:

$$u = -\frac{\partial \psi}{\partial y}, \quad v = \frac{\partial \psi}{\partial x},$$

where u is the eastward component and v is the northward component of the velocity field. Then, the trajectory of a device floating in the flow can

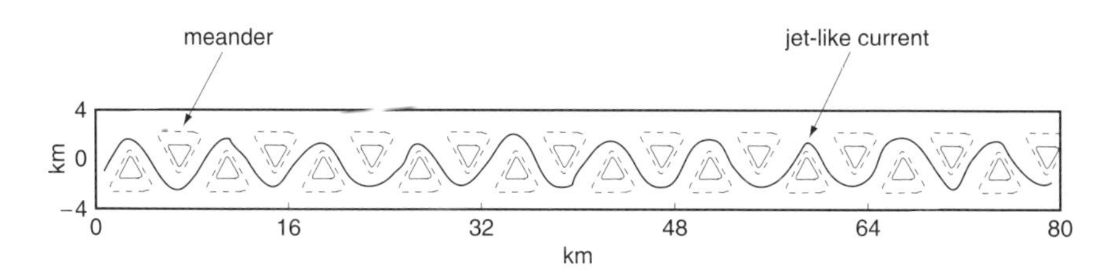

Figure 15.5 Jet-like current meandering between recirculating vortices.

be computed as the solution of the following system of ordinary differential equations:

$$\dot{x} = -\partial_y \psi(x, y, t), \quad \dot{y} = \partial_x \psi(x, y, t).$$

So, the trajectory of a floating device can be readily computed once the stream function describing the behavior of the flow is defined. Caruso et al. (2008) use a widely studied stream function first defined in Bower (1991) and later generalized by Samelson (1992), which has been designed to capture the main features of a typical ocean flow, namely, currents and vortices. The stream function, known as the *meandering jet model* since it represents a jet-like current meandering between recirculating vortices (see Figure 15.5), is defined as follows:

$$\psi(x, y, t) = -\tan\text{h}\left[\frac{y - B(t)\sin(k(x - ct))}{\sqrt{1 + k^2 B^2(t)\cos^2(k(x - ct))}}\right],$$

where $B(t) = A + \epsilon\cos(\omega t)$, k is a parameter corresponding to the number of meanders per unit length, c is the phase speed with which they shift downstream, and B is a time-dependent function modulating the width of the meanders; parameter A determines the average meander width, ϵ is the amplitude of the modulation, and ω is its frequency.

Caruso et al. (2008) suggest using the meandering jet model with the following parameters: $A = 1.2$, $c = 0.12$, $k = 2\pi/7.5$, $\omega = 0.4$, and $\epsilon = 0.3$. With these parameters, and by setting the length and time unit to 1 km and 0.03 days, respectively, we have that the average size of meanders is about 7.5 km, the peak speed inside the jet is about 0.3 m/s, and the modulation period is about half a day (corresponding to the main tidal period).

The motion patterns of a group of 30 sensor nodes initially deployed in a 4×4 km area are shown in Figure 15.6. As can be seen from the figure, the sensor nodes are initially very close to each other, but, as time goes by, they become more and more dispersed along the flow; after three days, the distance between the eastern and western sensor node in the flow is more than 60 km. Figure 15.7 shows a typical sensor path, with an alternation between phases of fast downstream motion when the sensor node is in the jet current,

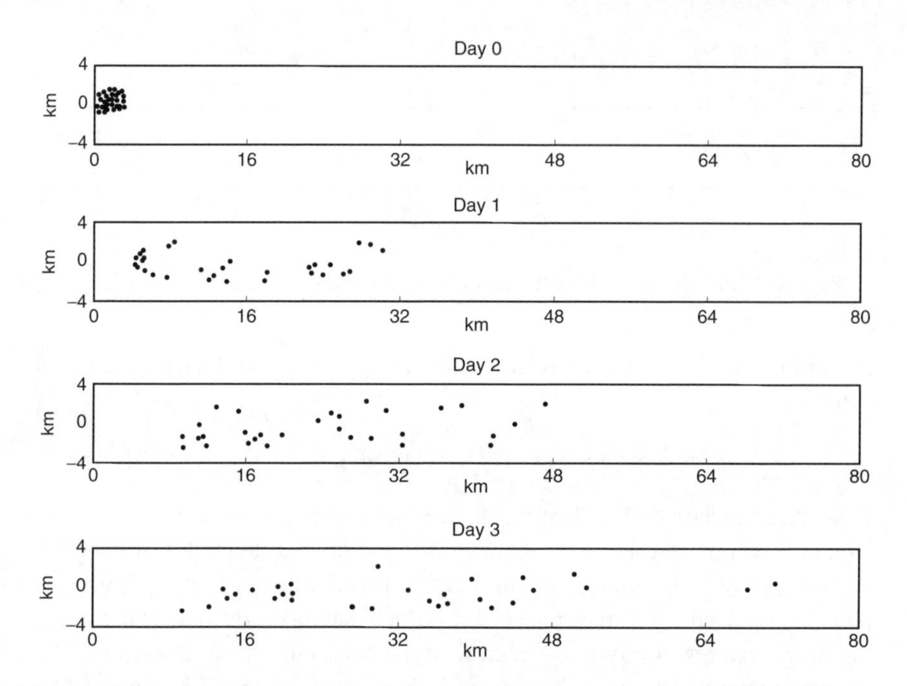

Figure 15.6 Motion patterns of a group of 30 sensor nodes initially deployed in a square of side 4 km.

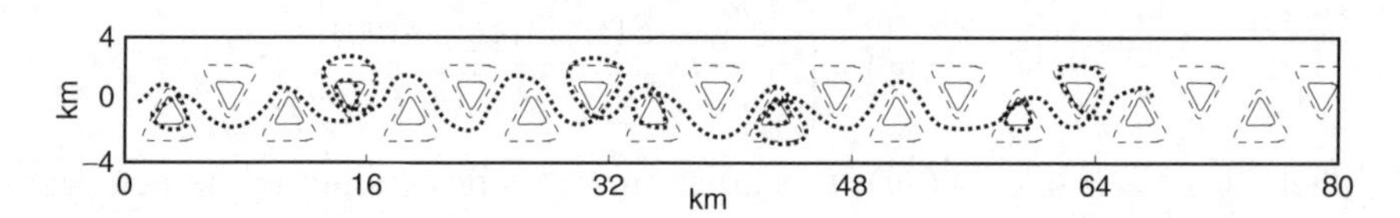

Figure 15.7 Trajectory of a typical sensor node path in the meandering current mobility model.

and looping motion when the sensor node is captured in a vortex. Notice that sensor node trajectories are highly correlated when they are trapped in the same vortex, while they are almost independent when they travel in the jet stream.

References

Altmann J 1974 Observational study of behavior: Sampling methods. *Behavior* **49**, 227–267.

Bower A 1991 A simple kinematic mechanism for mixing fluid parcels across a meandering jet. *Journal of Physical Oceanography* **21**(1), 173–180.

Caruso A, Paparella F, Vieira L, Erol M and Gerla M 2008 The meandering current mobility model and its impact on underwater mobile sensor networks. *Proceedings of IEEE Infocom*, pp. 771–779.

Juang P, Oki H, Wang Y, Martonosi M, Peh LS and Rubenstein D 2002 Energy-efficient computing for wildlife tracking: Design tradeoffs and early experiences with ZebraNet. *Proceedings of ACM ASPLOS*, pp. 96–107.

Martonosi M, Lyon S, Peh LS, Poor V, Rubenstein D, Sadler C, Juang P, Liu T, Wang Y and Zhang P 2004 *http://www.princeton.edu/mrm/zebranet.html*.

Samelson R 1992 Fluid exchange across a meandering jet. *Journal of Physical Oceanography* **22**(4), 431–440.

Small T and Haas Z 2003 The shared wireless infostation model - a new ad hoc networking paradigm (or where there is a whale, there is a way). *Proceedings of ACM MobiHoc*, pp. 233–244.

16

Wireless Sensor Networks: Active Mobility Models

In the previous chapter, we described models of passive mobility in WSNs. This chapter is devoted to the other type of possible mobility in sensor networks, that is, active mobility in which the mobile entity (typically, the sensor node, or the sink) can decide the direction, speed, and time of movement. Active mobility can be exploited in WSNs for different purposes, for instance, to improve sensing coverage after an initial, random sensor node deployment, or to optimize the data collection task. The former is an example of sensor node active mobility, while the latter is an example of sink node active mobility.

From a technological viewpoint, both sensor and sink node active mobility can be realized at reasonable costs with current technology. The feasibility and cost-effectiveness of sensor node mobility were first shown with the introduction of the Robomote platform (Sibley et al. 2002); more recently, sensor nodes have been mounted on cheap mobile robots such as the Roomba vacuum cleaner and Lego Mindstorm toy robots. Sink mobility can be realized in a similar way, possibly using more powerful robots, which is economically feasible given the limited number of sink nodes, or human operators.

In contrast to most types of mobility considered so far in this book, active mobility in WSNs is seen as a means of optimizing certain network-level properties, such as coverage, energy consumption, and so on. Given this, active mobility in WSNs is typically not described through a *model* – describing, for example, the position, direction, and velocity of the mobile entity at a certain time – but through the definition of *motion control algorithms*, whose purpose is, given the current network "state" and the optimization goal, to compute the best possible target location and/or trajectory of the mobile entities.

Mobility Models for Next Generation Wireless Networks: Ad Hoc, Vehicular and Mesh Networks, First Edition. Paolo Santi.

In this chapter, we will present representative examples of motion control algorithms for sensor and sink nodes. Prior to that, we will briefly illustrate a theoretical result showing the benefits of active mobility in terms of increased sensing coverage.

16.1 Active Mobility of Sensor Nodes

As discussed above, applications in which sensor nodes (or a fraction of them) are installed on cheap, mobile robots are becoming feasible with current technology. Sensor node active mobility can then be exploited for different purposes, such as improving sensing coverage of the monitored area and/or reducing energy consumption after an initial, random node deployment, repairing connectivity holes in the network topology, providing better monitoring of ongoing events in the sensed field, and so on. In this section, we will first illustrate a theoretical result showing that even random sensor node movements can be exploited to improve sensing coverage. We will then present two classes of motion control algorithms, the first aimed at increasing sensing coverage, and the second at improving monitoring of relevant events occurring in the sensed field.

16.1.1 Active Mobility and Sensing Coverage

Consider a WSN whose sensor nodes are deployed uniformly at random (e.g., by an aircraft) in a monitored region R. Assume, as commonly done in the literature, that each sensor node is able to monitor and cover a circular region of radius r – the *sensing range* – centered at its current position. Let us define a point $(x, y) \in R$ to be *covered* if there exists at least a sensor node at distance less than or equal to r from (x, y), and let $C(R)$ be the subregion of R formed by the covered points. We define *sensing coverage* as the ratio of the area of $C(R)$ to the area of R. In general, the network designer's goal is to maximize sensing coverage, which ideally should be 1. The reasons for that are quite intuitive. For instance, if the WSN is used for intrusion detection, an intruder can be detected only if she/he is located within $C(R)$, hence sensing coverage can be used as a measure of the likelihood of detecting an intruder.

Unfortunately, achieving complete sensing coverage with random sensor node deployment is possible only if a very large number of sensor nodes are deployed in R, which is often economically not feasible. Active sensor node mobility can be exploited to increase sensing coverage, if the temporal dimension is brought into the picture. In fact, if sensor nodes are mobile, and if we define *mobile sensing coverage* $MC(R) = MC(R, t)$ as the set of points in R which are covered by at least one sensor node in at least one time instant in the interval $[0, t)$, then we have $MC(R) \geq C(R)$, with $MC(R)$

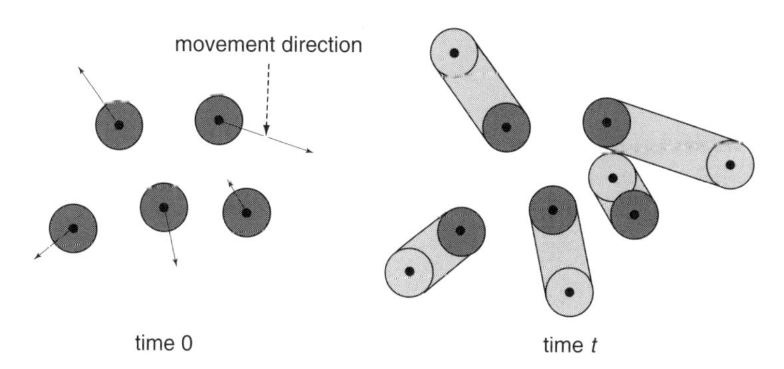

Figure 16.1 Effect of mobility on sensing coverage: the covered region increases in proportion to the movement speed and to the observed time interval.

possibly much larger than $C(R)$ if t is relatively large and/or movement speed is relatively high. The intuition behind this observation is pictorially represented in Figure 16.1: if the sensor nodes are stationary, $C(R)$ is formed by the union of the shaded disks centered at the sensor nodes; in the presence of mobility, the region covered is much larger, and is formed by the union of the shaded regions also including the disks centered at the initial sensor node positions. Note that, with fixed time interval $[0, t]$, a larger speed results in the coverage of a relatively larger region. Note also that, in the presence of linear trajectories, neglecting the effect of the boundary of R, and with fixed sensing range r, the area covered by a sensor node depends only on its velocity and not on the movement direction.

Clearly, mobile sensing coverage is not equivalent to stationary coverage, since region $MC(R, t)$ is covered only for a *fraction* of the time interval $[0, t]$. Yet, mobile sensing coverage is a useful measure of the likelihood of detecting an intruder, at least in the following situations: (i) the intruder is stationary; and (ii) the intruder is mobile but not aware of the presence and position of the sensor nodes.

Liu et al. (2005) investigate the relationship between sensing coverage and mobile sensing coverage in the presence of continuous sensor node mobility. More specifically, sensor nodes are assumed to move according to the following mobility model. Initially, at time 0, sensor nodes are distributed uniformly at random in R. After initial deployment, the sensor nodes immediately start moving, independently of each other, according to the following rules. The direction θ of movement is chosen in the $[0, 2\pi]$ interval according to a distribution $f_\Theta(\theta)$, while the movement speed v_s is chosen in the interval $[0, v_{max}]$ according to a distribution $f_V(v)$. The same direction and speed of movement are maintained by the sensor nodes for the entire time interval of interest $[0, t]$. This implies that, for the sake of analysis, the sensor nodes are assumed to be sufficiently far away from the boundary of R, so that the border effect can be ignored.

The main theoretical findings of Liu et al. (2005) are summarized in the following theorem:

Theorem 16.1 *Let λ be the average sensor node density in R, assume that sensor nodes move according to the mobility model described above, and let $\bar{v}$ denote the expected node speed as dictated by the speed distribution $f_V(v)$. Then,*

$$C(R) = 1 - e^{-\lambda \pi r^2}$$

and

$$MC(R, t) = 1 - e^{-\lambda\left(\pi r^2 + 2r\bar{v}t\right)}.$$

Thus, mobility increases sensing coverage by a factor that is proportional to the sensing range r, the average speed $\bar{v}$, and the duration t of the time interval. It is interesting to observe that an increase in sensing coverage is *independent* of the movement direction distribution.

To give the reader an idea of the benefits of active mobility in sensing coverage, Figures 16.2 and 16.3 show $C(R)$ and $MC(R)$ as a function of t and $\bar{v}$, respectively. As can be seen from the figures, mobility can considerably increase sensing coverage from about 0.27 in the case of stationary networks to nearly 1 if either t or $\bar{v}$ is sufficiently high.

Liu et al. (2005) also investigate another interesting theoretical question: under the assumption that movement speed is fixed to $\bar{v}$, what is the best possible mobility strategy (namely, the choice of the movement direction distribution $f_\Theta(\theta)$) if the purpose is to minimize the expected detection time of a mobile target? Interestingly, the authors find that the optimal movement strategy is when the movement direction is chosen uniformly in $[0, 2\pi]$, that is, when $f_\Theta(\theta)$ is the uniform distribution in $[0, 2\pi]$.

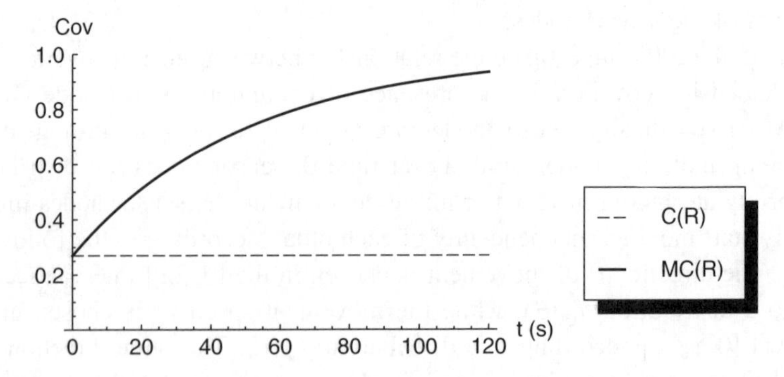

Figure 16.2 Effect of mobility on sensing coverage for different values of t. Sensor node density is $\lambda = 0.001$, sensing range r is 10 m, and $\bar{v} = 1$ m/s.

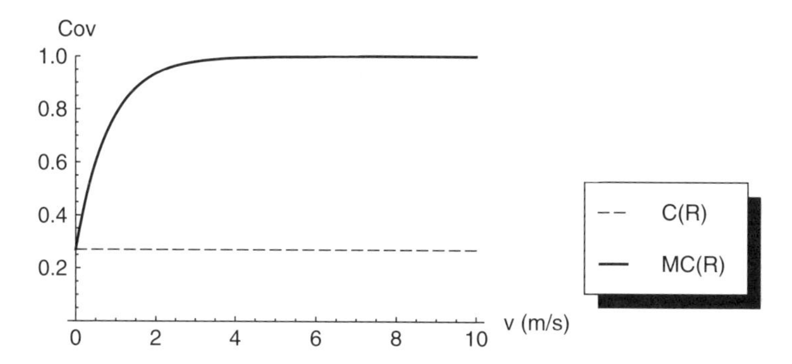

Figure 16.3 Effect of mobility on sensing coverage for different values of $\bar{v}$. Sensor node density is $\lambda = 0.001$, sensing range r is 10 m, and $t = 60$ s.

16.1.2 Motion Control for Sensing Coverage

An important class of motion control algorithms for sensor nodes is that aimed at improving sensing coverage after an initial, random sensor node deployment. The underlying idea is that, after initial deployment, sensor nodes can exchange information about each other's position (assumed to be known to sensor nodes through, for example, GPS or other forms of localization); then, a motion control algorithm is executed, with the purpose of identifying possible coverage holes and a new set of sensor node positions aimed at reducing or eliminating coverage holes. The new positions are communicated to the sensor nodes, which then autonomously move toward the new location. Typically, the motion control algorithm is executed in several iterations, until it converges to an "optimal" set of sensor node locations.

Motion control algorithms in this class can be classified as either centralized or fully distributed, and based on the optimization criteria used to find the optimal set of sensor node locations. For instance, the optimization goal can be increasing coverage only, increasing coverage combined with connectivity requirements, increasing coverage combined with minimizing sensor node movements, and so on. In the following, we present two representative approaches for motion control aimed at improving sensing coverage, the former belonging to the class of centralized algorithms and the latter to the class of fully distributed ones.

Zou and Chakrabarty (2003) present a centralized motion control algorithm based on virtual forces, called the virtual force algorithm (VFA). The basic idea underlying the algorithm is that each sensor node in R behaves like a "source of force" for all other sensor nodes. The intensity of the exerted force, as well as whether the force is attractive or repulsive, is determined by the distance between the sensor nodes and by a parameter called the *distance threshold* that can be used to optimize the resulting sensing coverage. Intuitively speaking, the force exerted by sensor node i on sensor node j is repulsive when the two nodes are "too close" to each other, while it is attractive when the two nodes are "too far" from each other. By suitably combining these forces, sensor node locations as computed by the VFA should converge to a quasi-regular pattern with optimal coverage.

The VFA is assumed to be executed at a centralized location (either a sink node or a cluster head in a clustered sensor network topology), to which all sensor nodes in the field initially communicate their location. Given these locations, and the locations of preferential areas and obstacles in R (see below), the VFA computes for each sensor node the combined force exerted by all other sensor nodes and preferential areas/obstacles, which results in a set of new locations for the sensor nodes. Instead of immediately communicating these new locations to the sensor nodes and letting them move, only "virtual movements" are performed, and a new iteration of the VFA is executed with the new set of "virtual" node locations. This process is iterated until a termination condition is met, at which point the final locations are communicated to the sensor nodes, which start with the actual movement phase. A pseudo-code description of the VFA is given in Figure 16.4.

The choice of performing only "virtual movements" during VFA execution is motivated by the need to reduce the average distance covered by sensor nodes which, in turn, is motivated by the need to reduce energy consumption. In fact, performing movements is a task that consumes a significant amount of the energy available at the sensor nodes, an amount orders of magnitude larger than that needed for sensing, computational, or communication tasks. As a further energy optimization, Zou and Chakrabarty (2003) suggest disabling the exertion of virtual forces (and, hence, movement) on a sensor

The VFA:

1. Set $loops = 0$
2. **While** $loop <$ MAXLOOPS
3. Evaluate current coverage
4. **if** coverage requirements are met
5. **return** final sensor positions; **terminate**
6. **for** each sensor s_i
7. compute resultant virtual force $\mathbf{F}_i$ on s_i
8. **for** each sensor s_i
9. move s_i to its next position determined by $\mathbf{F}_i$
10. Set $loops = loops + 1$
11. **return** final sensor positions; **terminate**

Figure 16.4 The VFA motion control algorithm.

node when the current distance between the initial sensor node position and the current "virtual" one reaches a pre-specified distance limit.

The coverage requirements at step 4 of the VFA are verified as follows. An $n \times m$ point grid is superimposed on the monitored region R, and the coverage requirements are assumed to be satisfied if at least a certain fraction f of the grid points is covered by at least one sensor. As in the theoretical study reported in the previous subsection, sensor nodes are assumed to cover a circular area of radius r.

The resultant force $\mathbf{F}_i$ exerted on sensor node s_i is computed at step 7 of the VFA as follows:

$$\mathbf{F}_i = \mathbf{F}_{iO} + \mathbf{F}_{iA} + \sum_{j \neq i} \mathbf{F}_{ij},$$

where $\mathbf{F}_{iO}$ is the resultant (repulsive) force exerted by obstacles located in R on s_i, $\mathbf{F}_{iA}$ is the resultant (attractive) force exerted by preferential coverage areas (i.e., areas within R that require higher coverage) on s_i, and $\mathbf{F}_{ij}$ is the force exerted by sensor node s_j on s_i. In turn, force $\mathbf{F}_{ij} = (r, \theta)$ expressed in polar coordinates is defined as follows:

$$\mathbf{F}_{ij} = \begin{cases} (w_A(d_{ij} - d_{th}), \alpha_{ij}) & \text{if } d_{ij} > d_{th} \\ 0 & \text{if } d_{ij} = d_{th} \\ (w_R/d_{ij}, \alpha_{ij} + \pi) & \text{otherwise} \end{cases},$$

where d_{ij} is the distance between s_i and s_j, d_{th} is the threshold distance, α_{ij} is the angle of the segment connecting s_i to s_j, w_A is a measure of the attractive force, and w_R is a measure of the repulsive force.

The distance threshold d_{th} can be used to control how close sensor nodes get to each other in the final, quasi-regular lattice obtained at the end of VFA execution and thus, indirectly, the provided degree of coverage. Smaller values of d_{th} result in relatively denser deployments, while larger values of d_{th} generate relatively sparse deployments. This can be seen from Figure 16.5: with $d_{th} = 2r$, sensor nodes tend to converge to the lattice on the left, which leaves a small portion of the area uncovered; by setting $d_{th} = \sqrt{3}r$, sensor

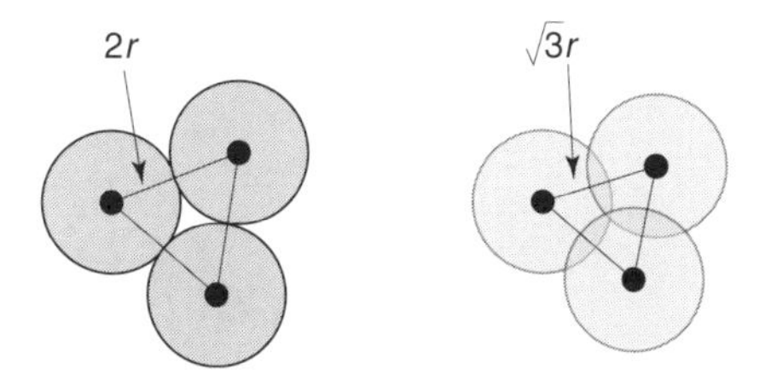

Figure 16.5 Effect of varying distance threshold d_{th} on sensor node deployment in the VFA.

nodes tend to converge to the lattice on the right, which is relatively denser and leaves no coverage holes.

Wang et al. (2006) present a framework for distributed motion control based on the use of Voronoi diagrams. The *Voronoi diagram* is the tessellation of a space composed of a set of non-overlapping Voronoi cells, each centered at a point in the space given as input – corresponding to sensor node locations in this case. The *Voronoi cell* centered at sensor node s_i is a convex polygon formed by all points in R closer to s_i than to any other sensor nodes s_j. Two sensor nodes are said to be *Voronoi neighbors* if the polygons delimiting their cells share an edge. An example of a Voronoi cell and Voronoi neighbors is shown in Figure 16.6. For a more extensive definition of Voronoi diagrams and their properties, the reader is referred to Appendix B.

Voronoi diagrams have interesting properties when applied to the sensing coverage problem. First of all, the Voronoi cell of a sensor node can be computed locally, provided the communication range is large enough to reach Voronoi neighbors. This property enables a fully distributed computation of the Voronoi diagram of the sensor network. Second, and most importantly, if a subset of the points located in the Voronoi cell of s_i cannot be covered by s_i (again, a sensor node is assumed to have a circular sensing range of radius r), they cannot be covered by any other sensor node in the network, given the property of Voronoi cells. Thus, coverage holes can be detected locally by finding points in the Voronoi cells that cannot be covered by the respective sensor node.

The pseudo-code of the motion control framework presented in Wang et al. (2006) is given in Figure 16.7. Sensor nodes repeatedly compute whether their own Voronoi cell has coverage holes; if there are no coverage holes, the sensor node remains stationary, otherwise it computes its next target location according to one of the three algorithms VEC, VOR, and Minmax described below. Before performing a movement, a movement-adjustment heuristic is applied. Basically, the sensor node checks whether the local coverage (i.e., coverage of its own Voronoi cell) is increased by its movement. If local

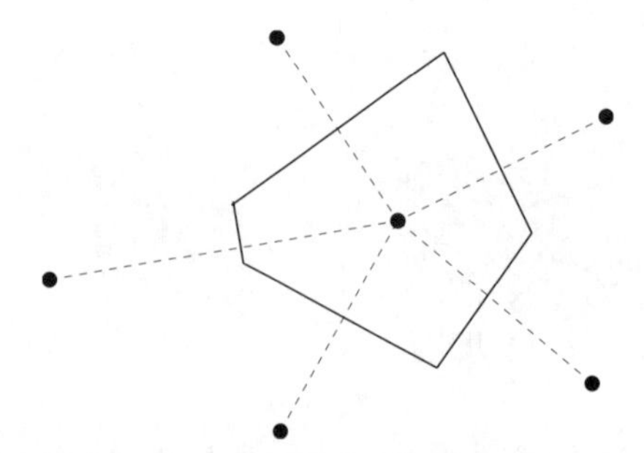

Figure 16.6 Example of Voronoi cell and Voronoi neighbors.

Voronoi-based motion control
(Algorithm for sensor node s_i)
1. Set $Terminate$ = **False**
2. **While** *not* $Terminate$
3. Evaluate coverage of own Voronoi cell
4. Compute next location
5. Apply movement adjustment
6. **If** local coverage is not increased **then**
7. Set $Terminate$ = **True**
8. **else**
9. Move to next location

Figure 16.7 Voronoi-based motion control.

coverage is increased, the sensor node actually moves to the next location, otherwise it remains at the current location.

The next position of a sensor node is computed according to one of the following algorithms. The first algorithm is called VEC, and it computes the next position using a virtual force approach similar to the one at the basis of the VFA. Similar to the VFA, VEC is a proactive algorithm, which attempts to arrange sensor nodes in a quasi-regular lattice.

Denoting by d_{ij} the distance between sensor nodes s_i and s_j, and with d_{ave} the average distance between sensor nodes as if they were evenly spread in the monitored region (this information is assumed to be known to all sensor nodes), we have that the virtual force between s_i and s_j will push them away from each other with intensity $(d_{ave} - d_{ij})/2$. In case one sensor node entirely covers its Voronoi cell, the other sensor node will be pushed $(d_{ave} - d_{ij})$ away. The virtual force exerted by sensor node s_j on s_i is denoted $\mathbf{F}_{ij}$.

In addition to the virtual forces generated by sensor nodes, also the boundary of R exerts a force on s_i, denoted $\mathbf{F}_b$, of intensity $d_{ave}/2 - d_b(i)$ and direction toward the center of R, where $d_b(i)$ denotes the distance between s_i and the boundary of R. Summarizing, the next location of sensor node s_i is obtained by taking the vectorial sum of vectors $\mathbf{F}_{ij}$, for each s_j which is a Voronoi neighbor of s_i, and of the vector $\mathbf{F}_b$.

The second algorithm, called VOR, is a heuristic algorithm that attempts to cover the largest coverage holes in each Voronoi cell. In VOR, if a sensor node detects the existence of a coverage hole in its Voronoi cell, it will move toward its farthest Voronoi vertex V_f, and stop when V_f can be covered – that is, when it is at distance r from V_f, see Figure 16.8.

Note that algorithm VOR may cause oscillating behavior if new holes are generated due to a sensor node leaving its current position. To solve this problem, an oscillation control heuristic is introduced: before a sensor node

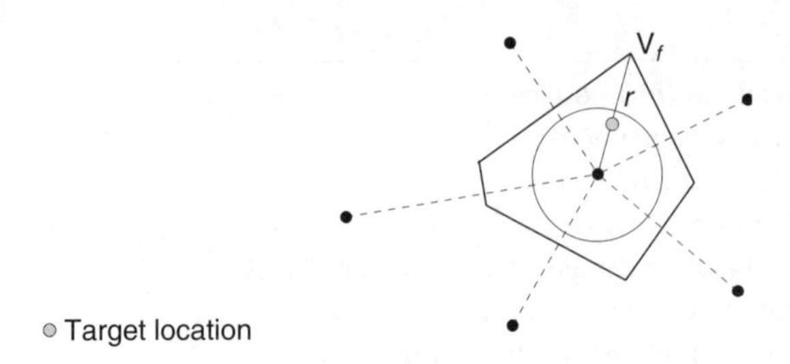

Figure 16.8 Computation of next location with algorithm VOR.

moves, it first checks whether its moving direction is opposite to that of the previous round. If so, it stops for one round to prevent oscillations.

Similar to VOR, algorithm Minmax fixes coverage holes by letting sensor nodes move toward the farthest Voronoi vertex; however, Minmax is designed to avoid situations in which a Voronoi vertex that was originally covered becomes uncovered after movement. Thus, in Minmax the next target location is chosen as the point inside the Voronoi cell whose distance to the farthest Voronoi vertex is minimized.

Some heuristics are proposed in Wang et al. (2006) to improve the performance of the proposed motion control algorithms. In particular, a variant of the algorithms is described in which "virtual movement" is performed at each step – similar to what was done in the VFA – instead of physical sensor node movements, and actual movements occur only upon convergence of the motion control algorithm. However, performing "virtual movements" in this case is more complex because of the distributed nature of the approach, and in particular because, if the communication range is not large enough, supposed Voronoi neighbors after virtual movement might not be within communication range. Wang et al. (2006) describe optimizations of the algorithms designed to solve this problem. The interested reader is referred to Wang et al. (2006) for details.

16.1.3 Motion Control for Event Tracking

Consider a situation in which the specific areas of interest within a larger monitored area are unknown at the deployment stage. Hence, initial sensor node deployment is random. After initial deployment, events occur in the monitored area, and the network designer would like to be able to carefully track such events, with a relatively denser sensor node deployment in the subregions of R where events are more likely to occur. Active sensor node mobility can be exploited for this purpose, by letting sensor nodes react to events occurring in R, with the goal of obtaining a final sensor node deployment resembling the distribution of events in R.

Butler and Rus (2003) present two distributed motion control algorithms designed to track the occurrence of events in the monitored region R. Besides these tracking events, the algorithms proposed in Butler and Rus (2003) are designed to minimize the amount of required computation, memory, and communication. Furthermore, heuristics are proposed to maintain coverage of the whole region R and network connectivity while relocating the sensor nodes for event tracking – see Butler and Rus (2003) for details. In both algorithms, it is assumed that at least one sensor node in the network detects each event and broadcasts its location to the other sensor nodes, so that every sensor node quickly learns about the occurrence of each event.

The first motion control algorithm presented in Butler and Rus (2003) does not maintain any history of past events. The algorithm resembles "virtual force" approaches such as the ones described in the previous section. The main difference is that elements exerting force are events, instead of sensor node positions. More specifically, denoting by x_i^k the position of sensor node s_i after the kth event E^k occurs, we have

$$x_i^{k+1} = x_i^k + f(E^{k+1}, x_i^k, x_i^0).$$

The relocation function $f()$ is designed according to the following guidelines. First, after an event occurs, the sensor node should never move past that event location. Second, the motion of sensor nodes should depend on their distance to the event location, with motion tending to 0 as the distance increases. This guideline is fundamental to avoid all sensor nodes converging to the last observed event. Finally, a monotonicity property should be fulfilled: no sensor node should move past another one moving in the same direction in response to the same event.

To address the above guidelines, Butler and Rus (2003) suggest using a relocation function with the following properties:

$$\text{(i)}\ \forall d, 0 \leq f(d) \leq d \quad \text{(ii)}\ f(\infty) = 0 \quad \text{(iii)}\ \forall d_1 > d_2,\ f(d_1) - f(d_2) < (d_1 - d_2),$$

where d denotes the distance between the sensor node and the location of the event. As an example of a function satisfying properties (i)–(iii), Butler and Rus (2003) suggest using $f(d) = de^{-d}$.

While the above motion control algorithm is very simple and requires minimal memory for the sensor nodes, its capability of event tracking is limited, due to the lack of knowledge of the history of past events. The second algorithm presented in Butler and Rus (2003) exploits this knowledge to improve event tracking capability. In the following, due to its simplicity, we present the algorithm for the case in which R is a one-dimensional region (e.g., a road, a pipeline, a bridge, etc.), and we refer the reader to

History-based event tracking motion control
(Algorithm for sensor node s_i)
1. **for each** event E^k occurring at position e^k
2. Increment CDF bins for positions $\geq e^k$
3. Scale CDF by $k/(k+1)$
4. Find bins b_j, b_{j+1} such that $b_j \leq x_0 \leq b_{j+1}$
5. Compute position x_i^k by interpolation of values b_j, b_{j+1}

Figure 16.9 Voronoi-based motion control.

Butler and Rus (2003) for generalization of the algorithm to deal with two-dimensional regions.

In order to determine its location, each sensor node maintains and updates a discrete version of the cumulative density function (CDF) of past event locations. The idea underlying the motion control algorithm is to assign each segment of the CDF with a proportional number of sensor nodes, so that sensor node density closely tracks event density. Since sensor nodes are supposed to be initially uniformly distributed in R, this is accomplished by mapping each CDF segment to a proportional interval of the sensor nodes' initial positions.

Figure 16.9 shows the algorithm used by sensor nodes to compute their positions: each time a new event occurs, the corresponding bins in the CDF are increased by 1. The CDF is then rescaled to account for the observed number of events. Then, bins b_j, b_{j+1} enclosing the initial sensor node position x_0 are found, and the new position of the sensor node is computed by interpolation based on the values of b_j and b_{j+1}.

16.2 Active Mobility of Sink Nodes

In the previous section, we described motion control algorithms for the scenario in which active mobility is exploited by sensor nodes. Many situations exist in which active mobility can be exploited by sink nodes, typically with the aim of optimizing the data collection task in terms of reduced energy consumption. One of the first proposals of a W0SN network architecture where sink nodes are mobile is the Data MULE concept introduced in Shah et al. (2003), which will be described next. After presenting the Data MULE architecture and related analysis, we will describe a methodology introduced in Wang et al. (2005) to optimize sink mobility and routing in order to maximize network lifetime.

Note that the problem of optimizing the trajectories of mobile entities with the aim of optimizing network-level parameters has been studied also in other contexts. A notable example is the *message ferrying* concept of Zhao et al. (2004), which was proposed for delay-tolerant networks in general and, hence, has different optimization goals (e.g., reducing communication delay) with respect to those typical of WSNs.

16.2.1 The Data MULE Concept

Shah et al. (2003) introduce the notion of Data MULEs (Mobile Ubiquitous LAN Extensions) to optimize data collection in sparse sensor networks. The authors propose a three-tier network architecture, displayed in Figure 16.10. In the bottom tier there are a number of stationary and sparsely deployed sensor nodes collecting data, which are typically so far that they cannot communicate with each other. In the intermediate tier there are a number of mobile MULEs that store data collected from sensor nodes when they get into their communication range. MULEs can be thought of as entities moving in the environment where sensor nodes are deployed, such as cars, public transport vehicles, individuals, etc. Finally, in the top tier of the architecture there is a set of stationary access points, through which MULEs – when within communication range – download sensor data to remote collection points.

While in the original Data MULE proposal sink nodes correspond to the top tier of the network architecture and are, then, stationary, Shah et al. (2003) note that, depending on the application scenario, the intermediate and top tiers of the architecture can be collapsed into a single tier formed of mobile MULEs with direct WAN connection (e.g., through a cell phone). This situation corresponds exactly to the case of stationary sensor nodes and mobile sinks of interest in this section.

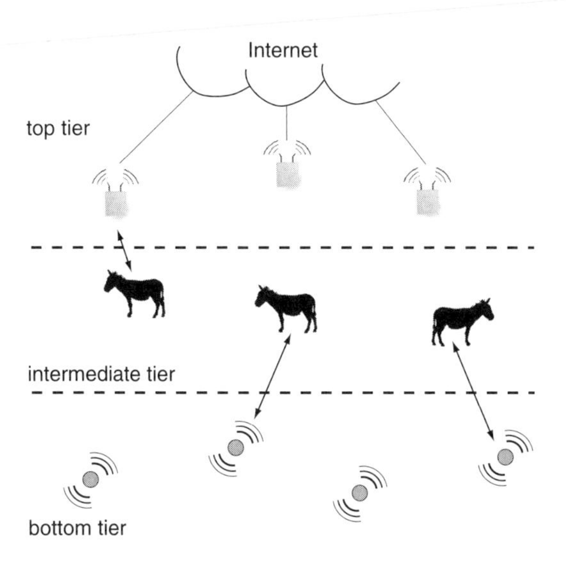

Figure 16.10 The three-tier Data MULEs architecture.

Shah et al. (2003) study the fundamental tradeoffs between the number of deployed sensor nodes, MULEs, access points, and parameters such as the data success rate (the fraction of generated data that reaches the access points) and the buffer size of the sensor nodes and the MULEs. The tradeoffs are studied under the following assumptions:

1. Time and space are discretized; in particular, possible sensor node, MULE, and access point locations correspond to the points of a two-dimensional square grid, with toroidal folding (to avoid boundary effects).
2. Access points are assumed to be evenly spaced on the grid, while sensor nodes are assumed to be distributed uniformly at random on the grid.
3. MULEs move according to independent random walks on the grid: at each time instant, a MULE moves with equal probability to one of the four neighboring points on the grid.
4. MULEs can communicate with sensor nodes when they are located at the same grid point; similarly for communication with access points. MULE-to-MULE communication is not possible; that is, even if two MULEs are located at the same grid point, they cannot exchange data.

With these assumptions, Shah et al. (2003) prove a number of theoretical results, from which the following observations are derived:

1. Buffer requirements on the sensor nodes are inversely proportional to the density ρ_m of MULEs per unit area.
2. Buffer requirements on MULEs are inversely proportional to both ρ_m and the density ρ_a of access points per unit area.
3. When the buffer on the sensor nodes is relatively large, the buffer size on the MULEs can be traded off against the number of MULEs to maintain the same data success rate.
4. The relative increase in sensor node buffer capacity needs to be larger than the increase in number of MULEs to keep the same data success rate.

16.2.2 Sink Mobility for Network Lifetime Maximization

Similar to what was done for the active mobility of sensor nodes, approaches have been proposed to *control* the movement of sink nodes in order to optimize a certain network-level performance parameter. Wang et al. (2005) introduce a linear programming-based approach for optimizing sink node mobility, with the aim of maximizing network lifetime.

The model considered by these authors is one in which sensor nodes are stationary and deployed on a two-dimensional square grid, while a single sink node can move freely on the grid. For simplicity, the sink node travel time between any two grid points is assumed to be negligible as compared to the sojourn times. With this assumption, the order with which grid points are visited by the mobile sink becomes irrelevant, and the sink node mobility optimization problem becomes one of determining sink node sojourn time at each grid point, with a sojourn time of 0 indicating that the respective grid point is not visited by the sink node.

During its sojourn time at a grid point, a sensor node can communicate with the sink, possibly in a multi-hop fashion. In fact, sensor nodes are assumed to have a communication range equal to the grid step; that is, each sensor node has at most four neighbors on the grid.

Each sensor node generates data packets at a fixed rate; if sensor node s_i is neither co-located with the sink, nor directly connected with it, which occurs when the sink node is located at a neighboring point on the grid, then the data generated by s_i will be sent to the sink through a multi-hop path, which is determined according to the following routing algorithm. If the sensor node lies on the same horizontal or vertical line as the current sink position, then the data is sent to the sink through the (unique) shortest path. Otherwise, the two routes on the perimeter of the rectangle with the sensor node and the sink node at opposite corners are chosen alternately – see Figure 16.11.

The goal of the sink mobility optimization problem considered in Wang et al. (2005) is to maximize sensor network lifetime, which is defined as the time for the first sensor node in the network to exhaust its energy budget.

The mobility optimization problem is stated as the following linear program, which can be efficiently solved for networks of relatively large size (as large as 256, as reported in Wang et al. (2005)):

$$\text{maximize} \quad \sum_{p \in G} t_p$$

$$\text{such that } \forall s_i, \ \sum_{p \in G} c_i^p t_p \leq e_0 \tag{16.1}$$

$$\forall p \in G, \ t_p \geq 0. \tag{16.2}$$

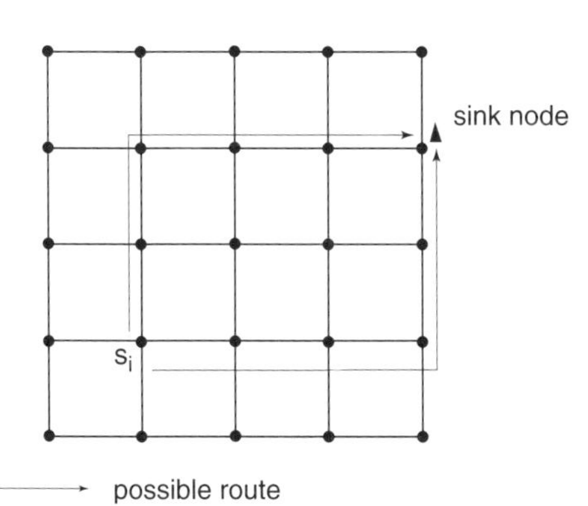

Figure 16.11 Multi-hop routing in the sink node mobility optimization approach of Wang et al. (2005).

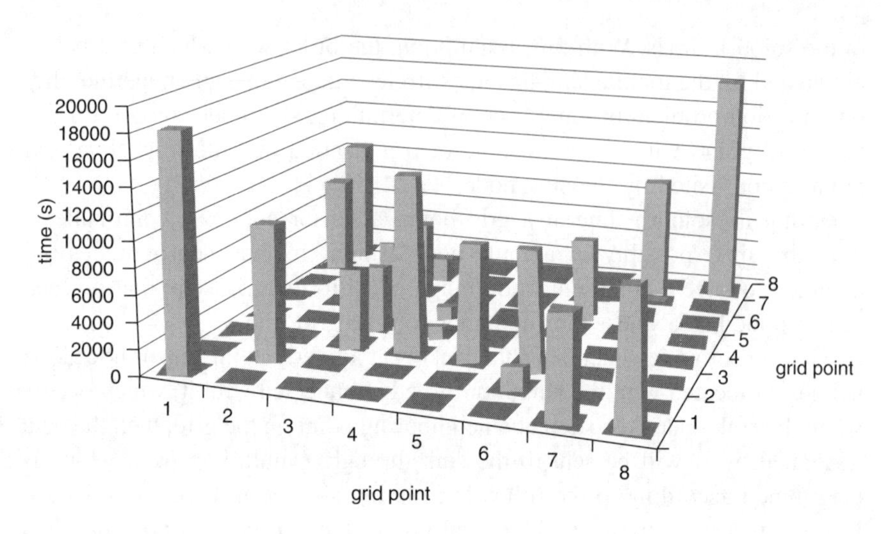

Figure 16.12 Optimal sink node sojourn times in an 8 × 8 grid.

In the linear programming formulation above, $p \in G$ represents a point in the two-dimensional grid G, t_p is the variable corresponding to the sink node sojourn time at point p, c_i^p is the power consumption at sensor node s_i for transmitting its data when the sink node is located at p (see Wang et al. (2005) for details on how this parameter can be computed), and e_0 is the initial energy budget at each sensor node (assumed to be the same for all sensor nodes). Constraint (16.2) imposes that the total energy consumed at each sensor node does not exceed the energy budget, while constraint (16.2) ensures non-negativity of sojourn times t_p.

Wang et al. (2005) compare the network lifetime obtained with sink mobility to that obtained when the sink node is stationary and located at the center of the grid, and find that up to five-fold improvements in network lifetime with respect to the stationary case can be achieved with active sink mobility. Furthermore, the authors observe that in the optimal solution the sink node tends to spend most of its time either at the corners or at the center of the grid. As an example, the optimal sink node sojourn times for an 8 × 8 network are shown in Figure 16.12.

References

Butler Z and Rus D 2003 Event-based motion control for mobile sensor networks. *IEEE Pervasive Computing* **2**(4), 34–42.

Liu B, Brass P, Dousse O, Nain P and Towsley D 2005 Mobility improves coverage of sensor networks. *Proceedings of the ACM International Conference on Mobile Ad Hoc Networking and Computing (MobiHoc)*, pp. 300–308.

Shah R, Roy S, Jain S and Brunette W 2003 Data MULEs: Modeling a three-tier architecture for sparse sensor networks. *Proceedings of the IEEE Workshop on Sensor Network Protocols and Applications (SNPA)*, pp. 30–41.
Sibley C, Rahimi M and Sukhatme G 2002 Robomote: a tiny mobile robot platform for large-scale sensor networks. *Proceedings of the IEEE Conference on Robotics and Automation (ICRA)*, pp. 1143–1148.
Wang G, Cao G and LaPorta T 2006 Movement-assisted sensor deployment. *IEEE Transactions on Mobile Computing* **5**(6), 640–652.
Wang Z, Basagni S, Melachrinoudis E and Petrioli C 2005 Exploiting sink mobility for maximizing sensor networks lifetime. *Proceedings of the IEEE Hawaii International Conference of System Sciences (HiCSS)*, p. 287.1.
Zhao W, Ammar Mand Zegura E 2004 A message ferrying approach for data delivery in sparse mobile ad hoc networks. *Proceedings of ACM MobiHoc*, pp. 187–198.
Zou Y and Chakrabarty K 2003 Sensor deployment and target localization based on virtual forces. *Proceedings of IEEE Infocom*, pp. 1293–1303.

[illegible], M., [illegible] and Trappe, W. (2007) [illegible] Modeling [illegible] for [illegible]. *[illegible] Networks [illegible]*, [illegible], pp. 40–51.

[illegible] (2003) [illegible]. *[illegible]*, pp. [illegible].

[illegible], L.S., [illegible] (2008) [illegible]. *[illegible] Mobile Computing*, [illegible], pp. [illegible].

[illegible] (2011) [illegible]. *[illegible]*, pp. [illegible].

Zhao, W., Ammar, M. and Zegura, E. (2004) A message ferrying approach for data delivery in sparse mobile ad hoc networks. *Proceedings of ACM MobiHoc*, pp. 187–198.

[illegible] and [illegible] (2004) [illegible]. *[illegible]*, pp. 184–190.

Part Six

Mobility Models for Opportunistic Networks

In this part of the book, we will focus our attention on an emerging class of next generation wireless networks, called *opportunistic networks*, which are attracting increasing interest among the research community. This class of networks is characterized by radically different topological properties with respect to the other classes of networks considered so far in this book, and by the fact that mobility of nodes is the prominent communication means within the network. Given this, mobility analysis and modeling has been and still is a very active research field in opportunistic networks, since a deep understanding of how network nodes move is a prerequisite for the effective design of opportunistic networking protocols.

After describing the features, state of the art, and perspectives of opportunistic networks, we will introduce the basics of how message routing is performed in opportunistic networks, introducing the well-known *store, carry, and forward* mechanism and some relevant routing protocols. We will then present the most relevant results concerning the analysis of mobility patterns observed in real-world opportunistic networks, which allow researchers to make important observations about the way people move and meet each other in daily life. We will then conclude this part of the book by describing representative mobility models proposed for opportunistic networks, whose goal is to resemble features of human mobility observed in real-world traces.

17

Opportunistic Networks

In this chapter, we will briefly describe the main features of an emerging class of next generation wireless networks, namely, opportunistic networks. After presenting the state of the art in opportunistic networking and related technologies, we will describe representative use cases. Finally, we will present a short overview of the main envisioned technological evolutions, and of the challenges to be faced by opportunistic network designers.

17.1 Opportunistic Networks: State of the Art

An opportunistic network is a short-range wireless network characterized by a very sparse network topology – see Figure 17.1: if one takes a snapshot of the network at an arbitrary time instant, what is typically observed is a large fraction of isolated nodes, and a small fraction of nodes having active links with a few other nodes in the network. From a networking perspective, what is lacking in the network topology is *connectivity*, that is, the possibility for a node in the network to establish a (possibly multi-hop) communication path with all the other nodes. However, network connectivity – or at least, as we will see, a weak form of connectivity – can be achieved by exploiting the temporal dimension and node mobility: since nodes in an opportunistic network move, isolated nodes can get in touch with other nodes as time goes by. Similarly, a node A which is currently in touch with node B might later get in touch with another node C, and so on – see Figure 17.1. So, if messages circulating in the network are stored in the node buffers for a long enough time interval, it is indeed possible to establish multi-hop communication paths between nodes in the network by exploiting these relatively seldom *communication opportunities* – whence the name of the network – achieving network-wide communications.

Mobility Models for Next Generation Wireless Networks: Ad Hoc, Vehicular and Mesh Networks, First Edition. Paolo Santi.

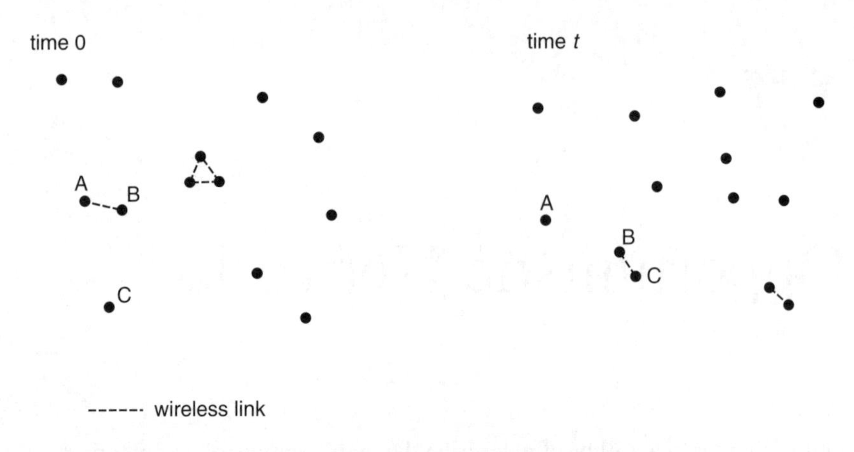

Figure 17.1 Typical opportunistic network architecture.

Since the communication mechanism described above – called *store, carry, and forward*, see Chapter 18 – relies on node mobility to physically carry around and forward messages within the network, and human/animal/vehicular mobility is several orders of magnitude slower than the speed of radio signal propagation in the air, which is comparable to the speed of light, it is clear that the delays in communicating a message from source to destination are much higher than those typical of other types of networks. Hence, applications running in opportunistic networks must be able to tolerate very large delays – in the order of minutes or hours – which explains why opportunistic networks are also called *delay-tolerant networks* (DTNs).

Unlike the other types of next generation wireless networks considered so far in this book, opportunistic networks are not related to a specific wireless technology, but rather are a class of mobile networks characterized by the above-described specific topological property of being extremely sparse. This means that, depending on the scenario, the radio technology used to establish wireless links between nodes can be different: it can be a WLAN technology such as WiFi, a PAN technology such as Bluetooth, or a vehicular or wireless sensor communication technology if some of the members of the opportunistic network are vehicles and/or wireless sensor nodes. It is indeed the case that in some scenarios different technologies coexist, and are alternately used to establish peer-to-peer wireless links: the most typical case in this respect is that of a smart phone with both WiFi and a Bluetooth interface and which can use either interface to establish a wireless link with another cell phone.

One type of opportunistic network that has recently attracted a lot of attention in the research community is the one which is dynamically formed by individuals carrying advanced personal communication devices (PCDs) capable of establishing direct wireless links, such as smart phones, PDAs, etc.

Table 17.1 Short-range wireless technologies for opportunistic network communications

Name	Domain	Typical range (m)	Max data rate (Mbps)
WiFi	WLAN	30–250	600 (802.11n)
Bluetooth	PAN	10–100	24 (v. 3.0)
ZigBee	WSN	10–75	0.25
DSRC	Vehicular	150–1000	54

This type of opportunistic network is known as a pocket-switched network (PSN) in the literature (Hui et al. 2005), and is considered very interesting under both the mobility analysis/characterization viewpoint and the potential application viewpoint. In fact, since members of a PSN are people, characterization of human mobility patterns can be used to improve the PSN performance. Furthermore, PSNs can potentially enable novel participatory and social networking applications based on opportunistic communications, as described in the next section.

It is important to observe that, while most nodes in an opportunistic network are mobile, there might exist application scenarios in which some of the nodes in the network are fixed and act as data collection and/or message relay points. An important example of this class of opportunistic networks is DieselNet, an opportunistic network composed of 40 radio-equipped public buses in the city of Amherst, Massachusetts (Group 2007). In DieselNet, some of the bus stops are equipped with wireless collection points called *throwboxes* that act as message relays, thus speeding up the message propagation process in the network. Another example is that of a very sparse wireless sensor network composed of mobile sensor nodes (e.g., sensor nodes attached to animals), where fixed data gathering stations can be installed in strategic locations (e.g., at a water source).

From a technological viewpoint, some of the short-range wireless technologies that can be used in opportunistic networks are mature (e.g., WiFi and Bluetooth for PSN applications), some are close to fully ripening (e.g., ZigBee for WSN-related applications), while others are in an advanced development stage (e.g., IEEE 802.11p/DSRC (Dedicated Short-Range Communications) for vehicular-related applications). A summary of the main features of wireless technologies for opportunistic network communications is given in Table 17.1.

17.2 Opportunistic Networks: User Scenarios

In this section, we report some representative user scenarios for opportunistic networks in the domain of PSN, WSN, and vehicular network applications.

17.2.1 User Scenarios in PSNs

Suppose traveler Bob has just arrived at an airport in a foreign country, and he would like to take a taxi to the Hotel Wonderland where he is staying. If possible, Bob would like to share the ride with other travelers staying at the same hotel, and thus to share the costs of the taxi ride. With PSNs, Bob could disseminate a message expressing his interest in going to the Hotel Wonderland by taxi. Due to the very dense population of PCD-equipped individuals typically found in an airport – highlighting that, in some cases, a part of an opportunistic network can indeed be relatively dense – Bob's expression of interest is likely to be disseminated quickly in the airport. So, in a few minutes since issuing the message, Bob could get a reply from other travelers interested in taking a taxi to the Hotel Wonderland, and who maybe are currently still waiting to claim their baggage. Another possible reply to Bob's expression of interest message might be a notification from a local shuttle service that a shared shuttle service to the Hotel Wonderland is available, together with directions on how to reach the shuttle stop. After getting the replies, Bob could decide to wait for the other travelers, or inform them that he was going instead to take the taxi by himself or the cheaper shuttle service.

Note that a smart transportation service like this one could be realized also through the Internet, for instance, by accessing a web server collecting travelers' posts about their transportation needs. However, this would require users to have Internet access enabled on their cell phones, which comes at a cost, especially considering that many subscription plans have rather high rates for accessing the Internet when roaming. If PSNs are used instead, no connection to the Internet is required, and also relatively cheap cell phones with a Bluetooth interface and no Internet connection can be used to access the smart transportation service.

Another possible use of PSNs is in the aftermath of disasters. Suppose an earthquake hit the city of Futuria, and the communication infrastructure – including the cellular network – is compromised. PSNs can be used during the hours/days needed to restore the communication infrastructure to enable exchange of information between survivors, members of the rescue teams, etc. For instance, a person living in Woodland Street who survived the earthquake could take some pictures with her cell phone to show the situation concerning the buildings and roads in the vicinity, and disseminate this information on the PSN. In a few minutes, the information could be delivered to a member of a rescue team, who would then share this information with other members of the team and local authorities to gain a better understanding of the situation. Alternatively, another person living in Woodland Street currently located in Central Avenue, where he works, might disseminate on the PSN an enquiry for any information relating to the situation in Woodland Street. After a few minutes, he might start to receive text messages and pictures reporting the situation in Woodland Street, from which he can see that his house survived the earthquake.

17.2.2 User Scenarios in WSNs

Suppose a WSN is used to monitor the health status and movement of wild animals. A small fraction of the monitored animals are equipped with radios. Furthermore, a set of fixed base stations with satellite Internet connection is located in selected feeding grounds and near water sources. When two radio-equipped animals move into each other's transmission range, they exchange summaries of their movement/health status in the last time period. This way, the data generated by an animal can reach the remote user monitoring the animals through the Internet in two ways: directly, when the tracked animal visits one of the regions covered by a base station; or through a multi-hop path, if the animal's movement/health status summary is delivered to a base station by another, recently met animal.

Note that this scenario, similar, for example, to that of the well-known ZebraNet experiment (Martonosi et al. 2004), is well within the realm of opportunistic networks when the density of radio-equipped animals is relatively low, and eventual communication of tracked data to the base stations is achieved through exploitation of animal movements.

17.2.3 User Scenarios in Vehicular Networks

Consider an advanced urban traffic monitoring system whose purpose is to give local authorities and vehicle drivers a real-time picture of the current traffic status in a city. Traffic status is tracked in real time both through fixed cameras at selected points and through traffic reports generated by the vehicles themselves. In particular, a set of traffic checkpoints is defined within the road network by the traffic authorities, and communicated to the traveling vehicles through the communication infrastructure – composed, for instance, of a set of sparsely deployed IEEE 802.11p RSUs. When traveling, vehicles automatically detect whether they have just passed a checkpoint and, if so, they compute the time elapsed since the passing time at the previous checkpoint. The computed traveling time, together with the ID of the two checkpoints, constitutes a traffic report, which is delivered to the traffic monitoring service through the communication infrastructure, for example, when the vehicle gets in touch with one of the RSUs. If the density of RSUs is not too sparse, or in case multi-hop forwarding of traffic reports is allowed, traffic reports reach the traffic monitoring service center in a few seconds, thus enabling a real-time assessment of the situation. A traffic monitoring service similar to the one described above has been recently proposed in the IPERMOB project (Consortium 2010).

The previously described scenario, at least the part concerning communication of traffic reports to the monitoring service center, displays all the features of opportunistic networks: a network composed of relatively sparse mobile nodes (we can expect that DSRC-equipped vehicles will be a minority of all circulating vehicles for several years to come), complemented with a few fixed collection points (RSUs); data generated by the vehicles is eventually delivered to the monitoring service center through a RSU, with a possibly faster data delivery process in case vehicle-to-vehicle forwarding of traffic reports is realized.

17.3 Opportunistic Networks: Perspectives

While opportunistic networks are considered a very promising type of next generation wireless network, technological and network design challenges are still to be satisfactorily addressed before these networks can be considered a feasible and practical solution.

In terms of technology, we are currently in a situation where some of the short-range wireless technologies exploited for opportunistic communications are mature, while others – especially in vehicular applications – are still to mature. However, where the radio technology is mature, the problem of designing energy-efficient solutions is still to be faced. Curiously, in those situations where energy consumption is not an issue, such as in vehicular applications, the communication technology is relatively less mature.

For instance, considering PSNs, it is well known that the energy drain of a smart phone when the WiFi or Bluetooth interfaces are active is very high, leading to a very short battery life if these interfaces are continuously kept active. Thus, applications for opportunistic networks should be designed taking into account that short-range radio interfaces are not necessarily active, possibly leading to missed communication opportunity detection. This means that energy efficiency should be carefully traded off against a lower likelihood of detecting communication opportunities, which are the fundamental building blocks of any opportunistic networking protocol. Another possible way of addressing the problem of energy efficiency is through improved hardware design of the short-range radio interfaces, possibly integrating them better with the other interfaces present in the smart phone.

It is important to observe that efficient use of the battery is a prerequisite for the success of opportunistic network-based applications: a user is likely to be willing to use these applications if the perceived "added value" they provide is not overwhelmed by the increase in energy consumption; if running an opportunistic network application – even a very exciting one – drains the battery in a few hours, it is likely that most users will end up not using it.

Another major challenge facing the opportunistic network designer is related to the design and realization of networking protocols explicitly designed for opportunistic networks. So far, most research efforts have been devoted to optimizing the performance of well-known networking primitives such as unicast, broadcast, multicast, and so on, in opportunistic networks. However, these primitives appear to be unsuitable for an effective design of opportunistic network applications such as those described in the previous section. Consider for instance the airport scenario. What networking primitive should be used to propagate Bob's message within the airport PSN? Unicast is not suitable, since there is no specific destination that Bob is aware of to which the message should be sent. Similarly for broadcast, since it is not even clear who the members of the airport PSN are at a given time; furthermore, most of the airport PSN members are likely not to be interested in Bob's message. Multicast could be a reasonable choice, but it is quite unlikely that a multicast group composed of "all the users staying at the Hotel Wonderland" could be distributively built and efficiently maintained in such a dynamic setting.

The example above highlights the need to design radically different networking primitives for opportunistic networks, which are designed, for instance, to deliver a message to all users sharing similar interests (staying at the Hotel Wonderland in the airport example) or to all users of a community (those living on Woodland Street in the earthquake example), and so on.

Finally, security and privacy issues are still to be addressed, especially in PSNs. Most existing opportunistic networking solutions exploit information about a user's mobility pattern, social ties, and so on, to optimize the spreading of messages and information within the network. While these solutions achieve their goal in terms of improved network performance, it is very unlikely that users will be willing to expose such sensitive information to potential strangers to run opportunistic network applications. Hence, techniques should be designed to enable a secure and privacy-preserving spreading of information within opportunistic networks.

17.4 Further Reading

This chapter is just a short introduction to opportunistic networks and related technologies. The reader interested in gaining a better understanding of this topic is referred to the books and surveys available in the literature, such as Zhang (2006), Farrell and Cahill (2006), Harras (n.d.), and Denko (2008).

References

Consortium I 2010 *http://www.ipermob.org*.

Denko M 2008 *Mobile Opportunistic Networks: Architectures, Protocols and Applications*. Auerbach Publications, Boca Raton, FL.

Farrell S and Cahill V 2006 *Delay- and Disruption-Tolerant Networking*. Artech House, London.

Group PR 2007 *http://prisms.cs.umass.edu/dome/umassdieselnet*.

Harras K n.d. *Challenged Networks: Protocol and Architectural Challenges in Delay and Disruption Tolerant Networks*. VDM, Saarbr.

Hui P, Chaintreau A, Scott J, Gass R, Crowcroft J and Diot C 2005 Pocket-switched networks and human mobility in conference environments. *Proceeding of the ACM Workshop on Delay-Tolerant Networks*.

Martonosi M, Lyon S, Peh LS, Poor V, Rubenstein D, Sadler C, Juang P, Liu T, Wang Y and Zhang P 2004 *http://www.princeton.edu/mrm/zebranet.html*.

Zhang Z 2006 Routing in intermittently connected mobile ad hoc networks and delay tolerant networks: Overview and challenges. *IEEE Communications Surveys & Tutorials* **8**, 24–37.

18

Routing in Opportunistic Networks

In the previous chapter, we described the distinguishing features of opportunistic networks, namely, sparse node density, lack of network-wide connectivity, and unique role of node mobility as communication means within the network. Given these features, the mechanisms governing message routing in an opportunistic network are very different from those typical of other types of wireless networks. In particular, lack of end-to-end communication paths between source and destination of a message forces the routing protocol to exploit the temporal dimension and node mobility to eventually deliver a message to the intended destination.

In this chapter, we will first describe the basic *store, carry, and forward* mechanism at the heart of any opportunistic network routing protocol, and present a few representative routing protocols for opportunistic networks. We will then carefully discuss the role of mobility in message routing, and define the most important mobility metrics used in the characterization and analysis of node mobility in opportunistic networks.

18.1 Mobility-Assisted Routing in Opportunistic Networks

Given the lack of end-to-end communication paths between source and destination of a message, the mechanisms at the basis of routing in opportunistic networks are radically different from those used in other types of wireless networks, such as route discovery/route reply messages, routing tables, and so on. In opportunistic networks, routing protocols are based on the *store, carry, and forward* principle, according to which a message M generated by a

Mobility Models for Next Generation Wireless Networks: Ad Hoc, Vehicular and Mesh Networks, First Edition. Paolo Santi.

source node S is *stored* in S's buffer and *carried* around, until a communication opportunity for node S with another node A arises. At that point, M can be *forwarded* to node A, either transferring or duplicating M into its buffer. This message propagation process is repeated until, at some point in time, a node carrying a copy of M in its buffer gets in touch with the destination node D, at which point the delivery of M to the destination takes place.

Although based on the same principle, routing protocols for opportunistic networks differ considerably due to the different design choices concerning: (i) the number of copies of a message circulating in the network; and (ii) the amount/type of network state information used by nodes to drive the message-forwarding process.

As regards the number of copies, routing protocols proposed in the literature can be classified into *single-copy* and *multi-copy* protocols. In the former case, only a single copy of the message is allowed to be present in the network at any point in time. The message is originally held by the source node S, which might either keep it till it directly gets in touch with the destination node, or decide to hand M to another node A. Decisions on whether to hand the message to another node are typically driven by metrics aimed at measuring the likelihood of a node to meet the destination node. While single-copy approaches induce minimal load in the network due to the lack of message replication, they typically fall short in terms of message delivery delay and successful message delivery rate, unless the mobility pattern of nodes is very predictable.

Multi-copy protocols have been introduced with the purpose of reducing delay and increasing success rate in delivering messages, at the expense of increasing message overhead. Multi-copy protocols allow replication of a message when communication opportunities arise. These protocols can be further divided into *controlled* and *uncontrolled* multi-copy approaches, depending on whether the message replication process is controlled or not. Control of the replication process is typically realized by either imposing a strict upper bound L on the number of copies of the message M circulating in the network, or putting an upper bound on the number of hops that a message can travel on its way to the destination, or both. While in uncontrolled multi-copy protocols the message overhead cannot be directly kept under control, in controlled approaches the tradeoff between delivery delay/success rate and message overhead can be carefully tuned. Note also that single-copy protocols can be considered as an instance of controlled multi-copy protocols if the upper bound L on the number of message copies is set to 1.

Besides the number of copies of a message circulating in the network, the other major design choice in opportunistic routing protocols concerns the way that forwarding decisions are taken when a communication opportunity arises. Suppose a node A currently holding a copy of M meets another node B, and that, given the rules on the number of M copies, A could potentially replicate M into B's buffer. What rule is used to decide whether to replicate M or not? In other words, what is node A's *forwarding strategy*? Note that this question is relevant only for single-copy or controlled multi-copy approaches, since in the case of uncontrolled multi-copy approaches M is always duplicated into node B's buffer.

Forwarding strategies presented in the literature can be divided into two broad classes, depending on whether or not they make use of *network state* (also known as *context*) information to undertake forwarding decisions. In the former class of strategies, the forwarding choice is based on a set of metrics aimed at predicting node B's likelihood of meeting the destination. These metrics are based on some network state information locally available at the node, which, depending on the specific routing protocol, can take the form of history of past encounters, profiles of recently encountered nodes, storage of a portion of a social graph, and so on. In the latter case, the forwarding strategy is very simple, and states that M is always duplicated into B's buffer provided rules on the number of message copies circulating in the network and/or number of hops are not violated. Clearly, network state-based protocols tend to have superior performance to network-oblivious protocols, since forwarding decisions are based on a richer amount of information. However, locally storing a portion of the network state on the nodes comes at a cost in terms of storage resources, which are used to store state information instead of additional network messages. Hence, whether network state-based approaches are superior to network-oblivious ones in the presence of limited storage capacity depends on the considered application scenario and on the node mobility pattern.

Table 18.1 presents a number of representative routing protocols for opportunistic networks, and their classification according to the criteria described above. In the rest of this section, we will describe in detail a few protocols belonging to different classes, focusing our attention mostly on network-oblivious routing protocols, whose performance is much easier to analyze formally. In fact, formal analysis of network state-based protocols would require accounting for the evolution of locally available network state information on the nodes, which contributes to making the analysis very complex.

18.1.1 Single-Copy Protocols

The simplest possible single-copy routing protocol is *Direct Transmission* (see, e.g., Spyropoulos et al. (2008b)), according to which the source node S keeps the single copy of M in its buffer until either M's time-to-live (TTL) counter expires, or S gets in touch directly with the destination D and delivers the message. Direct Transmission minimizes communication overhead – at most, a single transmission between S and D is performed – at the expense of very long average message delivery delays and low message

Table 18.1 Some representative routing protocols for opportunistic networks

Name	Reference	Copies	Forwarding strategy
Direct Transmission	Spyropoulos et al. (2008b)	Single-copy	Network-oblivious
Randomized	Spyropoulos et al. (2008b)	Single-copy	Network-oblivious
Seek and Focus	Spyropoulos et al. (2008b)	Single-copy	Network state-based
SimBet	Daly and Haar (2007)	Single-copy	Network state-based
Epidemic	Vahdat and Becker (2000)	Multi-copy, uncontrolled	Network-oblivious
2-Hops	Grossglauser and Tse (2002)	Multi-copy, controlled	Network- oblivious
Binary SW	Spyropoulos et al. (2008a)	Multi-copy, controlled	Network-oblivious
BubbleRap	Hui et al. (2008)	Multi-copy, uncontrolled	Network state-based

delivery ratios. Clearly, the performance of Direct Transmission depends on the node mobility pattern: if S and D meet each other relatively frequently, the delay and message delivery ratio can be satisfactory also with this simple routing protocol; however, in most cases Direct Transmission provides poor performance.

A less trivial example of a single-copy routing protocol is the Seek and Focus protocol proposed in Spyropoulos et al. (2008b). The protocol builds upon knowledge of network state locally stored at the nodes. More specifically, every node in the network is assumed to store a timer for every other node in the network, recording the time elapsed since the two nodes last encountered each other. Encounter timers are used by nodes to locally compute a utility function according to the following rules: let $U_i()$ be the utility function for node i; for each $j \neq i$, $U_i(j)$ is an arbitrary monotonically decreasing function of the encounter timer $\tau_i(j)$ between nodes i and j, satisfying the property that $U_i(i) \geq U_i(j), \forall i, j$.

The Seek and Focus protocol is a hybrid approach which alternates between randomized and utility-based forwarding. More specifically, the protocol alternates the following phases. In what follows, U_f denotes a predefined utility threshold, $0 < p < 1$ a predefined forwarding probability, and δ a predefined forwarding threshold.

1. *Seek phase*: if the utility of the current message holder is $\leq U_f$, then forward M to the encountered node with probability p.

2. *Focus phase*: if a node i with utility $> U_f$ just received M, then: (i) node i resets a timer t_{focus} to T_1 and starts counting down; and (ii) it performs utility-based forwarding, that is, it forwards M to another node j only if $U_j(D) > U_i(D) + \delta$.
3. *Re-seek phase*: if t_{focus} expires and node i still stores M, then: (i) node i sets timer t_{seek} to T_2 and U_f to the current utility value; and (ii) it performs randomized routing. If timer t_{seek} expires and node i still holds M, then reset U_f to its default value, and return to either the Seek (if current utility $< U_f$) or Focus (otherwise) phase.

Another interesting single-copy routing protocol is SimBet (Daly and Haar 2007), which attempts to exploit information related to the social relationships between network members to optimize the forwarding process. SimBet exploits two metrics to predict the likelihood of a node meeting the destination: *similarity* and *betweenness*. Similarity measures the number of common neighbors between two network nodes: denoting by L_A and L_B the list of nodes encountered by nodes A and B, respectively, the similarity between A and B is the number of nodes occurring in both lists, that is, $|L_A \cap L_B|$. Betweenness instead is a centrality metric aimed at measuring how well a node can facilitate communication to other nodes in the network – see the next chapter for a formal definition of centrality metrics. Note that both similarity and a version of the betweenness metrics computed on a local view of the network can be computed locally when two nodes meet each other if the two nodes exchange their contact lists. The similarity and betweenness metrics are used to compute a utility value in the [0, 1] interval, with higher utility values indicating a relatively higher likelihood of meeting the destination node. The utility metric is then used to drive the forwarding process: if node A holding message M destined for node D meets another node B, it hands over M to B only in case B has a higher utility (computed with respect to node D) than node A.

18.1.2 Multi-Copy Protocols

The simplest possible multi-copy routing protocol is *Epidemic* (Vahdat and Becker 2000) which, as the name suggests, resembles an epidemic spreading disease in a population. In Epidemic routing, initially only the source node S holds a copy of M and it is thus *infected*. Whenever an infected node A (i.e., a node holding a copy of M in its buffer) meets a non-infected node B, the infected node makes a duplicate of message M and copies it into the buffer of the non-infected node, thus making it infected.

Despite its simplicity, Epidemic routing is widely considered in opportunistic network performance evaluation. In fact, it has been shown that, as intuition suggests, Epidemic routing provides minimum message delivery delay if the buffer size on the nodes is assumed to be infinite. On the other hand, it is well known that Epidemic routing performance quickly degrades if node buffers have limited size: this is due to the very large message overhead induced by Epidemic routing, which quickly overloads node buffers leading to message losses.

Several controlled multi-copy routing protocols have been introduced in the literature with the purpose of reducing message overhead while at the same time providing acceptable performance in terms of message delivery delay.

A first example of this class of protocols is *2-Hops routing* (Grossglauser and Tse 2002; Spyropoulos et al. 2008a). In 2-Hops routing, the source node S holds $L \geq 1$ copies of the message M. Node S delivers a single copy of M to the first $L - 1$ nodes encountered – which become *relay nodes* – and keeps the last copy of the message for itself. Relay nodes are not allowed to forward their copy of M to other nodes, but they can deliver it only to the destination. Thus, the message is eventually delivered to the destination by either S, or one of the (up to) $L - 1$ relay nodes.

2-Hops routing controls both the number of message copies circulating in the network (L at most) and the number of hops in the source/destination path. In fact, since relay nodes can only deliver the message to the destination, the maximum length of the source/destination path is two hops, whence the name of this protocol.

Binary Spray and Wait (SW) (Spyropoulos et al. 2008a) can be regarded as an optimization of 2-Hops routing. In Binary SW, L is assumed to be a power of 2. Initially, the source node S holds all the $L = 2^q$ copies of the message. When a node with $h \geq 2$ copies of the message meets a node without message M, it hands the new node $h/2$ of its copies, and keeps the other $h/2$ for itself. When a node remains with a single copy of M, it can deliver this single copy only to the destination node. It is easy to see that the number of hops in the source/destination path is upper bounded also in Binary SW, and the upper bound is $q = \log_2 L$.

All the protocols described so far make no use of network state information. Several protocols have been proposed that exploit some form of locally stored network state information to optimize the message-forwarding process. Of particular interest in this class of protocols are those exploiting information about the social structure underlying the opportunistic network to optimize performance.

A representative example of this class of protocols is BubbleRap (Hui et al. 2008). BubbleRap builds upon two assumptions: (i) nodes are divided into communities, possibly overlapped, and possibly composed of a single node; and (ii) each node has a global ranking – computed through a centrality metric such as betweenness – across the whole network, as well as a local ranking within its local community(ies). Intuitively speaking, message routing is a hierarchical process performed as follows: when node S wants to send a message M to node D, it first attempts to *bubble* the message up the hierarchical ranking tree using the global ranking value, with the goal of delivering M to socially "well-connected" nodes. This process is repeated until a node B belonging to the same community as the destination is reached. Starting from this point, the local ranking value of the destination community is used to drive the routing process to the destination node. In order to reduce the message overhead, a heuristic is implemented requiring the original carrier of a message to delete M from its buffer when the message is delivered to a node in the destination community.

18.2 Opportunistic Network Mobility Metrics

We have seen in the previous section that mobility plays a fundamental role in the process of delivering a message from source to destination. It is then clear that mobility modeling on the one hand, and analysis of mobility properties observed in real-world traces on the other hand, have attracted considerable attention in the opportunistic networking community. In particular, a number of *metrics* related to node mobility have been defined and utilized in the literature to analyze the performance of opportunistic network routing protocols.

18.2.1 The Expected Meeting Time

A first important mobility metric is the *expected meeting time* (EMT), which is formally defined as follows.

Definition 18.1 *Let A and B be nodes in the opportunistic network, moving in a bounded region R according to a mobility model $\mathcal{M}$. Assume that at time $t = 0$ both A and B are independently distributed in R according to the stationary node spatial distribution of $\mathcal{M}$, and that A and B have a fixed transmission range r. The* first meeting time *T_{AB} between A and B is the random variable (r.v.) corresponding to the time interval elapsing between $t = 0$ and the instant when A and B first come into each other's transmission range. The* expected meeting time *is the expected value of the r.v. T_{AB}, where expectation is taken over all possible node pairs A, B.*

It is well known that some mobility models, such as RWP, give rise to a non-uniform node spatial distribution in stationary conditions. This is why, in the definition of EMT, it is required that nodes A and B are initially distributed according to the stationary node spatial distribution of the mobility model at hand.

EMT is especially important in characterizing the performance of network-oblivious routing protocols. In fact, in this class of protocols, a message held by node A is either forwarded to another node B upon their first encounter, or is not forwarded to B at all – for instance, because the upper bound L on the number of message copies and/or the upper bound on the number of hops has been reached. Thus, taking as $t = 0$ the instant of time when node A gets a copy of M, EMT is a good estimate of the time when node A is expected to encounter node B for the first time.

EMT for some well-known geometric mobility models has been derived by Spyropoulos et al. (2006). For instance, EMT in the presence of RWP mobility has been shown to be as follows:

$$\mathrm{EMT}_{\mathrm{RWP}} = \frac{1}{1.75 p_m + 2(1 - p_m)} \cdot \frac{A(R)}{2r\overline{L}} \cdot \left(\overline{T} + \overline{T}_{stop}\right),$$

where $A(R)$ is the area of the mobility region R, r is the transmission range, $\overline{L}$ is the expected length of a trip, $\overline{T}$ is the expected duration of a trip, $\overline{T}_{stop}$ is the expected pause time at waypoints, and $p_m = \overline{T}/(\overline{T} + \overline{T}_{stop})$ is the probability that a node is moving at any time.

Note that the above expression can be significantly simplified in case the pause time is set to 0. In fact, in this case we have $\overline{T}_{stop} = 0$ and $p_m = 1$, from which we can write

$$\mathrm{EMT}_{\mathrm{RWP}_0} = \frac{1}{1.75} \cdot \frac{A(R)}{2r\overline{L}} \cdot \overline{T} \approx \frac{A(R)}{3.5r\overline{V}},$$

where $\overline{V} \approx \overline{L}/\overline{T}$ is the expected speed in a trip.

A metric closely related to the meeting time is the *hitting time*, defined as the time needed for a node to move into the transmission range of a fixed, randomly chosen point in the movement region R, starting from the stationary node spatial distribution of $\mathcal{M}$. The hitting time for some well-known geometric mobility models has been derived in Spyropoulos et al. (2006).

18.2.2 The Inter-Meeting Time

Another metric that has been thoroughly studied in the opportunistic networking literature is the *inter-meeting time* – also called *inter-contact time* – which, informally speaking, is intended to measure the time elapsing between two consecutive meetings of the *same* pair of nodes. The inter-meeting time is important for studying the performance of network state-based routing protocols, since it can be used to estimate the time elapsing between successive communication opportunities of a node pair. Estimating this quantity is important in the evaluation of network state-based routing protocols, where a copy of message M is not necessarily forwarded from node A to B upon their first encounter after M has reached A – for instance, because the local view of the network state stored at A and/or B has changed in the time elapsing between the first and the second communication opportunity.

In the literature, two different definitions of the inter-meeting time can be found, which we report for completeness. In Groenevelt et al. (2005), the inter-meeting time is defined as the time elapsing between the *starting time* of two consecutive meetings of a node pair, where a meeting between nodes A and B is defined as a continuous-time interval during which nodes A and B are within each other's transmission range. Formally:

Definition 18.2 *Let $t_{AB}^s(k)$ be the starting time of the kth meeting between nodes A and B. The kth inter-meeting time is defined as $\tau_{AB}^s(k) = t_{AB}^s(k+1) - t_{AB}^s(k)$. The inter-meeting time τ^s can be considered as a random variable by considering repeated samples of $\tau_{AB}^s(k)$ for different values of k and node pairs A, B.*

The other definition of inter-meeting time, used for instance in Cai and Eun (2009) and Karagiannis et al. (2007), and more commonly used in the literature, defines the inter-meeting time as the time elapsing between the *ending time* of a meeting and the *starting time* of the next one. Formally:

Definition 18.3 *Let $t_{AB}^s(k)$ and $t_{AB}^e(k)$ be the starting time and the ending time of the kth meeting between nodes A and B, respectively. The kth inter-meeting time is defined as $\tau_{AB}(k) = t_{AB}^s(k+1) - t_{AB}^e(k)$. The inter-meeting time τ can be considered as a random variable by considering repeated samples of $\tau_{AB}(k)$ for different values of k and node pairs A, B.*

The inter-meeting time has been studied for synthetic mobility models, as well as estimated from observations made in real-world experiments. In the next chapter we will report extensively on existing studies of inter-meeting time characterization based on real-world data traces. As for the inter-meeting time in synthetic mobility models, it has been formally proven in Cai and Eun (2009) that well-known mobility models such as RWP and RW give rise to an inter-meeting time distribution whose tail decays at least exponentially with time.

18.2.3 Contact Duration

The third metric which is typically considered in opportunistic network performance evaluation is *contact duration*, which can be informally intended as the expected duration of a meeting between any two nodes A and B. Formally, contact duration is defined as follows:

Definition 18.4 *Let $t_{AB}^s(k)$ and $t_{AB}^e(k)$ be the starting time and the ending time of the kth meeting between nodes A and B, respectively. The kth* *contact duration is defined as* $\delta_{AB}(k) = t^e_{AB}(k) - t^s_{AB}(k)$*. The contact duration* δ *can be considered as a random variable by considering repeated samples of* $\delta_{AB}(k)$ *for different values of* k *and node pairs* A, B.

The metric defined above becomes relevant when the expected time needed to transfer the message M from A to B is very large – say, several seconds or minutes. Typically, this happens when M is a multimedia file such as a picture or a video. In these situations, it is very likely that the contact duration is not long enough to ensure complete transfer of M from A to B.

In general, estimating contact duration is necessary for understanding whether situations of incomplete message transfer between nodes such as the one described above are likely to happen. If, for instance, the contact duration is estimated to be in the range of 1–2 minutes and the average size of M is 1 MB, a few seconds are sufficient to transmit M from A to B even with the relatively slow Bluetooth technology; hence, it can be concluded that the event "message transfer between A and B during a contact is interrupted" is relatively unlikely. Under this assumption, contact opportunities between nodes can be well approximated by instantaneous "contact" events during which message M is entirely transferred, and the actual duration of the contact becomes irrelevant. On the other hand, if the expected message transfer time is comparable to the expected contact duration, the above simplifying assumption cannot be made, and the analysis of the message propagation process within the network becomes more complex.

A possible solution to avoid incomplete message transfer during a contact is to break down a relatively large message M into smaller pieces $M_1, \ldots, M_K$, and to transfer as many pieces as possible to the other node during a contact. This approach can be combined with network coding techniques to improve message delivery performance.

18.2.4 Relating the Three Metrics

The three mobility metrics defined above are clearly related to each other, as depicted in Figure 18.1. The difference between EMT and the inter-meeting time lies in the starting time of the observation interval: in the case of inter-meeting time, the observation of mobile nodes A and B starts at the moment when a contact between A and B has just ended. Thus, observation of A and B is done conditional on the knowledge of a recently past event, namely, a contact between A and B. On the contrary, in case of EMT no assumption on past contact between nodes is made, and the observation is assumed to start at a random point in time – that is, the observation is not conditioned on any past event. Note that the two quantities might be significantly different, depending on the mobility model. For instance, for a synthetic mobility model such as RWP, the condition that a meeting

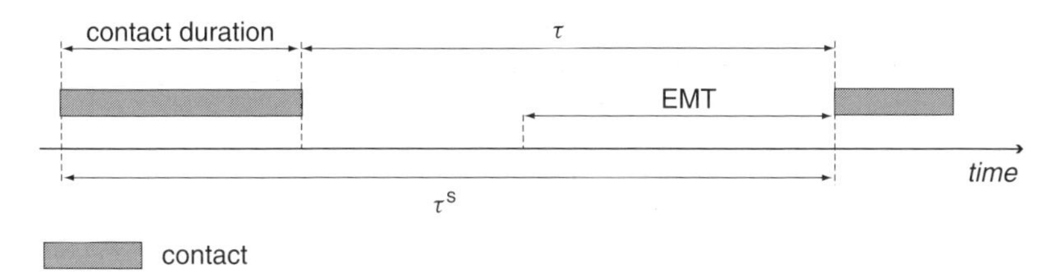

Figure 18.1 Relationships between expected meeting time, the two notions of inter-meeting time, and contact duration.

between A and B has just ended imposes strong constraints on the relative distance between A and B at the beginning of the observation period, which must be slightly larger than the transmission range r. On the contrary, for EMT, the expected distance between A and B at the beginning of the observation period equals the expected distance between two points in R distributed according to the node spatial distribution of the RWP model, which can be substantially different from r.

The two notions of inter-meeting time and contact duration are related as follows – see Figure 18.1:

$$\tau^s = \tau + \delta.$$

From this formula, we can see that the two notions of inter-meeting time become nearly equivalent when the contact duration is much shorter than the time elapsing between consecutive meetings; that is, $\tau^s \approx \tau$ whenever $\delta \ll \tau$.

Before concluding this chapter, it is worth observing that the mobility metrics defined above are based on a continuous-time model. They can be easily extended to a discrete-time model in those situations where the observation of the mobility and contact process occurs at periodic intervals, as is the case with real-world traces that have been studied in the opportunistic networking literature – see the next chapter.

References

Cai H and Eun D 2009 Crossing over the bounded domain: From exponential to power-law intermeeting time in mobil ad hoc networks. *ACM/IEEE Transactions on Networking* **17**, 1578–1591.

Daly E and Haar M 2007 Social network analysis for routing in disconnected delay-tolerant MANETs. *Proceedings of ACM MobiHoc*, pp. 32–40.

Groenevelt R, Nain P and Koole G 2005 The message delay in mobile ad hoc networks. *Performance Evaluation* **62**, 210–228.

Grossglauser M and Tse D 2002 Mobility increases the capacity of ad hoc wireless networks. *ACM/IEEE Transactions on Networking* **10**(4), 477–486.

Hui P, Crowcroft J and Yoneki E 2008 BUBBLE Rap: Social-based forwarding in delay tolerant networks. *Proceedings of ACM MobiHoc*, pp. 241–250.

Karagiannis T, Le Boudec JY and Vojnovic M 2007 Power law and exponential decay of inter contact times between mobile devices. *Proceedings of ACM Mobicom*, pp. 183–194.

Spyropoulos T, Psounis K and Raghavendra C 2006 Performance analysis of mobility-assisted routing. *Proceedings of ACM MobiHoc*, pp. 49–60.

Spyropoulos T, Psounis K and Raghavendra C 2008a Efficient routing in intermittently connected mobile networks: The multi-copy case. *ACM/IEEE Transactions on Networking* **16**, 77–90.

Spyropoulos T, Psounis K and Raghavendra C 2008b Efficient routing in intermittently connected mobile networks: The single-copy case. *ACM/IEEE Transactions on Networking* **16**, 63–76.

Vahdat A and Becker D 2000 Epidemic routing for partially connected ad hoc networks. *Technical Report CS-200006, Duke University*.

19

Mobile Social Network Analysis

In the previous chapters, we introduced opportunistic networks and described the store, carry, and forward mechanism at the basis of message routing in this kind of network. Furthermore, we introduced the main metrics that have been defined to characterize mobility in opportunistic networks. In this chapter, we will focus attention on a specific type of opportunistic network, called the *pocket-switched network* (see also Chapter 17), in which the nodes forming the network are individuals carrying portable communication devices such as smart phones. This class of networks is very interesting from a mobility modeling perspective, since network nodes are carried by people, hence characterization of node mobility in PSNs is equivalent to characterization of human mobility, which is a very interesting topic investigated in many fields such as sociology, transportation engineering, urban planning, and so on. What is particularly interesting is trying to understand the relationships between social interactions among individuals, and their mobility pattern. Understanding these relationships is fundamental to, for example, optimize the performance of social-aware routing protocols, design social-aware communication primitives, design and optimize innovative services for mobile social networking applications, and so on.

In this chapter, we will present the state of the art on the characterization of human mobility patterns, based on an analysis of data traces obtained from real-world experiments. We will start by presenting the notion of a social network graph, which is useful for studying and characterizing social interactions among a set of individuals. We will then introduce the most relevant metrics used to characterize the structure of social network graphs, which can be used to identify the most "important" nodes in a social network graph, or to identify communities within a larger social structure. Finally, we

Mobility Models for Next Generation Wireless Networks: Ad Hoc, Vehicular and Mesh Networks, First Edition. Paolo Santi.

will present the main findings of recent studies aimed at characterizing human mobility based on data trace analysis.

19.1 The Social Network Graph

A *social network* is a social structure made up of individuals and a set of connections between individuals representing some type of social interdependency, such as friendship, kinship, common interests, and so on. Social networks, which have been originally defined in field of sociology to, for example, measure social capital (i.e., the value that an individual gets from being part of a social structure), have now become a tool for explaining complex phenomena in different scientific fields such as anthropology, biology, economics, computer science, geography, and so on. For instance, social network analysis has been used in epidemiology to help understand how different human mobility patterns aid or inhibit the spread of infectious diseases such as HIV in a population. As another example, social networks are used to estimate the role of influential "opinion leaders" in spreading innovation within a population.

In a famous experiment performed in 1967, psychologist Stanley Milgram asked a sample of individuals in the USA to reach a particular target person by passing a message along a chain of acquaintances. Milgram then evaluated the average length of these message-passing chains, and established that five intermediaries – hence, *six degrees of separation* – were on average sufficient to connect an arbitrary pair of individuals. The fact that the chain of social acquaintances required to connect two arbitrary individuals in the world is in general short, and does not depend on the size of the population, is known as the *small-world phenomenon*, and was first noticed because of Milgram's experiment. Later on, small-world phenomena have been observed in many different types of social networks, such as the network formed by actors starring in the same movie, networks modeling hyperlinks in the World Wide Web, and so on.

The *social network graph* is a graph-based representation of a social network, where nodes correspond to social entities (e.g., individuals), and edges represent social ties between social entities. Depending on the specific social network considered, edges in the social network graph can be directed or undirected. In some cases, edges can be weighted, with the value of the weight representing the "intensity" of the social tie. In the following, we present some examples of social network graphs.

Based on Milgram's experiment, one can build a graph representing the social network underlying the experiment as follows: a node is assigned to each individual participating in the experiment, as the origin, target, or forwarder of a message; a directed edge is inserted between nodes u and v if node u passed a message to node v. The social network graph representing the Web is formed by adding a node for each `url` in the Web, and a direct edge between nodes u and v if page u contains a hyperlink to page v. A social network graph more relevant to PSNs of interest in this chapter is one in which nodes are the individuals forming the PSN, and there exists an edge between u and v if node u had at least a contact opportunity with node v during the period of interest. Alternatively, a weighted edge can be inserted, with a weight representing the number (or duration) of contact opportunities between the two nodes during the period of interest.

As the above examples highlight, in many cases social networks are very large, and are represented by graphs composed of several thousands (if not millions) of nodes and edges. For instance, the social network graph representing friendship relationships in an online social network application such as Facebook can be composed of up to several hundred million nodes and billions of edges. This explains why the field of social network analysis is closely related to the field of complex network analysis, an emerging research field at the intersection of many scientific disciplines such as mathematics, theoretical physics, computer science, biology, sociology, epidemiology, and so on. Complex networks are networks – represented as graphs – formed by a very large number of elements, and characterized by a non-trivial topology, which is neither regular nor entirely random.

The goal of social and complex network analysis is to provide a characterization of fundamental properties of the social (or complex) structure under study. For instance, typical goals of social/complex network analysis are the identification of "socially central" nodes (leaders) and the identification of "communities" within a larger network. To this end, several metrics have been defined on the social network graph, the most relevant of which are presented in the following section.

19.2 Centrality and Clustering Metrics

Several metrics have been defined on social network graphs to help understand and discover their structural properties. In the following, we introduce those metrics more relevant to the type of network under consideration in this part of the book, namely, PSNs. We first introduce a class of metrics aimed at measuring the "social centrality" of network nodes, and then a class of metrics aimed at measuring the degree of clustering observed in a network.

Before proceeding further, we observe that the metrics defined below can be applied to either the entire social network graph or a portion of it. Of particular importance is the so-called *ego social network*, that is, the portion of the social network as seen from a particular node in the network (the *ego* node). In graph-theoretic terms, the ego network of node u is the subgraph of the social network graph formed by node u itself, by all the edges incident to u, and by all nodes which are endpoints of an edge incident to u. Ego networks are used sometimes as proxies of the complete social network graph, since it has been observed that, at least in some cases, the structural properties of ego networks are similar to those of the complete network (Marsden 2002).

19.2.1 Centrality Metrics

Centrality metrics are aimed at measuring the importance of a node within a large graph. When applied to social networks, centrality metrics are used to identify socially influential nodes. In the following, we introduce some relevant centrality metrics, and discuss how they have been used to optimize the performance of social-aware routing protocols for PSNs. Furthermore, we use $G = (V, E)$ to denote a social network graph with node set V and (undirected) edge set E, and we assume G is connected.

A very simple centrality metric is *node degree*, that is, the number of edges in E incident to a node. Intuitively, nodes with higher degree have relatively more social ties, hence are socially better connected than nodes with relatively lower degree. However, it has been observed that, often, node degree is not sufficient to capture the "social importance" of a node within the network. For instance, a node which is part of a relatively isolated and tightly knit community within a larger network might have a relatively high degree because it has many connections with other members of its community, but nevertheless display a limited "social influence" due to the isolation of the community it belongs to. Justified by this observation, other metrics have been defined to measure node centrality.

A widely used centrality metric is *betweenness centrality*, which measures the fraction of shortest paths between any two nodes in the network that pass through a specific node. Formally, we have

$$C_B(u) = \sum_{s \neq u \neq t \in V} \frac{\delta_{st}(u)}{\delta_{st}},$$

where δ_{st} is the number of shortest paths from node s to node t in G, and $\delta_{st}(u)$ is the number of these paths passing through node u. The above value can be normalized by dividing it by $(n-1)(n-2)/2$, which represents the total number of node pairs in G which do not include u (n is the cardinality of the node set V). Betweenness centrality can be used, for instance, to measure how important a node is in spreading information within the social network: if a node u with high betweenness is removed from the network, the information will take much longer to spread throughout the network, since many shortest paths will no longer be available after node u's removal.

Another centrality metric is based on the topological notion of closeness, which, intuitively speaking, measures how "close" a node is to the other nodes in the network. Within the context of social networks, distance is defined in terms of number of hops in the social network graph, and it is called *geodesic distance*. Formally, the geodesic distance $d_G(u, v)$ between any two nodes in G is defined as the hop length of the shortest path connecting u and v in G. *Closeness centrality* is defined as follows:

$$C_C(u) = \frac{1}{\sum_{t \in V - \{u\}} d_G(u, t)}.$$

Closeness centrality can be used, for instance, to measure how long it will take for a piece of information (message) generated at node u to reach all the other nodes in the social network.

An example of a social network – based on Marsden (2002) – is shown in Figure 19.1, and the corresponding centrality metrics for each node are given in Table 19.1. The more central node(s) according to the different metrics are highlighted in Figure 19.1.

Centrality metrics have been used in the design of social-aware forwarding protocols for PSNs. The underlying idea is that relatively more central nodes in the social network graph should be preferred for carrying a copy of a

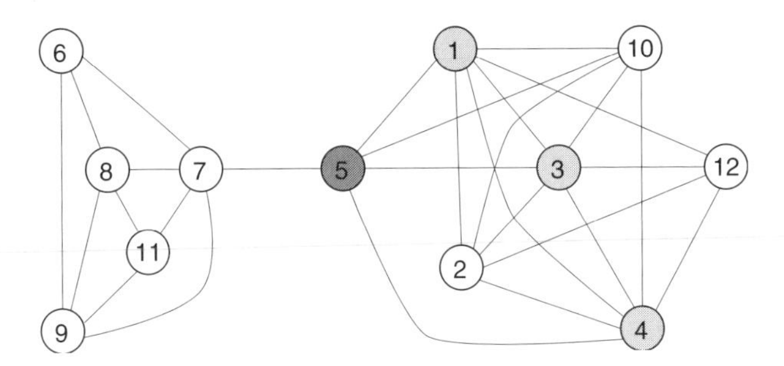

Figure 19.1 The bank wiring room social network (Marsden 2002). Nodes with the highest degree, betweenness centrality, and closeness centrality are highlighted in light gray (degree centrality) and dark gray (betweenness and closeness centrality).

Table 19.1 Centrality metrics of the social network displayed in Figure 19.1

Node	Degree	C_B	C_C	Node	Degree	C_B	C_C
1	6	3.75	0.050	7	5	28.33	0.053
2	5	0.25	0.038	8	4	0.33	0.038
3	6	3.75	0.048	9	4	0.33	0.038
4	6	3.75	0.050	10	5	1.5	0.048
5	5	30	0.059	11	3	0	0.037
6	3	0	0.033	12	4	0	0.037

message to relatively less central ones, since they have a higher likelihood of getting in touch with many other nodes within the PSN. Examples of protocols exploiting this idea are SimBet (Daly and Haar 2007) and BubbleRAP (Hui et al. 2008), which both use betweenness centrality (combined with other measures) to identify central nodes within the network.

19.2.2 Clustering Metrics

Clustering metrics are aimed at measuring the degree to which nodes in a graph tend to cluster together. In turn, clusters of nodes can be identified as communities within a larger social network.

In the following, we report the metric defined in Watts and Strogatz (1998), which is aimed at measuring the clustering as observed at a single node – that is, the clustering of a node's ego network. The *clustering coefficient* of node u quantifies how close its neighbors in G are to forming a clique (i.e., a complete graph). Formally, the clustering coefficient of node u is defined as follows:

$$CC(u) = \frac{2|E_u|}{deg(u)(deg(u)-1)},$$

where E_u is the set of edges in E such that both endpoints are neighbors of node u in G. In this formula, $deg(u)(deg(u)-1)/2$ represents the total number of possible edges between any two neighbors of node u, where $deg(u)$ is the degree of node u. Thus, the clustering coefficient represents the fraction of existing links between neighbors of u over the total number of possible such links. A clustering coefficient of 1 indicates that u's neighbors form a clique (maximal clustering), while a clustering coefficient of 0 indicates that u's neighbors are arranged as a star centered at u (minimal clustering).

Watts and Strogatz (1998) also define the *network average clustering coefficient* as the average of the clustering coefficients of all nodes in the network graph. The authors show that the network average clustering coefficient can be used to identify networks enjoying the small-world property: a network graph $G = (V, E)$ is considered to have the small-world property if it has an average clustering coefficient significantly higher than that of a random graph with the same edge density constructed on V, and if the network graph has approximately the same average shortest path length as compared to the corresponding random graph.

While clustering metrics are aimed at measuring the level of clustering in a social network, they cannot be directly used to *identify* clusters (communities) within the network. Finding communities in a social network is a very important and challenging task that has been (and currently is) the subject of intensive research. Challenges in addressing this problem are related to the definition of community itself, which is not uniquely defined, and to the computational complexity of community detection algorithms, which are **NP**-hard in most cases.

Intuitively speaking, a community is a group of social entities within a larger social network displaying a large number of social ties between themselves. However, how to turn this intuitive notion into a formal, graph-theoretic definition is not straightforward. As a matter of fact, there is no universally accepted formal definition of community in the social and complex network literature.

We start by observing that community detection is a relevant problem only in *sparse* social networks, that is, in social networks where the average node degree is $O(1)$, as is actually the case in most real-world social networks. If the number of edges in the social network graph is too large, then the distribution of edges among the nodes is likely to become too homogeneous for communities to exist.

To take a step forward toward a formal definition of community in a social network, we observe that, if we define as *internal* edges between community members, and as *external* edges joining a member of the community with a node outside the community, a community should satisfy the property that the number of internal edges is much larger than the number of external edges. Consider for instance the social network of Figure 19.1, where two communities can be clearly identified: a first community $\mathcal{C}_1$ formed by nodes $\mathcal{C}_1 = \{6, 7, 8, 9, 11\}$, and a second community $\mathcal{C}_2$ formed by nodes $\mathcal{C}_2 = \{1, 2, 3, 4, 5, 10, 12\}$. For community $\mathcal{C}_1$, the number of internal edges is 9, while the number of external edges is 1; for community $\mathcal{C}_2$, the number of internal edges is 16, while the number of external edges is 1. Thus, comparing the ratio of the number of internal edges to the number of external edges appears to be a good method for defining communities in a social network graph. Indeed, the above numbers should be normalized with respect to the maximum possible number of internal and external edges for a community $\mathcal{C}$, which is $|\mathcal{C}|(|\mathcal{C}| - 1)/2$ and $|\mathcal{C}|(n - |\mathcal{C}|)$, respectively. However, this is a heuristic criterion to identify communities, and there is no consensus in the literature on the minimum value of the ratio that a community should satisfy, nor on whether computing the above ratio is the only method that can be used to determine whether a subset $\mathcal{C}$ of the network nodes V forms a community. Given this lack of consensus on a definition of community, the set of community detection algorithms proposed in the literature is even more heterogeneous.

While surveying the many definitions of community introduced in the literature and the related detection algorithms is far beyond the scope of this chapter (see Section 19.4 for reading suggestions on this topic), in the following we mention a very simple definition of community that is often used in the opportunistic networking literature.

A set of nodes $\mathcal{C} \subseteq V$ can be defined as a community if the set of internal edges in $\mathcal{C}$ forms a clique. More formally, set $\mathcal{C}$ is a community if and only if the number of internal edges is $|\mathcal{C}|(|\mathcal{C}| - 1)/2$. Note that this definition accounts only for the number of internal edges to define a community, while the ratio between the number of internal and external edges is not considered. Thus, any subset of nodes $V' \subseteq V$ in a complete graph $G = (V, E)$ would be regarded as a community according to this definition.

The above definition is very strict in defining a community, since each member of the community is required to have a social tie with every other member. Furthermore, under this definition all members in a community have symmetric connections, that is, there is no role differentiation within the community. This is in contrast to evidence from real-world observations clearly hinting at the existence of a whole hierarchy of roles within social communities. Despite these limitations, the above is the simplest possible definition of community, and it is often used in the opportunistic networking literature.

19.3 Characterizations of Human Mobility

A great deal of attention has been recently devoted to the characterization of human mobility patterns based on real-world data traces. In fact, several wireless network data traces became available in the literature (mostly through the CRAWDAD website (Team 2011)) starting in the early 2000s, and researchers have deeply studied these traces to gain an understanding of human mobility patterns not only qualitatively, but also *quantitatively*.

Roughly speaking, two types of data traces have been used to this end: *location-based* and *contact-based* traces. In the former, the location of the user is traced either directly (e.g., by recording the GPS reading), or indirectly (e.g., by recording the user's association with a WiFi access point or with a cellular base station). In the latter type of trace, logs of pairwise contacts–established, typically, through a short-range wireless technology such as WiFi or Bluetooth–between network users are recorded, while the location of users is typically not recorded in the data trace.

Wireless network data traces can be used to study features of individual human mobility patterns (e.g., how frequently individuals visit certain locations), as well as their contact patterns. Observe that only location-based traces can be used to study individual mobility patterns, while both location-based and contact-based traces can be used to characterize contact patterns between individuals. In case location-based traces are used for this purpose, pre-processing of the data trace is needed, in order to derive contact logs starting from user locations. Typically, it is assumed that two users are in contact if they are associated with the same access point/base station, or if they are within a certain distance from each other. Observe, though, that the pre-processing phase introduces an approximation in defining contact between users. Hence, contact-based traces should be preferred if the goal is to study contact patterns between users.

19.3.1 Characterization of Individual Human Mobility Patterns

We first present recent findings concerning the characterization of individual human mobility patterns, which are mostly due to a series of contributions from A.-L. Barabasi and colleagues (Gonzales et al. 2008; Song et al. 2010; Wang et al. 2011). The authors in their analyses consider large data sets of mobile phone users, whose trajectories and phone calls are traced for a relatively long time period (up to one year). More specifically, each time a user initiates or receives a call or SMS (Short Message Service), the cellular base station with which the user is currently associated is recorded in the trace. This type of trace allows, on the one hand, tracking the location of users with a spatial granularity corresponding to the average radio coverage area of a cellular base station–around a few square kilometers. On the other hand, these traces also allow tracking phone calls between users, so potentially giving information not only on their mobility patterns, but also on their social interactions (more specifically, those occurring through the cellular network).

The data traces used by these authors contain massive amounts of data (about 6 million users are traced for several months); however, there is a major drawback concerning the characterization of human mobility patterns: that is, the location of a user is tracked with uneven temporal resolution, since user location tracking occurs only when the user initiates or receives a call/SMS. Thus, it might very well be the case that the location of a user remains untracked for a relatively long period of time (up to several hours), introducing inaccuracies in the location tracking process.

One of the data traces considered in Song et al. (2010), though, does not have this shortcoming, because it contains the location records of 1000 users who signed up for a location-based service; hence, their location is tracked with a one-hour granularity, independently of whether they initiate/receive phone calls/SMSs. What is reported in the following is mostly based on the analysis of this data trace.

The following metrics have been considered to characterize individual human mobility:

1. The *displacement* Δr at hourly intervals, defined as the distance between the location of a user at hour t and the location of the same user at hour $t - 1$. The location of a user is assumed to be the location of the cellular base station the user is registered with.
2. The *waiting time* Δt, defined as the time (expressed in multiples of a hour) a user spends at a particular location.

3. The *number of distinct locations* $S(t)$ visited by a user up to hour t.
4. The *visitation frequency* f, measuring the probability of a user to visit a given location.

Concerning displacement, it is observed that Δr follows a truncated power law with a cutoff at about 100 km, roughly corresponding to the maximum distance an individual can cover in one hour. More specifically, we have

$$Prob(\Delta r = k) \sim k^{-1-\alpha} e^{-k/100},$$

where Δr is expressed in kilometers and $\alpha \approx 0.55$.

A similar law governs the waiting time, which also obeys a truncated power law with a cutoff at about 17 hours, approximately corresponding to a user's daily awake period. More specifically, we have

$$Prob(\Delta t = h) \sim h^{-1-\beta} e^{-h/17},$$

where $\beta \approx 0.8$.

The number of locations visited within time t follows a power law with exponent 0.6, that is,

$$S(t) \sim t^{0.6}.$$

Finally, the visitation frequency closely resembles Zipf's law,

$$f_k \sim k^{-1.2},$$

where f_k represents the visitation frequency of the kth most visited location.

An interesting question to investigate is whether well-known mobility models such as random walks and Lévy flights are consistent with the above findings obtained from data traces, and whether they can be used to accurately model human mobility. Recall that Lévy flights in particular have been shown in the literature to faithfully reproduce animal movements. The analysis reported in Song et al. (2010) shows that random walks and Lévy flights *cannot* be used to faithfully reproduce human mobility patterns. In fact, if human mobility were to follow a random walk, we would have that the exponent of the law describing the number of locations visited with increasing time should be approximately equal to parameter β in the waiting time distribution. Based on the data traces, though, we have that this exponent is $0.6 \neq \beta = 0.8$. Similarly, if human mobility were to obey the Lévy flight model, the exponent of the law governing $S(t)$ should be 1, instead of 0.6 as observed from the data traces. Similar conclusions can be drawn for the visitation frequency, which would be constant under either random walk or Lévy flight mobility, while it is instead distributed according to Zipf's law based on data trace analysis.

From the previous observations, it can be concluded that, unlike animal mobility, human mobility cannot be accurately modeled by random walk-like models. This is essentially due to fundamental differences in how random walk-like models and human mobility deal with the following factors:

1. *Exploration*: random walk-like models assume memory-less exploration, that is, the destination of the next trip is independent of the locations already visited in the past; on the contrary, human mobility is characterized by an increasingly smaller probability of exploring new locations as time goes by. Individuals tend to repeatedly return to locations already visited in the past, and only occasionally visit new locations.
2. *Preferential return*: in random walk-like models, the probability of visiting a location is uniform in time and space; on the contrary, human mobility is characterized by a clear propensity to return to locations an individual visited frequently in the past, such as home and the workplace.

As we will see in the next chapter, the above observations have motivated researchers to define mobility models specifically aimed at faithfully reproducing individual human mobility patterns.

Before ending this subsection, we want to mention the study reported by Wang et al. (2011), who show that mobility patterns and social networks are correlated. In other words, if two users display similar mobility patterns in visiting similar locations at similar times, they are likely to be "close" to each other in the social network graph, where the social network graph in Wang et al. (2011) is built based on phone calls between users. Mobility models aimed at modeling the interplay between human mobility and social relationships will also be discussed in the next chapter.

19.3.2 Characterization of Pairwise Contact Patterns

While the results presented above are useful for gaining an understanding of human mobility in general, what is important in opportunistic network performance evaluation is characterizing the *pairwise contact pattern* between individuals, namely, the duration and frequency of contact opportunities between an arbitrary pair of nodes in the opportunistic network. The studies on individual human mobility patterns can only partially be exploited for this purpose; in fact, their focus is mostly on the mobility pattern of each individual, and information about pairwise mobility metrics (e.g., how often two individuals are registered with the same base station) is often disregarded. Furthermore, the spatial granularity of the traces based on cellular networks is in the order of a few square kilometers, so it is too coarse to accurately estimate whether two individuals are within each other's WiFi/Bluetooth communication range even if they are registered with the same cellular base station.

To gain a deeper understanding of the statistical properties of contact patterns between pairs of individuals, a series of experiments have been performed in the past, where short-range radio devices (mostly based on Bluetooth technology) have been given to a number of volunteers. The devices were programmed to periodically activate the radio interface and perform an inquiry to discover other devices within communication range. The typical inquiry period is in the order of a few minutes, and it is thus much shorter than that used in the location-based data trace analyzed in Song et al. (2010). Thus, not only is the spatial resolution of the data traces used to characterize contact patterns smaller than that used in the characterization of individual human mobility patterns, but also the temporal resolution is much smaller (a few minutes instead of hours). Note that using a smaller spatial and temporal resolution is a prerequisite for accurate characterization of contact patterns. In fact, given that the communication range in opportunistic networks is short (in the order of a few tens of meters at most), movements of even a few tens of meters–which are small movements at the spatial granularity considered in cellular networks–are very important to determine whether a contact opportunity arises. In turn, since even small movements can determine whether a contact opportunity arises or not, the temporal granularity used to produce the data trace should also be small enough to capture such small movements.

In this section, we report results on the characterization of pairwise contact opportunities mostly taken from Karagiannis et al. (2007), who conducted the first thorough study on this topic. The results are based on the analysis of six data traces, three of which were produced using Bluetooth devices. The features of these data traces are given in Table 19.2. The time granularity of the data traces is 300 s for the MIT BT trace, and 120 s for the Cambridge and Infocom data traces.

Karagiannis et al. (2007) were interested in characterizing the *inter-meeting time distribution*, where the inter-meeting time between nodes A and B is defined, we recall, as the time elapsing between the end of a contact between A and B and the beginning of the next contact between A and B.

The main finding of Karagiannis et al. (2007) is that the inter-meeting time distribution displays a dichotomy: initially, it behaves like a power law, but the tail of the distribution is exponentially distributed. In other words, the inter-meeting time distribution behaves like a truncated power

Table 19.2 Main features of the data traces analyzed in Karagiannis et al. (2007)

Name	Reference	Duration	Devices	Contacts
MIT BT	Eagle and Pentland (2006)	16 months	89	114 046
Cambridge	Chaintreau et al. (2006)	11.5 days	36	21 203
Infocom	Chaintreau et al. (2006)	3 days	41	28 216

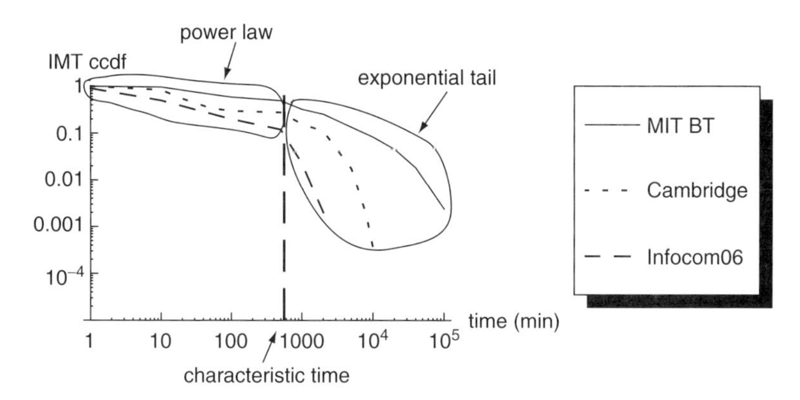

Figure 19.2 Complementary cumulative density function of the inter-meeting time distribution of the three data traces given in Table 19.2.

law. This can be seen from Figure 19.2 showing, for each data trace, the complementary cumulative density function of the inter-meeting time in log–log scale. The plots are initially linear, which, in log–log scale, corresponds to a power law, but, after a certain time, they display a sharp fall with time. Interestingly, the time at which the transition between power-law and exponential trend occurs (called the *characteristic time*) is approximately the same for the three data traces, being approximately equal to 12 hours. This fact clearly hints at a periodicity of human contact patterns similar to that displayed by individual movement patterns.

To investigate possible reasons for the displayed power-law and exponential tail dichotomy of the inter-meeting time distribution, Karagiannis et al. (2007) analyze the *return time* of a user to her/his *home location*, which is defined as the location where the user spends most of the time during the period of observation. To characterize return time, the authors use a set of location-based data traces containing GPS and cell phone tracks. The analysis reveals that the distribution of the return time to the home location displays the same dichotomy observed in the inter-meeting time distribution, with an initial power-law trend followed by an exponential drop-off. Driven by this observation, Karagiannis et al. (2007) formulate and validate the following hypothesis:

Inter-meeting time hypothesis: any two users in the opportunistic network tend to repeatedly meet at the same location, called the "meeting location."

The inter-meeting time hypothesis is validated on the data traces, verifying that for about 80% of all the possible user pairs in the trace, more than 90% of the meetings occurs in at most two locations.

Summarizing, we can empirically explain the observed power-law and exponential tail dichotomy of the inter-meeting time as follows: the initial power-law trend is dictated by occasional meeting opportunities in the "outside world"; as time goes by, the inter-meeting time complementary cumulative density function displays a sharp drop-off, which is explained by the fact that nodes, if not meeting earlier in the "outside world," tend to periodically meet at their "meeting location."

As we will see in the next chapter, the inter-meeting time hypothesis is at the basis of recently proposed mobility models for opportunistic networks aimed at capturing the unique features of individual human mobility and contact patterns described in this chapter.

19.4 Further Reading

The reader interested in gaining a better understanding of the field of social and complex network analysis is referred to the following books: Easley and Kleinberg 2010; Newman (2010), and Newman et al. 2006. See also recent surveys on selected topics, such as the exhaustive survey by Fortunato (2006) on community detection.

References

Chaintreau A, Hui P, Crowcroft J, Diot C, Gass R and Scott J 2006 Impact of human mobility on the design of opportunistic forwarding algorithms. *Proceedings of IEEE Infocom*, pp. 1–13.

Daly E and Haar M 2007 Social network analysis for routing in disconnected delay-tolerant MANETs. *Proceedings of ACM MobiHoc*, pp. 32–40.

Eagle N and Pentland A 2006 Reality mining: Sensing complex social systems. *Journal of Personal and Ubiquitous Computing* **10**, 255–268.

Easley D and Kleinberg J 2010 *Networks, Crowds, and Markets: Reasoning about a Highly Connected World*. Cambridge University Press, New York, US.

Gonzales M, Hidalgo C and Barabasi AL 2008 Understanding individual human mobility patterns. *Nature* **453**, 779–782.

Hui P, Crowcroft J and Yoneki E 2008 BUBBLE Rap: Social-based forwarding in delay tolerant networks. *Proceedings of ACM MobiHoc*, pp. 241–250.

Karagiannis T, Le Boudec JY and Vojnovic M 2007 Power law and exponential decay of inter contact times between mobile devices. *Proceedings of ACM Mobicom*, pp. 183–194.

Marsden P 2002 Egocentric and sociocentric measures of network centrality. *Social Networks* **24**, 407–422.

Newman M, Barabasi AL and Watts DJ 2006 *The Structure and Dynamics of Networks*. Princeton University Press, New Jersey, US.

Song T, Wang P and Barabasi AL 2010 Modelling the scaling properties of human mobility. *Nature Physics* **7**, 713–718.

Team C 2011 *http://crawdad.cs.dartmouth.edu/index.php*.

Wang D, Pedreschi D, Song C, Giannotti F and Barabasi AL 2011 Human mobility, social ties, and link prediction. *Proceedings of the ACM Conference on Knowledge Discovery and Data Mining (KDD)*, pp. 1100–1108.

Watts D and Strogatz S 1998 Collective dynamics of small-world networks. *Nature* **393**, 440–442.

20

Social-Based Mobility Models

As seen in the previous chapter, human mobility – and the resulting pairwise contact patterns – is characterized by features that make it substantially different from mobility patterns generated by simple mobility models such as random walks. In particular, human mobility has been recently shown to be influenced by, and to influence, social interactions between individuals. Given the above observations, researchers have recently proposed mobility models specifically designed to take the interplay between human mobility and social interactions into account.

In this chapter, we will present a set of recently proposed social-based mobility models aimed at faithfully reproducing individual human mobility patterns, and the resulting pairwise contact patterns. We will start by presenting a weighted version of the popular random waypoint model aimed at modeling different degrees of popularity in waypoint selection, and a mobility model where node movement is biased toward sub-areas (communities) specific to each node. We will then introduce a community-based mobility model, where an input social network graph is used to shape the mobility pattern of nodes in a network. The fourth model considered builds upon the notion of "home location" resulting from recent studies on human mobility characterization, and proposes a tradeoff between distance from home and popularity of the destination to determine individual mobility patterns. Another model, inspired by the principle of least action walking, is aimed at reproducing a set of features observed in human mobility traces concerning length of trips, pause time duration, inter-contact time, etc. Finally, we will present a contact-based mobility model which, instead of modeling the location of network nodes in a metric space, and then inferring contact patterns based on distance between nodes, is aimed at synthetically

Mobility Models for Next Generation Wireless Networks: Ad Hoc, Vehicular and Mesh Networks,
First Edition. Paolo Santi.

generating pairwise contact patterns directly. By exploiting the notion of "home location," this contact-based mobility model is shown to be able to faithfully reproduce contact patterns observed in real-world data traces.

Before we present the social-based mobility models, the reader might wonder about the actual need for defining synthetic human mobility models, given the availability of a relatively large collection of wireless network traces in the literature. Although a number of mobility traces are actually available in the literature, and despite the extensive use of these traces in the opportunistic networking literature, synthetic mobility models still provide benefits to the network designer by allowing the variation of many network parameters (number of nodes, density of nodes, average number of contacts, and so on), which is only partially possible with real-world data traces. Thus, a quite common methodology in opportunistic network performance evaluation is to run the protocol under test on a few real-world data traces, as well as on synthetically generated traces typically obtained for larger network sizes than those of the networks used to produce the real-world traces.

20.1 The Weighted Random Waypoint Mobility Model

The weighted random waypoint mobility model was introduced in Hsu et al. (2005) with the aim of more faithfully reproducing human mobility. In particular, the authors focus on pedestrian mobility in a typical university campus.

The ideas underlying the weighted random waypoint mobility model are as follows:

1. *Uneven waypoint popularity*: in a typical university campus, different locations (identified as buildings on a campus) display different degrees of popularity. Hence, waypoint selection should be weighted based on location popularity.
2. *Time-varying popularity*: location popularity changes with time – some locations are more popular in the morning (e.g., classrooms), while others are more popular at lunchtime (e.g., cafeterias), and so on.
3. *Location-dependent pause time distribution*: the pause time distribution at each location is different. For instance, pause time in classrooms tends to be highly concentrated around typical class durations, while pause time in cafeterias is more evenly distributed.

To faithfully calibrate the mobility model for location popularity, pause time distributions, etc., Hsu et al. (2005) conduct a mobility survey targeted at a random set of 268 students on the campus. In the survey, students are requested to supply the following information: current location (building), previous visited location, next location to be visited, and pause time at each of these three locations.

The data collected from the survey is then processed to obtain popularity and pause time distributions as follows. First, the campus buildings are grouped into four categories: classrooms, libraries, cafeterias, and others. Further, a fifth category is introduced to model off-campus mobility. The pause time distributions for each location category are directly obtained from the survey data. The movement area is then divided into different regions as shown in Figure 20.1, which are modeled based on the University of Southern California campus where the survey was conducted. Movement between locations is driven by popularity, as determined by the mobility survey. In particular, a Markov chain (see Figure 20.1) is used to model transition probabilities between the five location categories. It is important to observe that, in accordance with what was observed in the survey data, the transition matrix of the Markov chain is time dependent. In particular, two transition matrices are defined: one for modeling movements in the morning (from 9 AM to 1 PM), and one for modeling movements in the afternoon (from 1 AM to 5 PM).

Summarizing, in the weighted random waypoint mobility model a node is given an initial location based on popularity (computed as the stationary distribution of finding a node at a certain location in the underlying Markov chain); then, the node remains in the selected location for a pause time drawn from the location-dependent pause time distribution; when the pause time elapses, a new location is selected based on the Markov chain transition matrix, and so on.

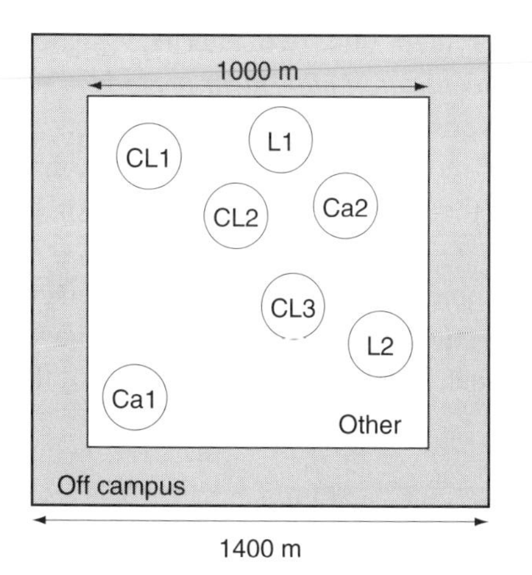

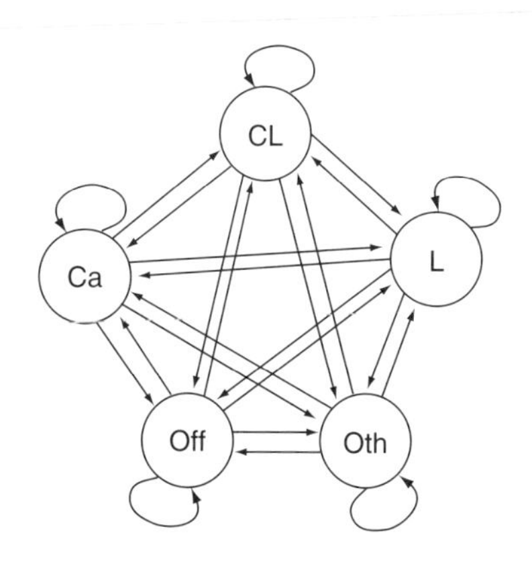

Figure 20.1 The weighted waypoint mobility model. The movement region is divided into five location categories (left): classrooms (CL), cafeterias (Ca), libraries (L), others on campus locations, and off campus. Movement between locations is governed by a five-state Markov chain (right), with transition probabilities derived from survey data.

20.2 The Time-Variant Community Mobility Model

The weighted waypoint mobility model presented in the previous section introduces some realistic aspects into the waypoint mobility model, such as the fact that different waypoints display different degrees of popularity, that waypoint popularity changes with time, and that pause time distribution is waypoint dependent. However, similar to the original RWP model, each mobile node in the weighted waypoint mobility model behaves identically (from a stochastic perspective). This is an unrealistic aspect of the mobility model described in the previous section, since in the real world different individuals display stochastically different mobility patterns.

The time-variant community (TVC) mobility model was introduced in Hsu et al. (2007) to account for heterogeneous individual mobility patterns. The model is designed also to account for other features observed in human mobility patterns, such as skewed location popularity distribution, time-variant mobility patterns, and periodical reappearance of nodes at the same location.

In order to create skewed location visiting preferences, each node in the network is randomly assigned to a subregion of the mobility region R, called its *community* – similar to the notion of "home location" used in other mobility models. A node tends to visit its community relatively more often than regions residing outside the community. The notion of community is also useful to model heterogeneous node mobility patterns: since communities are chosen randomly, and in general are different for different nodes, the resulting mobility patterns clearly have different stochastic properties.

In order to reproduce periodical reappearance of a node at the same location, time in the TVC model is divided into two alternate phases, called *normal movement periods* and *concentration movement periods* – see Figure 20.2. During concentration movement periods, the likelihood of performing movements within a node's community is even higher than in normal movement periods, implying that the node is likely to visit its community periodically.

In TVC, communities are defined as square subregions of the movement region R, with a different side during the normal and concentration movement periods – see Figure 20.3. Notice that, depending on the scenario, the

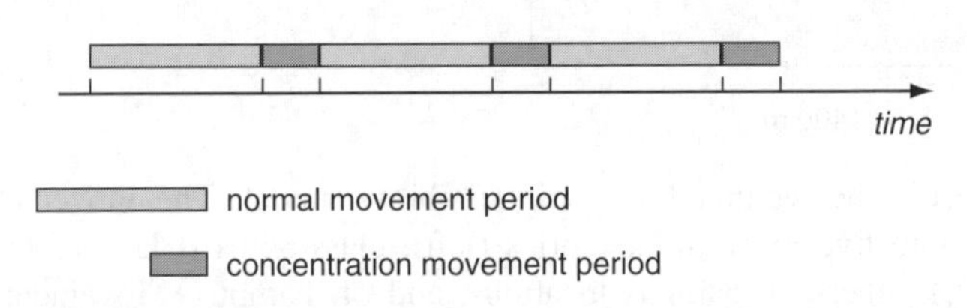

Figure 20.2 The TVC mobility model alternates between normal and concentration movement periods.

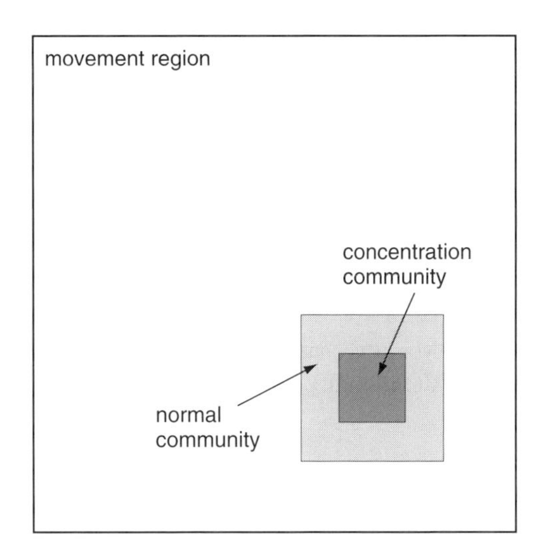

Figure 20.3 Normal and concentration communities in the TVC mobility model.

concentration community can be smaller than (as in Figure 20.3), larger than, or have the same size as the normal community.

During each time period, a node has two different movement modes, called *local epoch* and *roaming epoch*. During a local epoch, node movement is confined within the current community. In a roaming epoch, the node is free to move in the entire movement region R. The rule used to select speed is the same as in the RWP model: before initiating a new trip, a node chooses a speed uniformly at random in an interval $[v_{min}, v_{max}]$, and the speed is kept constant for the entire trip. The length of a trip is drawn from an exponential distribution with average trip length equal to a parameter $\bar{L}$, while trip direction is chosen uniformly at random in the $[0, 2\pi]$ interval (random direction mobility). The toroidal wrap-around rule is used in case a node hits the boundary of the allowed movement region (the community in local epochs, or the entire movement region R in roaming epochs). When the trip ends, the node pauses at the current location for a time drawn uniformly at random in a $[0, T_{max}]$ interval before starting a new trip.

Alternation between local and roaming epochs during a movement period is governed by the two-state Markov chain depicted in Figure 20.4, with probabilities p_l and p_r determining transition into local and roaming epoch state, respectively. Notice that these transition probabilities can be different in the normal and concentration movement periods.

Hsu et al. (2007) derive formulas for approximating two important mobility metrics of the TVC mobility model, namely, the expected hitting time and the expected meeting time – see Chapter 18 for a formal definition of these metrics. The expressions approximating the expected hitting and meeting

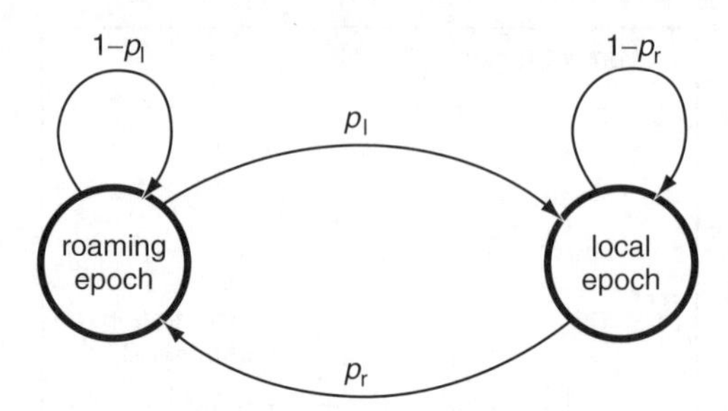

Figure 20.4 Two-state Markov chain governing transitions between local and roaming epochs in the TVC mobility model.

times with TVC mobility are cumbersome, and for this reason we do not report them here. The interested reader is referred to Hsu et al. (2007).

Besides providing explicit formulas for approximating the expected hitting and meeting times, Hsu et al. (2007) also provide a validation of the TVC mobility model based on a comparison to real-world traces, showing that the TVC mobility model achieves the design goals of reproducing skewed location visiting preferences and periodical reappearance of a node at the same location.

20.3 The Community-Based Mobility Model

Musolesi and Mascolo (2007) define a mobility model which explicitly takes into account social relationships between users to define mobility patterns. The model is based on the observation that human mobility is influenced by the social relationships between individuals. Thus, the authors suggest starting the definition of the mobility model from an input social network graph, which describes the social relationships between the individuals forming the network.

The block diagram of the community-based mobility model is shown in Figure 20.5. Mobility modeling is a five-step process:

1. *Social graph definition*: the first step consists of defining a social network graph modeling social relationships between individuals. In principle, any (undirected) social network graph can be used for this purpose. The graph can be obtained from real-world data; alternatively, the authors endow the model with an algorithm for producing synthetic, yet realistic, social network graphs (see further on in this section).
2. *Community detection*: after the social network graph is produced, a community detection algorithm is run to identify (non-overlapping) communities in the graph. In general, any community detection algorithm can be used for this purpose. In the implementation of their model, Musolesi

and Mascolo (2007) use the algorithm proposed in Newman and Girvan (2004), which repeatedly removes edges incident to nodes with maximal betweenness centrality to partition the social network graph into disjoint communities.

3. *"Home location" generator*: the next step consists of assigning a "home location" to each community in the social network graph. To this end, the movement area is divided into a number of square cells (assumed to be much larger than the number of communities in the social network graph), and a community is assigned a square cell chosen uniformly at random as its "home location," with the constraint that no two communities can be assigned to the same cell.
4. *Initial node position generator*: then, initial node positions are generated as follows: a node belonging to community $\mathcal{C}$ is positioned uniformly at random in the square cell corresponding to the "home location" of community $\mathcal{C}$.
5. *Movement generation*: after the four steps above have been performed, actual movement of nodes can start. Node movement obeys a weighted waypoint mobility rule, where the weight of waypoints is determined by the strength of social ties (see below for details).

The social network graph defined in step 1 is assumed to be a weighted, undirected graph $G = (V, E)$, where the weight of edge (u, v) represents the intensity of the (symmetric) social relationship between node u and node v. Before computing the community structure, graph G is turned into

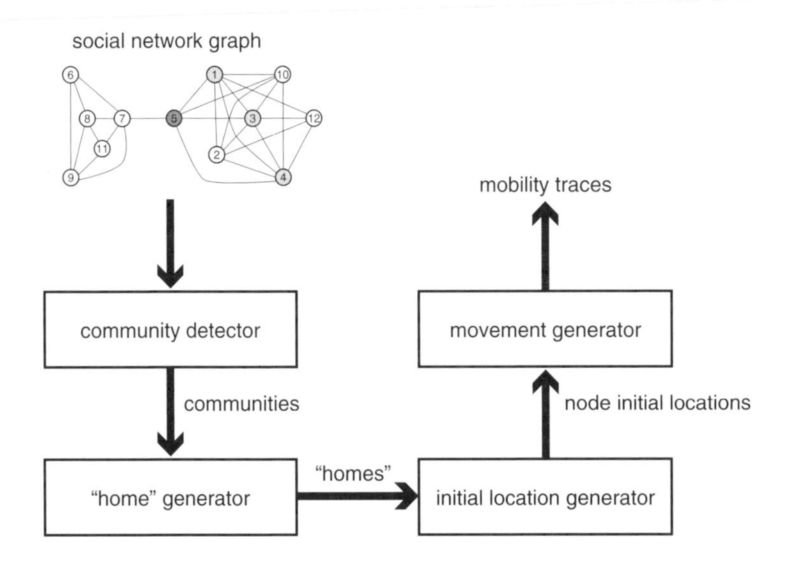

Figure 20.5 Block diagram of the community-based mobility model.

an unweighted graph G_σ by defining a threshold σ: in G_σ, only edges in G with weight $\geq \sigma$ are retained.

As mentioned above, besides allowing use of an arbitrary social network graph G as input, the community-based mobility model provides the user with an algorithm for producing synthetic, yet realistic, social network graphs. The algorithm used to generate synthetic social network graphs is based on the *caveman model* introduced in Watts (1999): initially, K fully connected graphs (corresponding to communities, called *caves* in this model) are added to the social network graph; then, every edge of this initial graph is rewired to point to another cave with a certain, fixed probability p. This simple algorithm has been shown by Watts (1999) to produce realistic social network graphs, characterized by a relatively high average clustering coefficient and a relatively low average path length.

Node movement after initial deployment obeys the following rules. Each node is assigned a *goal*, roughly corresponding to the popular notion of waypoint. More formally, node u is said to be associated with a certain cell $C_{i,j}$ (i and j are the cell indexes) if its current goal is located within $C_{i,j}$. Note that the fact that node u is associated with cell $C_{i,j}$ does not necessarily imply that it is currently located in $C_{i,j}$, but instead means that its current movement destination is within $C_{i,j}$.

Goals are assigned to nodes according to the following rule. Assume node u has reached its previous goal, and needs to select a new goal at time t; let $N_{i,j}(t)$ be the number of nodes associated with cell $C_{i,j}$ at time t. The following metric is used to define the *social attractivity* a certain cell exerts on node u:

$$SA^u_{i,j} = \frac{\sum_{v \in C_{i,j}} w_{uv}}{N_{i,j}(t)},$$

where w_{uv} is the weight of the edge connecting u and v in G – set to 0 if there is no edge between u and v in G.

The movement generator allows for two possible alternative mechanisms to select the next goal:

1. *Deterministic*: the next goal is chosen uniformly at random in the cell with highest social attractivity according to the above-defined metric.
2. *Random*: the next goal is chosen uniformly at random in a cell selected according to a weighted probability distribution, where the probability $Prob(c = C_{i,j})$ of selecting cell $C_{i,j}$ is defined as follows:

$$Prob(c = C_{i,j}) = \frac{SA^u_{i,j} + d}{\sum_{h,k} SA^u_{h,k} + d},$$

where d is a parameter greater than 1 used to ensure that the probability of selecting a goal in a cell is always positive, and to control the amount of "socially oblivious" randomness in the mobility model.

After the next goal is selected, the node starts moving toward the goal along a straight-line trajectory, with a fixed speed chosen uniformly at random in a certain interval – this part of the model is identical to RWP mobility.

An interesting feature of the community-based mobility model is that it allows the social network graph to be periodically changed, so as to induce different mobility patterns at different times of the day – which is a typical feature of human mobility, as discussed in the previous chapter. When the input social network graph changes, so does the community structure, and the mapping of nodes to home locations; hence, node mobility patterns will gradually tend to change to reflect the different structure of the underlying social network graph.

To validate the model, Musolesi and Mascolo (2007) compares sample synthetic traces produced by the community-based model to the Cambridge data trace (Chaintreau et al. 2006). The metrics considered in the comparison are the inter-meeting time and the contact duration. For the sake of comparison, the authors consider also the well-known RWP model. Synthetic and RWP traces are obtained by placing 100 nodes in a square movement region of side 5 km, and choosing movement speeds uniformly at random in the $[1, 6]$ m/s interval. For the sake of "home location" computation in the community-based model, the movement region is divided into 625 squares of side 200 m. Contact patterns between nodes are generated under the assumption that nodes have a transmission range of 250 m.

The cumulative distribution of inter-meeting and contact duration obtained from simulations is shown in Figure 20.6. As can be seen from the figure, by suitably tuning parameter p used in the generation of the social network graph with the caveman model, the community-based mobility model can be made to produce contact patterns quite similar to those encountered in the Cambridge data trace, especially for the inter-meeting time. This is evidently not the case for the RWP mobility model, which produces contact patterns very different from those displayed by the data trace.

20.4 The SWIM Mobility Model

Mei and Stefa (2009) introduced the SWIM (Small World In Motion) mobility model, which is designed to be a simple mobility model able to faithfully reproduce contact patterns between individuals. The intuition behind SWIM is as follows: people tend to visit more often places not too far from their home, and where they can meet lots of other people. In other words, Mei and Stefa (2009) identify two major factors influencing individual human mobility: proximity to home and popularity of the destination. In particular, the authors assume a tradeoff between these two factors when designing SWIM, observing that an individual typically travels a long distance only to reach a very popular location.

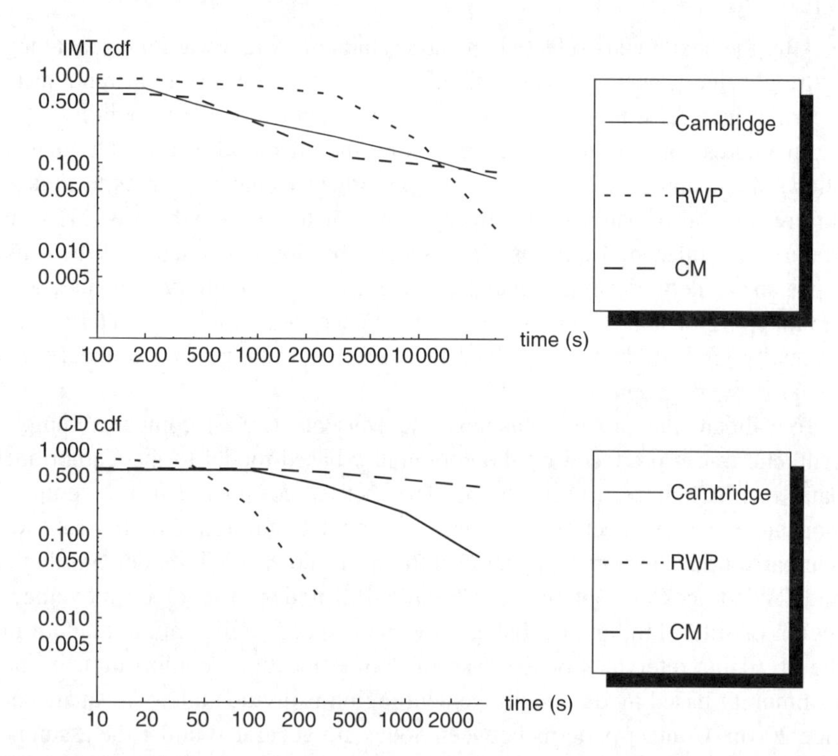

Figure 20.6 Inter-meeting time (top) and contact duration (bottom) distributions of the Cambridge data trace, RWP mobility model, and community-based mobility model. Rewiring probability p in the community-based mobility model is set to 0.2.

Other important observations on which SWIM is based are the following:

1. *Speed selection*: in real life, the speed of a trip tends to be proportional to the distance to be covered. For instance, if the destination is only a few hundreds meters away from an individual's current location, it is very likely that the distance will be covered by walking, that is, at very low speed. On the other hand, if one needs to travel several kilometers, one will likely use a car, that is, traveling speed will be much higher in this case. As we will see, in SWIM this observation is used to considerably simplify the speed selection process.
2. *Skewed waiting time distribution*: in real life, people tend to stay a long time in only a few locations (e.g., home and workplace), while spending considerably less time in other places (bank, shops, etc.). In accordance with this observation, waiting times at destinations in SWIM are chosen according to a skewed probability distribution, namely, a power law (see further on in this section for details).

The SWIM mobility model operates in a number of steps, described below:

1. *Initialization*: before starting the simulation, the movement region is divided into a number of square cells, which are used to assign "homes" and destinations to mobile nodes. In particular, the authors assume the side of a cell to be small enough to enable direct communication between any two nodes located in the same cell. In other words, if the node's transmission range is set to r, the side of square cells is set to $r/\sqrt{2}$. After square cell partitioning, each node u is assigned a "home" cell $C(u)$ uniformly at random, and its initial position is chosen uniformly at random in cell $C(u)$.
2. *Destination selection*: for each node u, the next destination (waypoint) is chosen uniformly at random in the *destination cell*, which is selected uniformly at random according to a weighted probability distribution. More specifically, for each cell $C_{i,j}$ in the movement region, the following metric relative to node u is computed and used to probabilistically select the destination cell:

$$w(C_{i,j}) = \alpha \cdot dist(pos(u), C_{i,j}) + (1 - \alpha) \cdot seen(C_{i,j}),$$

   where $dist(pos(u), C_{i,j})$ is a function that decays as a power law with the distance between the position $pos(u)$ of node u and the center of cell $C_{i,j}$, $seen(C_{i,j})$ is the number of nodes seen at $C_{i,j}$ the last time node u visited $C_{i,j}$ – initialized to 0 for all the $C_{i,j}$ at the beginning of the simulation – and α is a constant in the [0, 1] interval.
3. *Waiting time selection*: the waiting time at the current location is chosen according to a power law with slope 1.45 and upper bound of 4 hours.
4. *Movement phase*: as the waiting time elapses, the node starts moving toward the next destination along a straight-line trajectory, with a speed proportional to the distance to be traveled. More specifically, to simplify the model definition as much as possible the authors assume that if the distance to be covered is d, the node speed is d/k meters per second, implying that each trip in SWIM has the same duration of k seconds.

As can be seen from the above description, SWIM is designed to be very simple, with only a few parameters available to tune its mobility characteristics. In particular, parameter α has a major effect on mobility patterns, allowing the "home proximity" vs. "popularity" tradeoff to be tuned: the larger the value of α, the more the nodes tend to repeatedly visit locations close to home; on the contrary, smaller values of α give rise to mobility patterns with nodes tending to visit quite distant, yet popular, locations.

Note that, unlike the community-based mobility model described in the previous section, social interactions between individuals are *indirectly* accounted for in SWIM: in particular, a node's mobility pattern is influenced by other nodes' mobility patterns through function *seen*(), which records the number of nodes encountered in a cell during the most recent visit.

Mei and Stefa (2009) evaluate SWIM ability to faithfully reproduce human contact patterns both theoretically and through simulation. In particular, the authors formally prove that the *tail* of the inter-meeting time distribution generated by SWIM is exponentially distributed, and use simulations to assess that the head of the distribution is a power law, thus showing that the traces generated by SWIM display the power-law and exponential tail dichotomy of the inter-meeting time distribution observed by Karagiannis et al. (2007).

Mei and Stefa (2009) compare the contact patterns generated by SWIM to those of three data traces, including the Cambridge and Infocom traces given in Table 19.2. In Figure 20.7, we show the complementary cumulative density function of the inter-meeting time and contact duration distributions obtained with SWIM and with the Cambridge and Infocom traces. The SWIM traces have been produced as follows. In order to closely match the conditions under

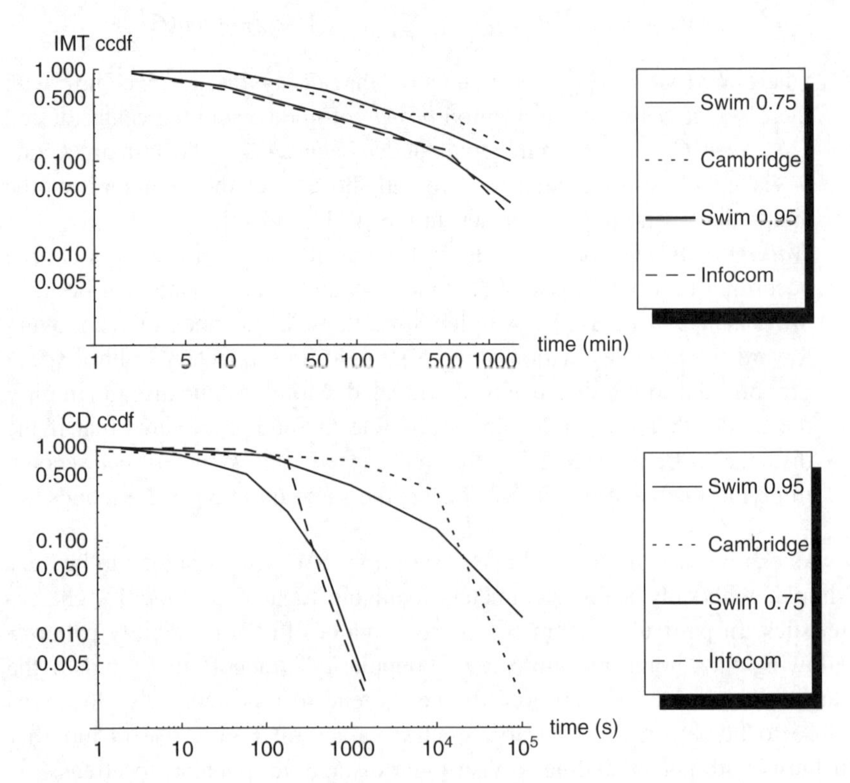

Figure 20.7 Inter-meeting time (top) and contact duration (bottom) distributions of the Cambridge and Infocom data traces, and SWIM mobility model with parameter α set to 0.95 and 0.75, respectively.

which the real-world data traces were produced, a square movement region of side 300 m is considered, and nodes are assumed to have a transmission range of 30 m. Function *dist*() is set as follows:

$$dist(pos(u), C_{i,j}) = \frac{1}{(1 + 0.05 \cdot \|pos(u) - c_{i,j}\|)},$$

where $c_{i,j}$ is the point corresponding to the center of cell $C_{i,j}$. Parameter α is optimized to reproduce contact patterns of real-world traces as faithfully as possible, and it is set to 0.75 in the SWIM trace resembling the Infocom trace and to 0.95 in the other trace. As seen in the plots displayed in Figure 19.7, by properly tuning parameter α the SWIM model is able to reproduce real-world contact patterns quite accurately.

In a recent paper Mei et al. (2011) propose a variation of the SWIM model aimed at explicitly taking into account the similarity of user interests when associating "home" location to nodes. More specifically, each node in the network is associated with an *interest profile*, which is an m-dimensional vector in the $[0, 1]^m$ *interest space* modeling a node's interest along m possible dimensions. An interest dimension in Mei et al. (2011) is intended as modeling interest in a certain topic (e.g., music, books, cinema, etc.), as well as individual habits (e.g., living in a certain neighborhood, working in a certain place, etc.). A similarity metric is defined on the interest space to express similarity between individual interests, thus expressing "homophily" between individuals. In social science, the notion of "homophily" expresses the fact that two individuals share similar interests.

The goal of the modified SWIM model defined in Mei et al. (2011) is to produce contact traces where nodes with relatively more similar interests are relatively more likely to meet than nodes with diverse interests. This is motivated by sociological studies showing that there exists a positive correlation between "homophily" between individuals and the frequency/duration of their contacts (McPherson 2001). Positive correlation between similarity of individual interests and contact frequency/duration has been observed also by Mei et al. (2011) and Noulas et al. (2009) based on the Infocom data trace, for which not only logs of pairwise contacts are available, but also anonymized profiles of the individuals who participated in the experiment.

In Mei et al. (2011), it is proposed to change the "home" location assignment in SWIM as follows. Instead of assigning the "home" location to nodes uniformly at random in the movement region, allocation is performed based on the similarity of user interests, with the goal of assigning nodes with relatively similar interests to relatively close "home" locations. This way, given the SWIM mobility pattern according to which nodes return to their home quite frequently, nodes with relatively similar interests tend to meet relatively more often than nodes with diverse interests. These authors propose to allocate nodes to "home" locations as follows. First, two out of the

m possible interest dimensions are chosen. The specific values $i_u, i'_u \in [0, 1]$ of node u along these directions give the location of node u's home, after suitable normalization of the movement region.

20.5 The Self-Similar Least Action Walk Model

The mobility models described so far have the common property of selecting a new destination at a time, subject to some random rule. In real life, however, human movements are often *planned*: given a set of locations to visit in a forthcoming period of time (e.g., workplace, the shopping mall, a fitness center, a school, etc.), an individual plans an itinerary to visit all locations, with the goal, typically, of reducing distance traveled or trip time. The ability of an individual to plan an itinerary across a set of locations to visit is not captured by models where trips are selected one after the other, as in the models described so far in this chapter.

The self-similar least action walk (SLAW) model was introduced in Lee et al. (2009) with the aim of modeling the above-described trip planning ability of humans. The model is inspired by the least action principle, according to which all objects (humans in this context) move in the direction of minimizing their discomfort (traveled distance in this context). The intuition is that, when planning an itinerary, people tend to reduce the distance traveled by visiting all nearby locations before visiting farther destinations, unless some high-priority events such as appointments induce them to overrule the above strategy.

The SLAW model which we describe in this section is shown by Lee et al. (2009) to fulfill several properties typical of human mobility patterns, as derived from analysis of real-world mobility traces (recall Chapter 19):

1. *Truncated power law of trip length and pause time*: by defining a trip as a continuous trajectory without directional changes or pauses, the trip length distribution has been observed to obey a power law with exponential cutoff; similar behavior has been observed for the pause time distribution.
2. *Heterogeneous mobility areas*: people have been observed to move in confined mobility areas, which are in general different for different individuals.
3. *Truncated power law of inter-contact times*: this is the power-law and exponential time dichotomy of pairwise inter-contact time distribution thoroughly discussed in Chapter 19.
4. *Fractal destinations*: based on GPS traces of human walks reported in Lee et al. (2008), it has been observed that the destinations of an individual's movement can be modeled as fractal points. Fractal points are characterized by the *self-similarity* property, which, intuitively speaking, means that the features of the set of points holds independently of the

spatial scale of observation. In particular, Lee et al. (2008) observe that human destinations tend to form hotspots in the movement region, and that hotspots can clearly be observed independently of the spatial granularity of the observation.

It is interesting to observe that the above properties are closely interrelated: Lee et al. (2009) show that fractal destinations (4) induce truncated power-law trip length (1). Hong et al. (2008) show that mobility patterns confined in a restricted area (2) give rise to the power-law and exponential time dichotomy of inter-contact times (3).

The block diagram of the SLAW mobility model is shown in Figure 20.8. Initially, a large population of possible destinations (waypoints) is generated. Waypoints are generated so as to satisfy the self-similarity principle (see Lee et al. (2008) for details). Alternatively, waypoints can be extracted from an existing GPS trace. These waypoints constitute the universe W of all possible

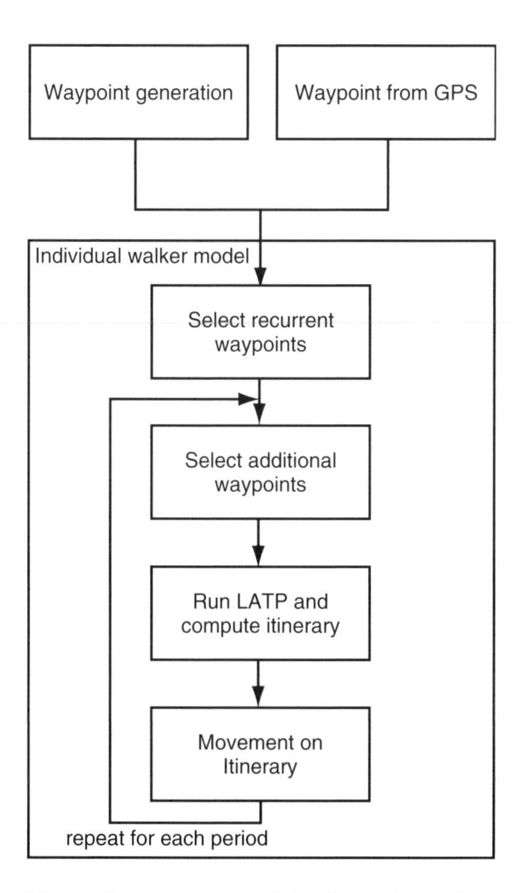

Figure 20.8 Block diagram of the SLAW mobility model.

destinations for *all* nodes in the network. In order to satisfy property 2 of heterogeneous node mobility patterns, an *individual walker model* is then implemented. In order to diversify individual mobility patterns, a subset W' of waypoints is randomly extracted from W, corresponding to the recurrent destinations of a specific node. Note that different nodes in general have different recurrent destinations, although overlapping between sets W' is possible. After selecting set W', a set of waypoints is chosen to form an itinerary for the current *period*. The period is defined in SLAW to be approximately 12 hours. The destination set for the period is composed of the set W' or recurrent waypoints, augmented with a set W'' of additional waypoints chosen from $W - W'$ in order to increase randomness in movement patterns and mixing of individual trajectories. The set $W' \cup W''$ of waypoints for the current period is then given as input to the least action trip planning (LATP) module, which randomly computes the itinerary for the current period satisfying the least action principle (on average). Once the itinerary is computed, the set of movements for the current period is performed; each time a node reaches a destination, it pauses for a random time drawn from a truncated power law.

It is important to observe that particular care should be taken in selecting the individual set of recurrent waypoints in step 1 of the individual walker model. In fact, recurrent waypoints should be selected so as to satisfy the self-similarity principle, which is not necessarily guaranteed by the fact that the set W of all possible destinations from which set W' is chosen does satisfy self-similarity. To preserve self-similarity also on subset W', the following random selection procedure is implemented. First, waypoints in W are clustered by connecting together waypoints that are less than 100 m apart. Denoting by $\mathcal{C} = \{C_1, \ldots, C_h\}$ the set of clusters and by $|X|$ the cardinality of set X, we have that each cluster C_i in $\mathcal{C}$ is assigned weight $w_i = |C_i|/|W|$. Each individual walker then selects k clusters from $\mathcal{C}$ with probability proportional to the weights w_i, where k =3–5 is a parameter of the model. Then, for each selected cluster C_i, 5 to 10% (the exact percentage of selected waypoints is another parameter of the model) of the waypoints in C_i are chosen uniformly at random and added to the set W' of recurrent destinations.

The additional waypoints in step 2 of the individual walker model are selected as follows. A cluster C_j is selected uniformly at random from the clusters in $\mathcal{C}$ not used to extract waypoints in W'. Then, 5 to 10% of the waypoints in C_j are selected uniformly at random and constitute set W'' for the current period.

The LATP module of SLAW is shown in Figure 20.9. Given an unordered set WW of waypoints to visit and an initial waypoint w, the algorithm returns an ordered set WW_o of all waypoints in WW, starting from w. The ordered set is formed in an incremental way: given the current waypoint, the next point is randomly selected according to a distance-weighted probability distribution. Parameter α is used to tune obedience to the least action principle: for large

LATP Algorithm

Input: Set WW of waypoints in the itinerary
An initial waypoint $w \in WW$
Distance weight factor $\alpha \geq 0$
cw denotes the current waypoint

Output: An ordered set WW_o of waypoints

1. $WW_o = \{w\}$; $cw = w$
2. **while** $WW_o \neq WW$ **do**
3. calculate distance d_i for all $w_i \in WW - WW_o$
4. calculate probability p_i to move to waypoint w_i as
$$p_i = \frac{(1/d_i)^\alpha}{\sum_j (1/d_j)^\alpha}$$
5. choose a node w' in $WW - WW_o$ according to probabilities p_i
6. $cw = w'$; $WW_o = WW_0 \cup \{w'\}$
7. **return** WW_o;

Figure 20.9 The LATP algorithm in SLAW.

values of α, selection of the next waypoint becomes a quasi-deterministic process based on distance to the current waypoint; on the other hand, if $\alpha = 0$ the next waypoint is chosen uniformly at random.

20.6 The Home-MEG Model

In the previous sections, we introduced representative mobility models taking into account the social dimension of human mobility. A common characteristic of these models is that they model human mobility in a metric space, and then derive pairwise contact patterns – which are used to validate the accuracy of the mobility model through comparison to real-world data traces – based on node trajectories combined with a notion of transmission range. Since the main goal of a mobility model for opportunistic networks is to generate realistic pairwise contact patterns between network nodes, some authors have proposed models aimed at directly modeling the occurrence/disappearance of pairwise wireless links between individuals, without modeling physical individual mobility in an underlying metric space. This approach has the potential advantage of producing simpler models of contact patterns amenable for use in protocol performance analysis, while not sacrificing accuracy in the generated contact patterns.

Becchetti et al. (2011) introduce the Home-MEG (Markovian Evolving Graph) model to model the occurrence/disappearance of wireless links between pairs of individuals. The Home-MEG model is a discrete-time model, according to which existence of the link between u and v can change state at time $t, t+1, \ldots$ and so on. The model is based on the well-known

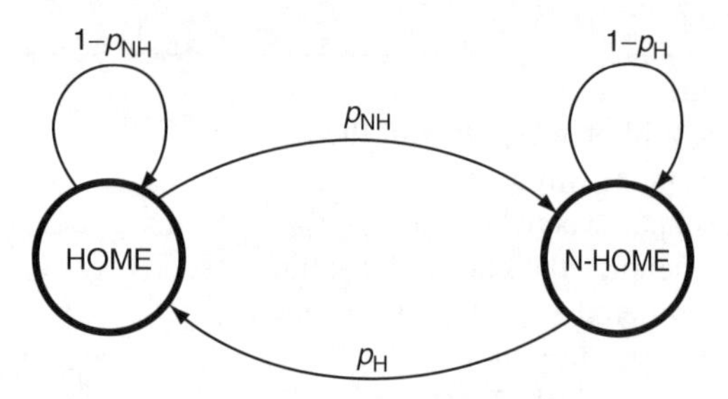

Figure 20.10 Pairwise link state transition diagram in the Home-MEG model.

observation made by Karagiannis et al. (2007) that pairs of individuals tend to repeatedly meet in a few locations, known as "home" (or "meeting") location. Thus, taking an arbitrary pair u, v of individuals in the network, the wireless link between these nodes can be modeled according to the two-states transition diagram shown in Figure 20.10. The wireless link between u and v can be in one of two possible states: the "Home" state corresponds to the situation in which both nodes are in the "home" location, while the "Not-Home" state corresponds to the situation in which one of the two nodes (or both) is not in the "home" location. In accordance with Karagiannis et al. (2007), the instantaneous probability for link (u, v) to exist at any time t is p_h, with $0 < p_h < 1$, when the link is in state "Home," while it is p_l, with $0 \leq p_l < p_h$, when the link is in state "Not-Home." Besides p_h and p_l, the Home-MEG model has two more parameters, p_H and p_{NH}, determining the transition probabilities between the two possible link states at each time instant.

It is easy to see that the existence/non-existence of any pairwise link in the Home-MEG model can be modeled with a four-states Markov chain, where the states are HC and NC modeling the existence of the link (u, v) when the link is in state "Home" and "Not-Home," respectively, and HD and ND modeling the non-existence of the link when in state "Home" and "Not-Home," respectively. The corresponding transition matrix is given in Table 20.1.

Table 20.1 Transition matrix of the Markov chain used to model existence/non-existence of a pairwise wireless link in the Home-MEG model

	HC	HD	NC	ND
HC	$(1 - p_{NH})p_h$	$(1 - p_{NH})(1 - p_h)$	$p_{NH}p_l$	$p_{NH}(1 - p_l)$
HD	$(1 - p_{NH})p_h$	$(1 - p_{NH})(1 - p_h)$	$p_{NH}p_l$	$p_{NH}(1 - p_l)$
NC	$p_H p_h$	$p_H(1 - p_h)$	$(1 - p_H)p_l$	$(1 - p_H)(1 - p_l)$
ND	$p_H p_h$	$p_H(1 - p_h)$	$(1 - p_H)p_l$	$(1 - p_H)(1 - p_l)$

In order to reproduce pairwise contact patterns in a network with n nodes, it is sufficient to generate $n(n-1)/2$ identical and independent copies of the above Markov chain (one for each possible pairwise link) in the network.

It is important to observe that the notion of "Home" state in the Home-MEG model is referred to a *specific* node pair, and, in particular, the transitive property does not hold. In other words, the fact that the pairs of nodes u, v and u, w are in the "Home" state does not imply that pair v, w are also in the "Home" state. This is a consequence of the fact that the $n(n-1)/2$ copies of the Markov chain modeling a single link in the network are *independent*. Thus, the "Home" state in the Home-MEG model must not be intended as a single popular location within the network where *all* the nodes regularly meet, but rather as a location where a specific pair of nodes regularly meet, where meeting locations are assumed to be different for each pair of nodes. Admittedly, this is an approximation of reality, where meeting locations are likely to be shared by several pairs of nodes. However, in the following we show that, despite this approximation, the Home-MEG model is able to accurately reproduce the *aggregate* inter-contact time distribution observed in several real-world traces.

Summarizing, the Home-MEG model has four parameters: parameters p_H and p_{NH} used to tune the transition between the "Home" and "Not-Home" states, and parameters p_h and p_l used to set the probability of link existence in the two states.

Becchetti et al. (2011) verify the accuracy of the contact traces generated by the Home-MEG model by comparing the synthetic traces produced by the Home-MEG model to the six data traces used in Karagiannis et al. (2007). In particular, Becchetti et al. (2011) consider the inter-meeting time distribution to assess the accuracy of the Home-MEG traces. The four parameters of the Home-MEG model are optimized to match the real-world data traces through an iterative search process over the four-dimensional parameter space. The inter-meeting time complementary cumulative density function of the Home-MEG traces vs. the three traces given in Table 19.2 is shown in Figure 20.11, and the corresponding optimal parameters for the Home-MEG model are given in Table 20.2. As can be seen from the plots, the Home-MEG model, if suitably tuned, is able to accurately reproduce pairwise contact patterns

Table 20.2 Optimal parameters of the Home-MEG model for the three data traces shown in Figure 20.11

Trace	p_H	p_{NH}	p_h	p_l
Mit BT	$4.5 \cdot 10^{-5}$	$1.5 \cdot 10^{-4}$	$1.2 \cdot 10^{-3}$	$8.6 \cdot 10^{-7}$
Cambridge	$2.5 \cdot 10^{-4}$	$8.3 \cdot 10^{-3}$	$4.7 \cdot 10^{-2}$	$4.6 \cdot 10^{-4}$
Infocom	$3 \cdot 10^{-3}$	$2.5 \cdot 10^{-2}$	$7 \cdot 10^{-2}$	$3 \cdot 10^{-4}$

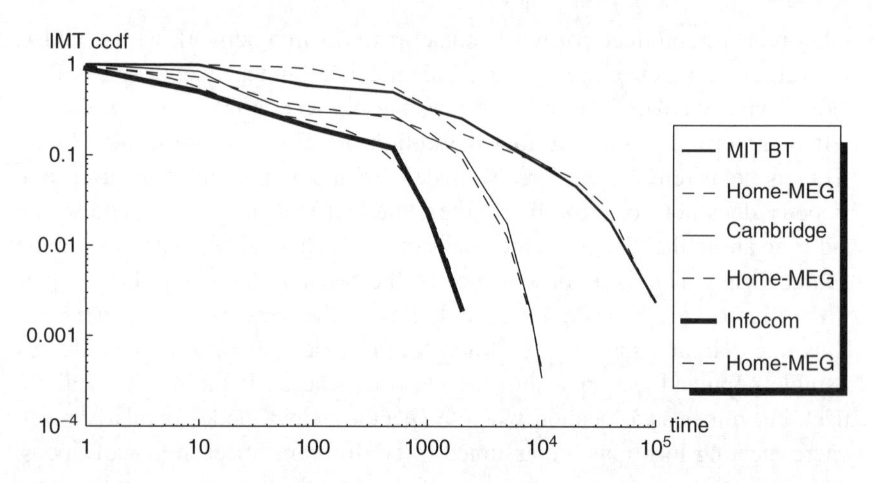

Figure 20.11 Inter-meeting time complementary cumulative density function of the MIT BT, Cambridge, and Infocom data traces, and of the synthetic traces produced with the Home-MEG model. The time step in the Home-MEG model is set to 82 s.

observed in real-world traces. It is also interesting to observe that, while optimal Home-MEG parameters for the three real-world traces are quite different in absolute terms, their relative values are quite similar, hinting at the following guidelines for Home-MEG model parameter tuning:

1. p_H is much smaller than p_{NH}, indicating that pairs of individuals tend to spend most of their time in "Not-Home" state;
2. p_h is much larger than p_l, in accordance with the intuition that motivated the Home-MEG model;
3. $p_H + p_{NH}$ is much smaller than 1, indicating that link states are persistent and transition events are rare.

In Chapter 22, we will see how to exploit these guidelines in the analysis of asymptotic broadcast time in opportunistic networks through the Home-MEG model.

20.7 Further Reading

This chapter presented a non-exhaustive survey of mobility models aimed at modeling individual human mobility, including human "social" behavior. Other models have been proposed in the literature, such as the Truncated Levy Walk model (Rhee et al. 2008), the Working Day Movement model (Ekman et al. 2008), and the Home-cell Community-based Mobility Model (Boldrini and Passarella 2010). We refer the reader to the respective papers for details.

References

Becchetti L, Clementi A, Pasquale F, Resta G, Santi P and Silvestri R 2011 Information spreading in opportunistic networks is fast. *Internet draft: available at http://arxiv.org/abs/1107.5241*.

Boldrini C and Passarella A 2010 HCMM: Modeling spatial and temporal properties of human mobility driven by users' social relationships. *Computer Communications* **33**, 1056–1074.

Chaintreau A, Hui P, Crowcroft J, Diot C, Gass R and Scott J 2006 Impact of human mobility on the design of opportunistic forwarding algorithms. *Proceedings of IEEE Infocom*, pp. 1–13.

Ekman F, Keranen A, Karvo J and Ott J 2008 Working day movement model. *Proceedings of the ACM Workshop on Mobility Models*, pp. 33–40.

Hong S, Rhee I, Kim S, Lee K and Chong S 2008 Routing performance analysis of human-driven delay tolerant networks using the truncated Levy walk model. *Proceedings of the ACM Workshop on Mobility Models*, pp. 25–32.

Hsu WJ, Merchant K, Shu HW, Hsu CH and Helmy A 2005 Weighted waypoint mobility model and its impact on ad hoc networks. *Mobile Computing and Communications Review* **9**, 59–63.

Hsu WJ, Spyropoulos T, Psounis K and Helmy A 2007 Modeling time-variant user mobility in wireless mobile networks. *Proceedings of IEEE Infocom*, pp. 758–766.

Karagiannis T, Le Boudec JY and Vojnovic M 2007 Power law and exponential decay of inter contact times between mobile devices. *Proceedings of ACM Mobicom*, pp. 183–194.

Lee K, Hong S, Kim S, Rhee I and Chong S 2008 Demystifying the Levy walk patterns in human walks. *Technical Report, North Carolina State University*, pp. 1–14.

Lee K, Hong S, Kim S, Rhee I and Chong S 2009 SLAW: A new mobility model for human walks. *Proceedings of IEEE Infocom*, pp. 855–863.

McPherson M 2001 Birds of a feather: Homophily in social networks. *Annual Review of Sociology* **27**(1), 415–444.

Mei A and Stefa J 2009 SWIM: A simple model to generate small mobile worlds. *Proceedings of IEEE Infocom*, pp. 2106–2113.

Mei A, Morabito G, Santi P and Stefa J 2011 Social-aware stateless forwarding in pocket switched networks. *Proceedings of IEEE Infocom*, pp. 251–255.

Musolesi M and Mascolo C 2007 Designing mobility models based on social network theory. *Mobile Computing and Communications Review* **11**, 1–11.

Newman M and Girvan M 2004 Finding and evaluating community structure in networks. *Physical Review E*.

Noulas A, Musolesi M, Pontil M and Mascolo C 2009 Inferring interests from mobility and social interactions. *Proceedings of the Workshop on Analyzing Networks and Learning with Graphs*.

Rhee I, Shin M, Hong S, Lee K and Chong S 2008 On the Levy walk nature of human mobility. *Proceedings of IEEE Infocom*, pp. 924–932.

Watts D 1999 *Small Worlds:The Dynamics of Networks between Order and Randomness*. Princeton University Press, Princeton, NJ.

Part Seven

Case Studies

In the last part of this book, we will present two case studies aimed at illustrating the importance of mobility modeling in two fundamental methodologies for wireless network performance evaluation: *simulation* and *theoretical analysis*. In the first case study, we will discuss how the standard implementation of the random waypoint model in wireless network simulation can lead to inaccurate results due to the border effect and speed decay phenomena. Then, we will present methodologies recently developed to remove the border effect and speed decay and, hence, the resulting simulation inaccuracy. In the second case study, we will consider the problem of deriving asymptotic performance bounds for opportunistic networks, and, through discussion of recently derived theoretical results, illustrate the profound impact of the mobility model used to perform the analysis on the derived performance bounds.

21

Random Waypoint Model and Wireless Network Simulation

In Chapter 5, we introduced the most popular mobility model for ad hoc networks, the random waypoint (RWP) model, and studied its properties. In particular, two phenomena generated by RWP mobility have been illustrated: the *border effect* and *speed decay*. The border effect is present whenever RWP mobile nodes move in a bounded region: due to the presence of the border, and since waypoints are uniformly selected in the movement region, we know that the node spatial density generated by RWP mobility is not uniform, but it is relatively higher in the center than at the border of the movement region. The speed decay phenomenon is due to the fact that, when RWP mobile nodes choose trip velocity in a bounded interval $[v_{min}, v_{max}]$ with $v_{min} < v_{max}$, the average nodal speed observed at the beginning of the simulation is different (higher) from that observed when stationary conditions are reached.

As we will see, the border effect and speed decay, if not adequately accounted for when performing simulations, can cause significant inaccuracies in the simulation results obtained when estimating the performance of, say, a routing protocol for ad hoc networks.

In this case study, after discussing why the border effect and speed decay can impair the accuracy of simulation results, we will present recently developed methodologies aimed at completely removing the border effect and speed decay in RWP mobile networks. At the end of the chapter, we will describe how these methodologies can be combined to build a "perfect" simulation framework.

Mobility Models for Next Generation Wireless Networks: Ad Hoc, Vehicular and Mesh Networks, First Edition. Paolo Santi.

21.1 RWP Model and Simulation Accuracy

The RWP model is the most popular mobility model for ad hoc networks, and it is included in the vast majority of wireless network simulators. However, it is important to understand that, if not adequately used, the RWP mobility model can lead to inaccurate simulation results due to the border effect and speed decay phenomena.

The border effect, we recall, is due to the fact that waypoints are selected uniformly at random in the movement region R implying that, if R is a bounded region, the likelihood of a trajectory crossing an infinitesimal area located near the center of R is higher than that of crossing an infinitesimal area located near the border of R. This implies that the node spatial distribution generated by the RWP mobility model in a bounded region is bell-shaped – see Chapter 5.

What are the implications of the border effect on wireless network simulation accuracy? To answer this question, consider a quite common situation in which the network designer is interested in evaluating the performance of a certain networking protocol, say, a routing protocol for ad hoc networks. One of the first steps in setting up simulation experiments is network dimensioning, in which parameters such as the movement area, number of nodes, transmission range, etc., have to be set. Suppose the network designer is interested in evaluating the performance of the routing protocol in a mostly connected network, that is, a mobile network in which the communication graph is connected most of the time; in order to generate a mostly connected network, the node transmission range should be carefully set. A typical approach with respect to this is to choose a value of the transmission range close to the smallest possible value giving rise to a connected network – known as the *critical transmission range* (CTR) in the literature. This is motivated by well-known foundational results showing that the minimal transmission range ensuring network-wide connectivity maximizes network transport capacity (Gupta and Kumar 2000).

The value of the CTR for connectivity has been characterized in a number of papers (Gupta and Kumar 1998; Santi and Blough 2003), most of which are based on the assumption that the node spatial distribution is uniform. Unfortunately, due to the border effect, the node spatial distribution of a RWP mobile network is *not* uniform, hence the value of the CTR for connectivity in a RWP mobile network is different from that predicted by theoretical results based on the uniformity assumption. As shown in Santi (2005), the CTR with RWP mobility is *larger* than in the case of uniformly distributed nodes, and the difference between the two values tends to increase with the number n of nodes in the network. This can be seen from Table 21.1, listing the CTR with uniformly distributed nodes and with RWP mobile nodes for increasing values of n (data from Mitsche et al. (2011)). In the table, the

Table 21.1 The critical transmission range for connectivity with uniform node spatial distribution and distribution generated by RWP mobility for networks of increasing number n of nodes

n	CTR (uniform)	CTR (RWP)
25	0.7397	0.7378
100	0.3857	0.4152
500	0.1789	0.2466

critical transmission range is estimated as the 99% quantile of the empirical distribution of minimum transmission range for connectivity obtained by performing 10^5 experiments for each value of n. The movement region is assumed to be a square of side 1, the CTR values are normalized with respect to side length, and the pause time in the RWP model is set to 0.

As can be seen from the table, when n is very small, the CTR for connectivity with RWP mobility is a little smaller than that with uniform node distribution, which is due to the fact that, with few nodes in the network, the probability of finding at least one node close to the border (which is the cause of the higher CTR value with RWP mobility) is very small, and node concentration in the center of R tends to favor a relatively small CTR value. However, as n increases, the probability of finding at least one node close to the border with RWP mobility increases as well, leading to relatively higher CTR values. When $n = 500$, the difference between the CTR with uniform node distribution and with RWP mobility is substantial.

Suppose the network designer is interested in evaluating the performance of a large network composed of $n = 500$ nodes. According to theoretical studies on the CTR for connectivity with uniformly distributed nodes, a value of the transmission range in the order of 0.18 would be a reasonable choice. However, if nodes move according to the RWP mobility model, this value of the CTR is likely to generate networks that are *disconnected* most of the time. Hence, the result of the simulation experiments is very likely to be an unsatisfactory routing performance, which, however, could be due to the topological properties of the mobile network (lack of connectivity most of the time), instead of being due to shortcomings in the routing protocol. In other words, the network designer, if not aware of the border effect and of its consequences on network dimensioning, could draw the wrong conclusions about a networking protocol performance in a mobile environment.

Let us now consider the speed decay phenomenon, which, we recall, is due to the fact that in the RWP model the distance and the velocity of a trip are chosen independently. Hence, when computing the average nodal speed, nodes with relatively low speed are more likely to occur than nodes with relatively high speed, due to the fact that a trip performed at low speed lasts longer than a trip performed at high speed. Summarizing, and denoting by v_{RWP} the average nodal speed as time $t \to \infty$, we have that $v_{RWP} < v_0$ whenever the speed of a trip is chosen uniformly at random in an interval $[v_{min}, v_{max}]$ with $v_{min} < v_{max}$. In other words, the initial average nodal speed $v_0 = (v_{max} + v_{min})/2$ is *larger* than the stationary average nodal speed v_{RWP}.

The main consequence of the speed decay phenomenon on simulation accuracy is that the "amount of mobility" in the network – summarized by the average nodal speed – changes as the simulation proceeds, and requires some time to stabilize. Thus, if simulation results are collected before stationary conditions are reached, they can turn out to be inaccurate since they refer to a mobile system where the "amount of mobility" varies with time.

A possible way of improving simulation accuracy is to "warm up" the simulation, and start collecting simulation results only after the "warm-up" period is over. This, coupled with a careful tuning of the CTR for connectivity, would solve both the issues with simulation accuracy mentioned above. However, this methodology is difficult to apply in practice for the following reasons. First of all, "warming up" a simulation wastes considerable computational resources, especially considering that wireless network simulators are computationally intensive programs. Second, even if wastage of computational resources were not an issue, estimating the duration of the "warm-up" period is not easy, since it depends on mobility parameters, and different periods to allow settling of the average nodal speed and spatial distribution could be needed. Finally, using a CTR for connectivity with RWP mobile networks that is different from the one used with uniformly distributed nodes should be avoided, since important topological parameters (e.g., average number of nodes within transmission range) change with respect to the case of uniformly distributed nodes if a different CTR value is chosen. Thus, if the goal of a network designer is to compare, say, a routing protocol performance in stationary, uniformly distributed networks to that of the same protocol executed in a mobile network, it is important that the transmission range is not changed in the two considered scenarios.

For all the reasons discussed above, it would be desirable to redesign the RWP mobility model so as to completely *remove* the border effect and speed decay, thus solving in a satisfactory way the issues with simulation inaccuracy mentioned in this section. Methodologies aimed at this will be presented in the remainder of this chapter.

21.2 Removing the Border Effect

Mitsche et al. (2011) present two variants of the RWP model in which the border effect is almost completely removed. Otherwise stated, the node spatial distribution generated by these versions of RWP is near uniform, despite the fact that nodes move in a bounded region. The RWP variants introduced in Mitsche et al. (2011) exploit either the *temporal* or the *spatial* dimension to remove the border effect. In the first variant, called *temporal-RWP*, the speed of nodes along a trajectory is continuously changed in order to reduce the *time* that nodes spend in the center of the movement region; in the second variant, called *spatial-RWP*, the spatial distribution of waypoints is no longer uniform, but is suitably shaped with the aim of removing the border effect. Since the border effect is maximum when the pause time is set to 0, the authors consider this case in their analysis. Furthermore, in what follows the movement region is assumed to be the unit disk, denoted R.

21.2.1 The Temporal-RWP Model

The temporal-RWP model builds upon the observation that what determines the node spatial distribution of a mobility model is the amount of *time* a node spends in an infinitesimal area $\partial I(x, y)$ of the movement region centered at $(x, y) \in R$. Since in the standard RWP model the speed of a node can be different across trips, but remains fixed in a single trip, the amount of time a node spends in $\partial I(x, y)$ is proportional to the length of a trajectory's segment crossing $\partial I(x, y)$ – recall Figure 5.2. In other words, under the assumption of fixed speed during a trip, the temporal and spatial dimensions of mobility are equivalent, and the node spatial distribution of the RWP – which is determined by the amount of time a node spends in $\partial I(x, y)$ – can be equivalently estimated as the density of *trajectories* crossing $\partial I(x, y)$, as has been done, for example, in Bettstetter et al. (2003) and Hyytia et al. (2006).

Based on the observation above, results such as those presented in Bettstetter et al. (2003) and Hyytia et al. (2006) have shown that, when waypoints are uniformly chosen in R, the density of trajectories in R is bell-shaped. Let us denote by f_t this bell-shaped density function as characterized, for example, by these authors. In order to obtain a uniform node spatial distribution of the RWP mobility model, it is sufficient to ensure that the amount of time a node spends in $\partial I(x, y)$ is independent of its position within R. Since the density f_t of trajectories crossing $\partial I(x, y)$ is location-dependent, in order to remove the dependence on location of the crossing time it is necessary that the speed of a node when crossing $\partial I(x, y)$ is location-dependent. More specifically, the goal is to find a value $v(x, y)$ of the speed such that the (expected) time $t(x, y)$ that a node spends when crossing $\partial I(x, y)$ is constant and does not depend on (x, y). More formally,

$$t(x, y) = \frac{E[L_{x,y}]}{v(x, y)} = k,$$

for some constant $k > 0$, where $E[L_{x,y}]$ represents the expected length of the intersection between a random trajectory and $\partial I(x, y)$, and $v(x, y)$ is the location-dependent velocity of a node when located in $\partial I(x, y)$.

In the above formula, $E[L_{x,y}]$ corresponds, up to constant factors, to the density of trajectories in (x, y), that is, $E[L_{x,y}] = c \cdot f_t(x, y)$ for some $c > 0$. Thus, in order to obtain a location-independent value of $t(x, y)$, it is sufficient to set $v(x, y) = c' \cdot f_t(x, y)$, for some $c' > 0$.

Summarizing, in the temporal-RWP model the speed of a node is continuously changed along its trip to a destination, in a way which is proportional to the density of trajectories in R, that is, $v(x, y) = c' \cdot f_t(x, y)$, for some $c' > 0$. This way, regions of R where the trajectory density is higher are crossed at a higher speed, while regions with lower trajectory density are crossed at lower speed, resulting in a uniform node spatial distribution.

It is important to observe that, while the spatial-RWP model presented in the next section yields a uniform node spatial distribution only when the pause time in the RWP model is set to 0, the temporal-RWP model is guaranteed to generate a uniform node spatial distribution with any setting of the pause time. This is because, in the temporal-RWP model, both the mobility and pause component of the node spatial distribution (recall Chapter 5) are uniform, hence their weighted sum remains uniform independently of the value of the weights – which is determined by the pause time distribution.

Mitsche et al. (2011) suggest a methodology to implement the temporal-RWP model in simulations. Note that, in the temporal-RWP model, the minimal node speed is always 0 at the border – in fact, function f_t is 0 on the border of R, meaning that in the (extremely unlikely) case where a waypoint lying exactly on the border of R is selected, the node will stop and turn into a stationary node when reaching the destination. While the minimum node velocity is fixed, the maximum node velocity (achieved when a node crosses the center of R) can be arbitrarily set by suitably setting constant c'. In particular, we observe that (see Hyytia et al. (2006)) the maximum of f_t is 0.7031. Thus, if the designer wishes to set a maximum node speed of, say, 2 m/s (pedestrian speed), it is sufficient to set $c' = 2/0.7031 = 2.846$.

In order to implement the temporal-RWP model in simulations, it is necessary to discretize time and use a sufficiently small time step to set node velocities. With a fixed value Δt for the time step, node speed is updated at (simulated) time $t, t + \Delta t, t + 2\Delta t, \ldots$ as follows. For each node u in the network, it is first checked whether the node is currently in the *move* or *pause* state. If the node is pausing, no action is taken. Otherwise, the speed for the next time interval is set to $f_t(x_u, y_u)$, where (x_u, y_u) is the current position of node u in R.

The next question to address is how small the time step Δt should be in order to have a time-discrete version of the temporal-RWP model that closely resembles the time-continuous version of the model – which is guaranteed

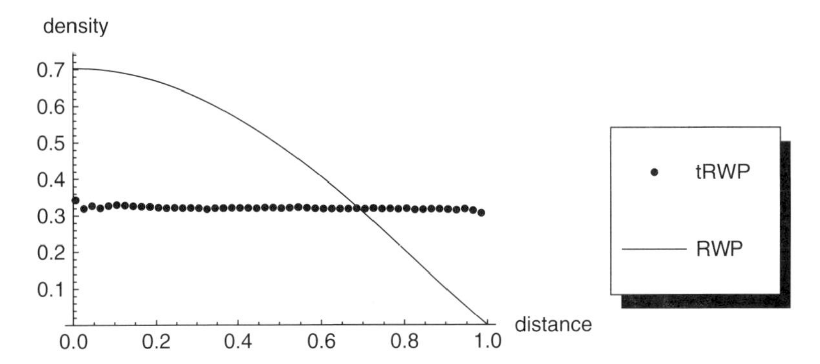

Figure 21.1 Node spatial distribution of the original RWP model, and of the temporal-RWP model with time step set to 10 s. The pause time in both models is set to 0.

to yield a uniform node spatial distribution. Note that a relatively large time step is desirable, since using a small value of the time step Δt would negatively impact the simulation running time. Mitsche et al. (2011) show that a relatively large time step, which only negligibly impacts running time, is sufficient to obtain a time-discrete implementation of temporal-RWP mobility, yielding a node spatial distribution indistinguishable from a uniform one. This can be seen from Figure 21.1, showing the cross-section of the node spatial distribution of the original RWP model, and of the temporal-RWP model with the time step set to 10 s in a scenario where the radius of R is 1 km and the maximum node speed is set to 2 m/s.

21.2.2 The Spatial-RWP Model

The temporal-RWP model presented above is very simple, and is guaranteed to yield a uniform node spatial distribution under quite general conditions: arbitrary convex shape of the movement region R, arbitrary pause time distribution, etc. However, the assumption of location-dependent speed selection for nodes is scarcely representative of real-world mobility. The spatial-RWP model was introduced in Mitsche et al. (2011) with the aim of generating a uniform node spatial distribution without changing the rules for speed selection. Thus, in the spatial-RWP model, nodes select a speed at random prior to initiating a trip, and maintain this speed for the entire duration of the trip. However, in contrast to the original RWP model, waypoints are not selected uniformly at random in R, but are selected according to a mix of probability densities, which is designed to yield a near-uniform node spatial distribution of the resulting trajectory density. We recall that when the speed during a trip is fixed, the node spatial distribution can be estimated as the density of trajectories crossing an infinitesimal area $\partial I(x, y)$, hence in the following the terms “node spatial distribution” and “trajectory density distribution” are considered equivalent.

To compute the “uniforming” mix of probability densities, the authors start by deriving the node spatial distribution resulting when waypoints are chosen uniformly at random among the points in R at a distance no less than x from the center, where $0 < x < 1$. More specifically, the probability density function according to which waypoints are chosen is as follows:

$$f_x(r) = \begin{cases} \dfrac{1}{[\pi(1 - x^2)]} & \text{if } r \geq x \\ 0 & \text{otherwise} \end{cases},$$

where the waypoint density function is expressed in terms of the distance r from the center of R.

The special cases $x = 0$ and $x = 1$ were covered by Hyytia et al. (2006): the case $x = 0$ corresponds to the standard RWP model on the unit disk, while the case $x = 1$ corresponds to the RWPB model, where waypoints are uniformly distributed on the border of R.

The node spatial distribution F_x resulting when waypoints are chosen according to density f_x for different values of x is shown in Figure 21.2. As can be seen from the figure, the node spatial density tends to be concentrated in the center of R for low values of x, but is instead concentrated on the border for high values of x. This suggests that it should be possible to design a version of RWP mobility yielding a uniform node spatial distribution by suitably mixing waypoint densities with different values of x when selecting waypoints.

In order to find the “uniforming” mix of waypoint densities, Mitsche et al. (2011) consider the family $\mathcal{F} = \{f_0, f_{1/h}, f_{2/h}, \ldots, f_1\}$ of waypoint densities, and the corresponding family of resulting node spatial distributions

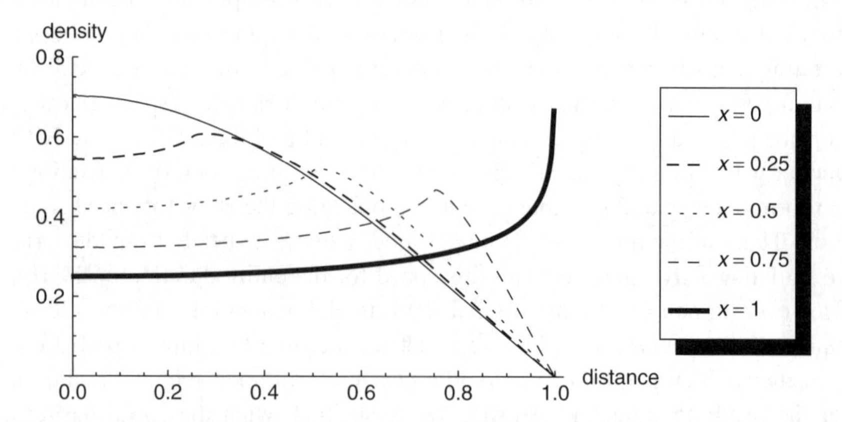

Figure 21.2 Node spatial distribution F_x of the RWP model when waypoints are selected according to density f_x, for different values of x.

$\mathbb{F} = \{F_0, F_{1/h}, F_{2/h}, \ldots, F_1\}$, for some value of h (set to 100 in Mitsche et al. (2011)). These authors observe that, if a set of weights $\alpha_0, \alpha_{1/h}, \ldots$ is chosen such that function $\hat{F} = \sum_i \alpha_i F_i$ is as close to uniform as possible – which, according to the diversity of shapes shown in Figure 21.2, is possible – then the same mixing weights $\alpha_0, \alpha_{1/h}, \ldots$ can be used in the selection of waypoints, where the weights are used to determine, each time a new waypoint needs to be chosen, the density function according to which the next waypoint will be selected. In fact, due to the linearity of the integral operator used in the computation of the node spatial distribution, we have that the node spatial distribution resulting when waypoints are selected as above is exactly $\hat{F}$; this is true with some technical tricks which we will describe shortly.

The mixing weights $\alpha_0, \alpha_{1/h}, \ldots$ are computed in Mitsche et al. (2011) through an iterative local search, performed on a constrained solution space obtained by imposing that only a small number s of the α values are not zero. As can be seen in Figure 21.3, having six non-zero weights is already sufficient to generate a node spatial density function $\hat{F}$ which is virtually indistinguishable from a uniform one. The corresponding mixing weights are as follows:

$$\alpha_{5/100} = \alpha_5 = 0.12, \quad \alpha_{64/100} = \alpha_{64} = 0.10, \quad \alpha_{86/100} = \alpha_{86} = 0.10,$$
$$\alpha_{95/100} = \alpha_{95} = 0.11, \quad \alpha_{99/100} = \alpha_{99} = 0.13, \quad \alpha_1 = \alpha_{100} = 0.44.$$

The spatial-RWP model is defined as follows. The rules for selecting speed are the same as in the original RWP model, that is, before initiating a trip, a node chooses the speed uniformly at random in an interval $[v_{min}, v_{max}]$. Waypoint selection is then performed as follows. The density functions used to

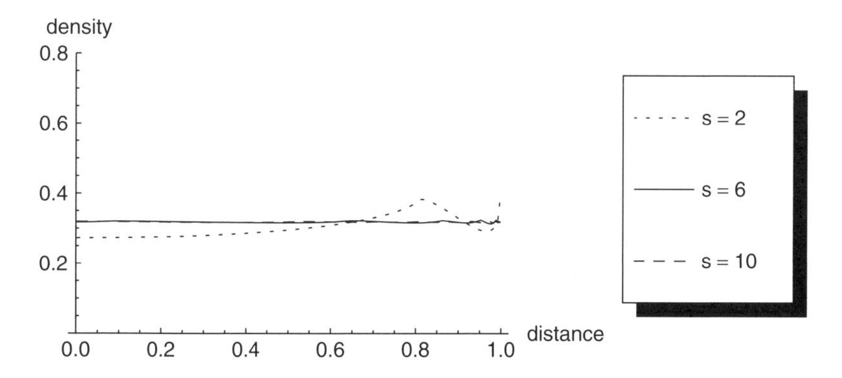

Figure 21.3 Node spatial distribution $\hat{F}$ of the RWP model with "optimal" mixing weights under the constraint that only s of the α values are non-zero.

select waypoints are $f_{5/100}$, $f_{64/100}$, $f_{86/100}$, $f_{95/100}$, and $f_{99/100}$, f_1 – renamed as f_5, f_{64}, f_{86}, f_{95}, f_{99}, and f_{100} below – which are weighted as above. Which specific density function of the mix is used to select waypoints is determined as follows:

1. A real number y is chosen uniformly at random in the [0,1] interval.
2. The value of the index j such that $\sum_{i<j} \alpha_i < y \leq \sum_{i<j+1} \alpha_i$ is found.
3. Density function f_j is chosen to randomly select the waypoint.

For instance, if the random number y generated at step 1 is 0.415, the waypoint will be selected according to density function f_{95}.

Mitsche et al. (2011) then observe that, in order to obtain a node spatial distribution closely resembling $\hat{F}$, it is not sufficient to select waypoints at each step using the above mixing weights. In fact, the "quasi-uniform" node spatial distribution $\hat{F}$ is computed as a weighted sum of spatial density functions F_x referring to different values of x. However, each one of the F_x has been derived under the assumption that *both waypoints* are selected according to the *same* density function f_x.

To better understand this, denote by $P_1, P_2, \ldots$ the sequence of waypoints selected by a mobile node, and call a *mix event* the following event: "waypoint P_i is chosen according to waypoint density f_j, and waypoint P_{i+1} is chosen according to waypoint density $f_{j'} \neq f_j$." Formally speaking, function $\hat{F}$ accurately approximates the node spatial distribution of the spatial-RWP model under the assumption that the probability of observing a mix event in the sequence of selected waypoints is much smaller than the probability of the complementary event "no mix occurs." Based on this observation, Mitsche et al. (2011) propose the following methodology to reduce the probability of observing a mix event: once a waypoint density function is chosen to randomly select waypoints, the *same* density function is used to generate a fixed number b of consecutive waypoints (batch); once the b waypoints have been selected, the waypoint density function is again chosen according to the above-described procedure, and this (possibly) different density function is used to select the next batch of b waypoints. It is easy to see that with this technique we have at most one mix event every b consecutive waypoint selections, implying that the node spatial distribution resulting from the spatial-RWP model is more and more accurately approximated by $\hat{F}$ as b increases. It should be observed, though, that increasing b also has the negative effect of increasing the time needed for the spatial distribution of the spatial-RWP model to settle to the stationary regime. Hence, a trade-off should be pursued in the choice of b. Based on extensive simulations, Mitsche et al. (2011) suggest using a batch size of $b = 4$ which, as shown in Figure 21.4, is sufficient to generate a node spatial distribution very close to a uniform one.

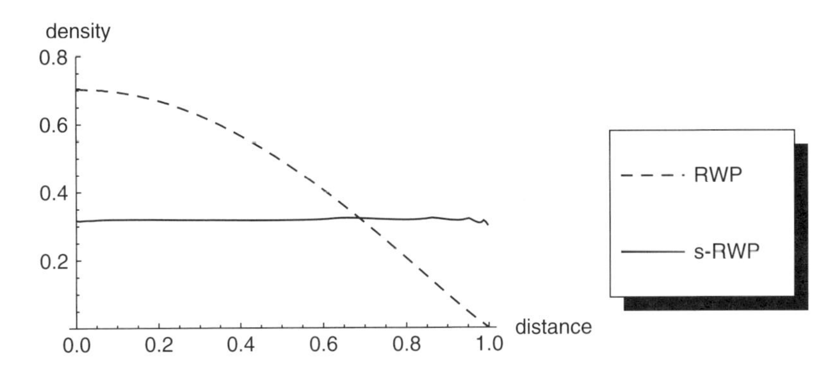

Figure 21.4 Node spatial distribution of the spatial-RWP model with "optimal" mixing weights and batch size $b = 4$. The node spatial distribution of the original RWP model is also shown for comparison.

Summarizing, the spatial-RWP model is defined as follows:

1. Define the candidate waypoint density functions $\{f_5, f_{64}, f_{86}, f_{95}, f_{99}, f_{100}\}$ and the respective weights $\{0.12, 0.10, 0.10, 0.11, 0.13, 0.44\}$;
2. Choose a waypoint density function f_j from the set of candidate functions according to the procedure described above.
3. Select a batch of $b = 4$ waypoints using density function f_j.
4. At the beginning of each trip, choose speed uniformly at random in the $[v_{min}, v_{max}]$ interval.
5. Once the last waypoint in the current batch is reached, return to step 2.

Note that, unlike the temporal-RWP model which is guaranteed to generate a uniform node spatial distribution with arbitrary pause time distributions, the spatial-RWP model yields a uniform node spatial distribution only when the pause time at waypoints is 0. In fact, waypoints are no longer uniformly distributed in the spatial-RWP model, and having a non-zero pause time introduces a non-uniform component in the node spatial distribution.

21.3 Removing Speed Decay

While the previous section was concerned with removing the border effect, in this section we will present a methodology introduced in Yoon et al. (2003) to completely remove the speed decay phenomenon.

These authors start by observing that the speed decay phenomenon arises because, in the RWP mobility model, speed and distance of a trip are chosen independently, and trip duration is a by-product of these random choices. When conditioning on a specific trip distance, trip duration is longer if a relatively smaller speed is chosen. Thus, if one observes the mobile system at a random (and large enough) time instant, the node speed distribution will be biased toward relatively small speed values.

Yoon et al. (2003) formally prove the above fact by deriving the node speed distribution of the RWP model in the stationary regime, and by comparing this distribution to the initial distribution – which, we recall, is the uniform distribution in $[v_{min}, v_{max}]$. The node speed distribution of the RWP model is described by the following pdf:

$$f_V(v) = p_{move} \cdot \frac{1}{v \ln\left(v_{max}/v_{min}\right)} + p_{pause} \cdot \delta(v),$$

where the function is defined for $v \in [v_{min}, v_{max}]$, and p_{pause} and $p_{move} = (1 - p_{pause})$ are the probabilities of finding a node in the pause and move states, respectively. In this expression, $\delta(v)$ denotes the Dirac delta function taking value 1 when $v = 0$, and value 0 otherwise. We recall that, in the RWP model, under the assumption that the pause time is set to $t_p \geq 0$, p_{pause} can be computed as follows:

$$p_{pause} = \frac{t_p}{t_p + E[L]\dfrac{\ln\left(v_{max}/v_{min}\right)}{v_{max} - v_{min}}},$$

where L is the random variable denoting the distance of a trip.

Figure 21.5 shows the stationary node speed distribution f_V as defined above and the node speed distribution f_0 at initial conditions, which can be derived in a straightforward way as follows:

$$f_0(v) = p_{move} \cdot \frac{1}{v_{max} - v_{min}} + p_{pause} \cdot \delta(v),$$

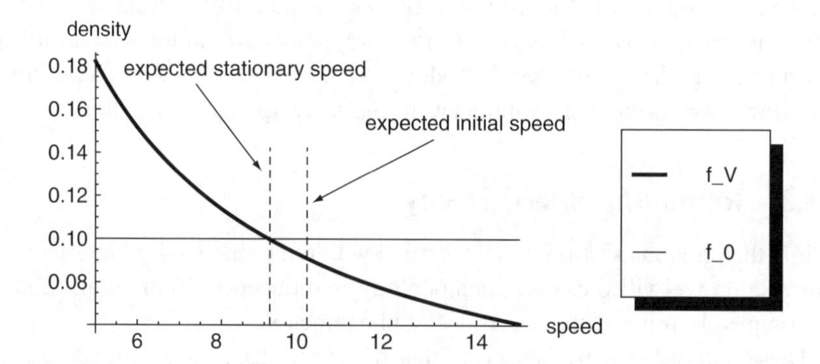

Figure 21.5 Node speed distribution of the RWP model in the unit square, with pause time set to 0, $v_{min} = 5$ m/s, and $v_{max} = 15$ m/s. Both the stationary distribution f_V and the initial uniform distribution f_0 are shown, along with the respective expectations.

with the assumption that nodes are initialized in the move or pause state according to probabilities p_{move} and p_{pause}, respectively.

As can be seen from the figure, the stationary node speed distribution has a probability mass shifted toward lower speed values, as witnessed also by the expected speed value at stationary conditions, which, as described and commented on in Chapter 5, is

$$E[V] = \frac{v_{max} - v_{min}}{\ln\left(v_{max}/v_{min}\right)};$$

it is always smaller than or equal to the initial expected speed $(v_{max} - v_{min})/2$.

Yoon et al. (2003) then exploit the characterization of speed density f_V to completely remove the speed decay phenomenon, which always occurs when $v_{min} < v_{max}$. Note that one possible way of eliminating the speed decay phenomenon is to set $v_{min} = v_{max}$, that is, to remove randomness in the choice of speed; however, this approach has the shortcoming of considerably reducing the range of possible scenarios that can be simulated with the RWP model. Thus, it is preferable to devise a methodology that eliminates speed decay also when $v_{min} < v_{max}$.

To achieve this, and defining

$$f_{V,move} = \frac{1}{v\ln\left(v_{max}/v_{min}\right)},$$

Yoon et al. (2003) suggest using the following approach, assuming a fixed pause time $t_p \geq 0$ at the waypoints:

1. Determine whether a node starts from the pause or move state according to probabilities p_{pause} and $p_{move} = 1 - p_{pause}$ as defined above.
2. If a node starts from the pause state, set its pause time to t_p.
3. If a node starts from the move state, use density function $f_{V,move}$ to select the travel speed.
4. After the first trip (either move or pause) of a node, select speed uniformly at random in $[v_{min}, v_{max}]$ for all subsequent trips.

The authors also show that the above methodology achieves the goal of completely removing the speed decay phenomenon when $v_{min} < v_{max}$. In particular, if the above methodology is used, the initial node speed distribution equals the stationary node speed distribution f_V.

21.4 The RWP Model and "Perfect Simulation"

In the previous sections, we saw how to modify the original RWP model in order to *separately* remove the border effect and speed decay. The next question to investigate is whether there exists a methodology that allows the *simultaneous* removal of both these effects, enabling so-called "perfect simulation" – that is, a simulation methodology where the mobile system is initialized directly in the stationary regime and the node spatial distribution is uniform.

The answer to this questions lies in the wealth of theoretical results presented in Le Boudec and Vojnovic (2006), which, besides the framework for demonstrating the existence and characterization of the stationary regime as described in Chapter 7, also include the definition of a "perfect simulation" methodology for random trip models.

It is important first to notice that the notion of "perfect simulation" considered in Le Boudec and Vojnovic (2006) accounts only for the existence of a stationary regime and for the initialization of the mobile network directly into this regime. Indeed, for the reasons explained in Section 21.1, we are interested here in a stronger notion of "perfect simulation" where, besides the above properties, the additional property that the stationary node spatial distribution is uniform must be fulfilled.

Notice also that, although the original RWP model belongs to the class of random trip models and, hence, according to Le Boudec and Vojnovic (2006), a RWP mobile network can be initialized directly in the stationary regime, it holds true that the node spatial distribution of the original RWP model *is not uniform*. Thus, the issues with wireless network simulation accuracy described in Section 21.1 are not completely solved even if the simulation methodology described in Le Boudec and Vojnovic (2006) is applied.

To completely solve the issues related to wireless network simulation when the RWP model is applied in a bounded domain (as is the case in most wireless networking studies), we need to use a version of the RWP model to generate a uniform node spatial distribution. In what follows, we describe how to devise a "perfect simulation" methodology using the spatial-RWP model presented in Section 21.2. We recall that in the spatial-RWP model presented in Section 21.2 the pause time at waypoints is assumed to be 0, and nodes are assumed to move in the unit disk.

We start by observing that the only difference between the original RWP model and the spatial-RWP model is in the choice of the waypoints, which is no longer uniform in the mobility region R. However, whether a mobility model belongs to the class of random trip models is determined by general conditions on the existence of non-degenerate distributions for choosing waypoints. Whether these distributions are uniform or not is not relevant for determining membership of a mobility model in the class of random trip models. Based on this, and observing that the Markov property of the stochastic process $Y = (I_n, P_n)$ (recall Chapter 7) is still fulfilled by suitably defining the set of phases, we can then conclude that the spatial-RWP model belongs to the class of random trip models. In particular, the set of phases is defined by a triplet (i, j, r), where i represents the waypoint distribution used to select the starting point of the current path, j the waypoint distribution used to select the ending point of the path, and r the index in the current batch of selected waypoints. The fact that the spatial-RWP model is a random trip model, coupled with the observation that the average trip duration in the spatial-RWP model, although different from that in the original RWP model, remains *finite*, implies that there exists a stationary regime for the spatial-RWP model (recall Theorem 7.1).

The stationary regime of the spatial-RWP model is given by stationary distributions (as $t \to \infty$) for the following quantities:

$$(P(t), S(t), U(t)),$$

where $P(t)$ denotes the *path* (i.e., a segment connecting two consecutive waypoints) along which a node is moving at time t, $S(t)$ is the duration of the trip (from starting to ending waypoint), and $U(t)$ is the fraction of elapsed time on the trip at time t. To initialize the system directly into stationary conditions, we then need to generate: (i) the *initial path*, which, in turn, is determined by the location of the starting and ending waypoints; (ii) the *duration* of a trip, which, in the spatial-RWP model, is determined by a randomly chosen *speed* of the trip; and (iii) the *fraction* of trajectory already covered at time t – see Figure 21.6.

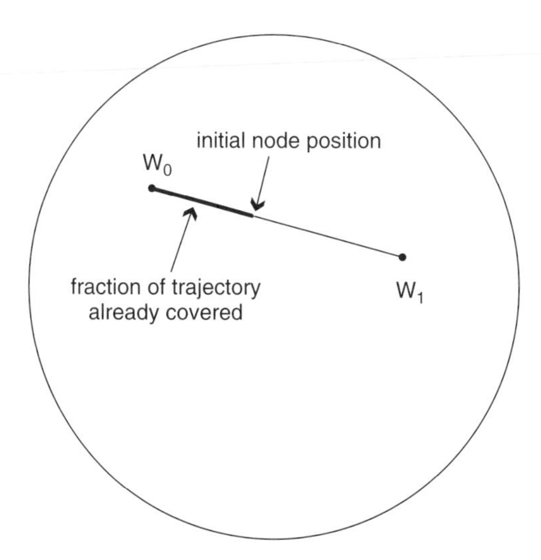

Figure 21.6 Initialization of the spatial-RWP model for "perfect simulation."

In the following, we present a version of the "perfect simulation" methodology introduced in Le Boudec and Vojnovic (2006) customized to the spatial-RWP model with zero pause time. The "perfect simulation" methodology for the spatial-RWP model is summarized below:

1. *Determine the initial trajectory*: first, with a fixed value for b (batch size) in the spatial-RWP model, determine whether the two initial waypoints belong to different batches (with probability $1/b$), or to the same batch (with probability $1 - 1/b$).
   1.1 If the two waypoints belong to different batches, select waypoints W_0 and W_1 according to the mix of waypoint distributions defined in Section 21.2; set *initialBatchSize* to b.
   1.2 If the two waypoints belong to the same batch, select one of the waypoint density functions f_j in the mix based on the α weights, and select both waypoints W_0 and W_1 according to f_j. Select an integer i uniformly at random in $\{1, \ldots, b - 1\}$, and set *initialBatchSize* to i.
2. *Determine the initial speed*: select a random speed v using density function $f_{V,move}$ defined in Section 21.3.
3. *System initialization*: select a number u uniformly at random in the $[0, 1]$ interval; initialize the node speed to v, the node's next waypoint to W_1, and the node's current position along the segment connecting W_0 with W_1, at distance $u \cdot dist(W_0, W_1)$ from W_0, where $dist(x, y)$ is the Euclidean distance between x and y.
4. *Movement phase*: after initialization, rules for selecting waypoints and speed are as in the spatial-RWP model, the only difference being that the first batch of selected waypoints has size i instead of b.

Based on the results presented in Le Boudec and Vojnovic (2006), Mitsche et al. (2011), and Yoon et al. (2003), we can conclude that the above methodology initializes the mobile system directly into the stationary regime, and that the stationary regime of the node spatial distribution is the uniform distribution on the disk. Thus, no "warm-up" period is needed when performing simulations, and important network parameters such as the node transmission range need not be changed with respect to the case of uniform node spatial distribution. This way, the issues related to simulation accuracy when the RWP mobility in a bounded region is used (discussed in Section 21.1) are completely solved.

References

Bettstetter C, Resta G and Santi P 2003 The node distribution of the random waypoint mobility model for wireless ad hoc networks. *IEEE Transactions on Mobile Computing* **2**, 257–269.

Gupta P and Kumar P 1998 Critical power for asymptotic connectivity in wireless networks. *Stochastic Analysis, Control, Optimization and Applications* pp. 547–566.

Gupta P and Kumar P 2000 The capacity of wireless networks. *IEEE Transactions on Information Theory* **46**, 388–404.

Hyytia E, Lassila P and Virtamo J 2006 Spatial node distribution of the random waypoint mobility model with applications. *IEEE Transactions on Mobile Computing* **5**, 680–694.

Le Boudec JY and Vojnovic M 2006 The random trip model: Stability, stationary regime, and perfect simulation. *IEEE/ACM Transactions on Networking* **14**, 1153–1166.

Mitsche D, Resta G and Santi P 2011 The random waypoint mobility model with uniform node spatial distribution. *Technical Report IIT-CNR, IIT-25-2011*.

Santi P 2005 The critical transmitting range for connectivity in mobile ad hoc networks. *IEEE Transactions on Mobile Computing* **4**(3), 310–317.

Santi P and Blough D 2003 The critical transmitting range for connectivity in sparse wireless ad hocnetworks. *IEEE Transactions on Mobile Computing* **2**(1), 25–39.

Yoon J, Liu M and Noble B 2003 Sound mobility models. *Proceedings of the ACM International Conference on Mobile Computing and Networking (MOBICOM)*, pp. 205–216.

22

Mobility Modeling and Opportunistic Network Performance Analysis

In the previous chapter, we discussed the effects of mobility modeling on the accuracy of wireless network simulation results. In this chapter, we will consider the effects of mobility modeling on another important performance evaluation methodology for next generation wireless networks, namely, theoretical analysis.

More specifically, we will focus our attention on recent results aimed at characterizing *asymptotic* opportunistic network performance, that is, performance trends observed when the number n of nodes in the network grows to infinity. The aim of this chapter is to show how different assumptions about the probability distribution of the inter-contact time, which is a by-product of the underlying mobility model, lead to radically different conclusions about performance trends for large networks. This observation applies to different communication schemes – *unicast* and *broadcast* – as well as to different forwarding strategies used to route a message from source to destination.

We will start this chapter by discussing the case of unicast communication, in which a single message M has to be delivered from a source node S to a destination node D, and then consider the case of broadcast communication, in which a message M originating at a source node S has to be delivered to all other nodes in the network.

22.1 Unicast in Opportunistic Networks

Unicast is one of the most important and widely used communication schemes, according to which a sender node S has to deliver a message M

Mobility Models for Next Generation Wireless Networks: Ad Hoc, Vehicular and Mesh Networks, First Edition. Paolo Santi.

to another node D in the network, called the destination node, which can be uniquely identified through a network address (e.g., the IP address in IP-based networks).

In order to characterize unicast performance in opportunistic networks, the following need to be defined:

1. The *forwarding strategy* that is used to deliver the message from source to destination.
2. The *mobility model* governing node mobility, from which statistical properties of the pairwise contact patterns can be derived; alternatively, these statistical properties can be directly modeled, leaving the underlying mobility model unspecified.

The characterization of unicast performance in opportunistic networks has been the subject of a number of recent papers (Chaintreau et al. 2006; Resta and Santi 2012; Spyropoulos et al. 2008; Zhang et al. 2007). In what follows, we report selected results that have been specifically chosen to show the importance of mobility modeling in performance analysis.

22.1.1 Network Model

We consider an opportunistic network composed of n nodes; a randomly selected source node S wants to send a message M to a randomly selected destination node D, with $S \neq D$. Two possible routing strategies can be used to deliver message M from S to D:

1. *Epidemic routing* (Vahdat and Becker 2000): source node S delivers a copy of M to all nodes it encounters, unless the encountered node already has a copy of M. The same process is repeated at all nodes holding a copy of M, until the message is eventually delivered to the destination.
2. *Two-hops multi-copy routing* (Grossglauser and Tse 2002; Groenevelt et al. 2005; Spyropoulos et al. 2008): as in Epidemic routing, source node S delivers a copy of M to all nodes it encounters; however, relay nodes holding a copy of M are only allowed to deliver M to the destination node, and they cannot further propagate the message.

Node mobility is not explicitly modeled in what follows; instead, assumptions are made for what concerns pairwise inter-contact patterns between nodes:

1. Pairwise contacts are assumed to be instantaneous, but sufficiently long to entirely transfer message M between node buffers if required by the forwarding strategy.

2. Let $T_{u,v}$ be the random variable denoting the time interval elapsing between two consecutive contacts between nodes u, v; random variables $T_{u,v}$ and $T_{i,j}$ are mutually independent, for any $(i, j) \neq (u, v)$.

Assumption 1 above models situations where the time scale of the analysis is much larger than the average contact duration, and messages circulating in the network are relatively short, so that they can be entirely copied from one node's buffer to another node's buffer during a typical contact. It is also worth observing that, under assumption 1 above, the two notions of inter-contact time defined in Chapter 18 become equivalent.

Three alternative assumptions are made in the following concerning the *probability distribution* of random variable $T_{u,v}$:

1. *Exponential distribution*: for any u, v, $T_{u,v}$ is an exponential random variable with parameter λ. Formally, we have that

$$Prob(T_{u,v} > t) = e^{-\lambda t}.$$

2. *Power law*: for any u, v, the probability distribution of random variable $T_{u,v}$ obeys a power law for parameter $\alpha > 0$. Formally, we have that

$$Prob(T_{u,v} > t) = t^{-\alpha}.$$

3. *Truncated Pareto distribution*: for any u, v, $T_{u,v}$ is a random variable with truncated Pareto distribution of parameters γ, a, b, with $\gamma > 0$ and $0 < a < b$. Formally, we have that

$$Prob(T_{u,v} > t) = 1 - \frac{1 - (a/t)^{\gamma}}{1 - (a/b)^{\gamma}}.$$

The assumption of exponentially distributed inter-contact time is justified by the fact that it has been proven that well-known mobility models such as random waypoint, random direction, random walks, etc., give rise to exponentially distributed inter-contact times – see, for example Groenevelt et al. (2005) and Spyropoulos et al. (2006). The power-law assumption is justified by the fact that, at least in a certain period of time, the complementary cumulative density function of the inter-contact time observed in real-world traces obeys a power law (Chaintreau et al. 2006). Indeed, as commented in Chapter 19, more correctly the inter-contact time complementary cumulative density function has been observed to obey a dichotomy, with an initial power law followed by an exponential tail. Thus, the power-law assumption can be considered to be accurate in modeling the inter-contact time distribution only up to the *characteristic time*, which, we recall, is defined as the time when the transition between the power law and exponential regime

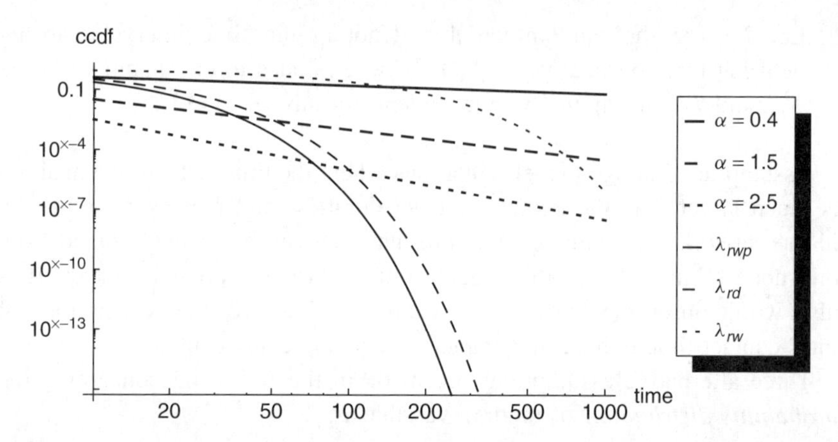

Figure 22.1 Trends of exponential distribution and power law for increasing values of t and different parameters. Notice that both the x- and y-axes are in logarithmic scale.

occurs. The assumption of truncated Pareto distribution of inter-contact time is motivated by the fact that this distribution has been shown (Hong et al. 2008) to be able to reproduce the power-law and exponential tail dichotomy observed in inter-contact time distributions derived from real-world traces.

Figure 22.1 shows the trends of the exponential distribution and of the power law for different parameters as t increases. In particular, the three values of parameter λ in the exponential distribution refer to three different mobility models – random waypoint, random direction, and random walk – applied in a square area of side 4 km, with speed chosen uniformly at random in the [4, 10] km/h interval – for random waypoint and random direction mobility, and transmission range set to 100 m. The resulting values of λ are $\lambda_{rwp} = 0.148$, $\lambda_{rd} = 0.115$, and $\lambda_{rw} = 0.014$, respectively (Groenevelt et al. 2005). Values of the α parameter in the power law are modeled after the trend observed in the Infocom data trace (Chaintreau et al. 2006), yielding $\alpha_1 = 0.4$. For reference, larger values of α ($\alpha_2 = 1.5$ and $\alpha_3 = 2.5$) are also reported.

As shown in Figure 22.1, the exponential distribution and power law have very different trends as time variable t increases. In particular, a faster decay with t is shown for the exponential distribution with respect to the power law. This implies that the probability for any two nodes to meet within a certain time $\bar{t}$ (i.e., $Prob(T_{u,v} \leq \bar{t})$) is much *larger* with an exponential distribution than with a power law, for large enough values of $\bar{t}$. Thus, better unicast performance can be expected under the assumption of exponentially distributed inter-contact times, with respect to the case of inter-contact times obeying a power law. This intuition is confirmed by the theoretical results reported in the following.

22.1.2 Epidemic Routing Performance

We first consider the case of Epidemic routing, and present results concerning the asymptotic trend of the *expected delivery delay*, which is formally defined as follows:

Definition 22.1 *Let T_D be the random variable corresponding to the time elapsing between the time instant at which message M is generated at node S and the time instant at which the message M is first delivered to node D. The* expected delivery delay *is the expected value of random variable T_D, and is denoted $E[T_D]$.*

The following theorem, derived in Groenevelt et al. (2005), characterizes the expected delivery delay with epidemic routing under the assumption of exponentially distributed pairwise inter-contact times:

Theorem 22.1 *The expected delivery delay with epidemic routing and exponentially distributed pairwise inter-contact times with parameter λ is*

$$E_{exp}[T_D^{epi}] = \frac{1}{\lambda(n-1)}\left(\log n + 0.577\,21 + O\left(\frac{1}{n}\right)\right).$$

Notice that, if λ is an arbitrary constant not depending on the network size n, the above theorem implies the following asymptotic trend for $E_{exp}[T_D^{epi}]$:

$$\lim_{n\to\infty} E_{exp}[T_D^{epi}] = \lim_{n\to\infty} \frac{\log n}{n} = 0.$$

The next theorem, which is a consequence of Theorem 16.1 in Chaintreau et al. (2006), characterizes the asymptotic trend of $E[T_D]$ with epidemic routing under the assumption of inter-contact times obeying a power law of parameter $0 < \alpha < 1$, as is the case in the real-world data traces considered in Chaintreau et al. (2006):

Theorem 22.2 *The expected delivery delay with epidemic routing and inter-contact times obeying a power law for parameter $0 < \alpha < 1$ grows unboundedly with n, that is,*

$$\lim_{n\to\infty} E_{pl}[T_D^{epi}] = +\infty.$$

Thus, comparing Theorem 22.2 to the asymptotic trend derived for the case of exponentially distributed inter-contact times, we observe that radically different conclusions about the expected delivery time of epidemic routing in large-scale networks can be drawn depending on the assumptions made about the inter-contact time distribution: the expected delivery time *converges to 0* as network size increases if *exponentially distributed* inter-contact times are assumed, while it *grows unboundedly with n* if inter-contact times obey a *power law* of relatively low exponent.

22.1.3 Two-Hops Routing Performance

Let us now consider the case of two-hops routing. The following theorem, derived in Groenevelt et al. (2005), characterizes the expected delivery delay with two-hops routing under the assumption of exponentially distributed inter-contact times:

Theorem 22.3 *The expected delivery delay with two-hops routing and exponentially distributed pairwise inter-contact times with parameter λ is*

$$E_{exp}[T_D^{2h}] = \frac{1}{\lambda}\left(\sqrt{\frac{\pi}{2(n-1)}} + O\left(\frac{1}{n}\right)\right).$$

If λ is an arbitrary constant, this theorem implies the following asymptotic trend for $E_{exp}[T_D^{2h}]$:

$$\lim_{n\to\infty} E_{exp}[T_D^{2h}] = \lim_{n\to\infty} \frac{1}{\sqrt{n}} = 0.$$

Since Theorem 22.2 holds also for the case of two-hops routing, by comparing the above asymptotic trend to that stated in Theorem 22.2 we can conclude that, also in this case, radically different conclusions about asymptotic performance trends can be drawn depending on the assumptions made on the inter-contact time distribution: we have an expected delivery delay converging to 0 as n increases with exponentially distributed inter-contact times, while the expected delivery delay grows unboundedly with n if inter-contact times are assumed to obey a power law.

From the results presented above, the reader might wonder which trends for expected delivery delay could be expected in real-world networks, where inter-contact times have been shown to obey a power-law and exponential tail dichotomy.

To answer such a question, Hong et al. (2008) studied the expected routing delay when the inter-contact time distribution is a truncated Pareto distribution, which they showed was able to reproduce the power-law and exponential tail dichotomy of inter-contact times observed in real-world traces. In particular, the authors consider the *single-copy* 2-Hops routing protocol, according to which the source node S delivers its copy of the message to the first node it meets (relay node R), and then node R is in charge of delivering message M to the destination node D. Note that, since the multi-copy version of two-hops routing is strictly faster than its single-copy counterpart in delivering a message to its destination, the expected routing delay characterization presented in the following theorem can be considered as an upper bound to expected routing delay under multi-copy two-hops routing. By a similar argument, the one presented below can be considered an upper bound to expected routing delay also for Epidemic routing.

Theorem 22.4 *The expected delivery delay with single-copy two-hops routing and inter-contact times with truncated Pareto distribution of parameters* $\gamma, 1, b$, *with* $0 < \gamma < 2$ *and* $1 < b$, *is*

$$E_{Par}[T_D^{s2h}] = \frac{1}{E_{Par}[T_{u,v}](1 - b^{-\gamma})}\left(\frac{b^{2-\gamma} - 1}{2 - \gamma} - \frac{b^2 - 1}{2b^\gamma}\right),$$

where $E_{Par}[T_{u,v}]$ *is the expected inter-contact time with truncated Pareto distribution and is defined as follows:*

$$E_{Par}[T_{u,v}] = \begin{cases} [\ln b/(1 - b^{-1})] & \textit{if } \gamma = 1 \\ (\gamma/\gamma - 1)[(1 - b^{1-\gamma})/(1 - b^{-\gamma})] & \textit{otherwise} \end{cases}.$$

Since $E_{Par}[T_D^{s2h}]$ does not depend on the number n of nodes in the network, we can conclude that

$$\lim_{n\to\infty} E_{Par}[T_d^{2h}] \leq \lim_{n\to\infty} E_{Par}[T_d^{s2h}] = c,$$

for some $c > 0$. Thus, the expected delivery delay with two-hops routing (and also with Epidemic routing) converges to a finite value as n increases under the (realistic) assumption that inter-contact times are distributed according to a truncated Pareto distribution.

22.2 Broadcast in Opportunistic Networks

In the previous section, we showed the effect of different assumptions concerning inter-contact time distribution on asymptotic bounds that can be obtained for the time needed to send a message M from a source node S to a single destination node D. In this section, we focus our attention on another important communication scheme, *broadcasting*, according to which message M originating at a random source node S must be delivered to *all* the other nodes in the network. In what follows, we present asymptotic bounds on the time needed to complete a broadcast operation in opportunistic networks – called *broadcasting time*, or *flooding time*, in the literature – obtained by making different assumptions about the pairwise inter-contact time distribution. Selected results are presented from a number of recent papers addressing this topic (Becchetti et al. 2011; Clementi et al. 2011; Peres et al. 2011; Pettarin et al. 2011).

22.2.1 Network Model

In what follows, we model opportunistic networks using the general notion of *dynamic graphs* introduced in a number of recent papers analyzing properties of dynamic networks (see, e.g. Kuhn and Oshman (2011) and references therein).

Formally speaking, a *dynamic graph* on a set V of n nodes is a time-evolving graph $G = (V, \{E_t\}_{t\geq 0})$, where V is the (time-stationary) set of nodes, and E_t is the set of edges in the graph at time t. Time in dynamic graphs is assumed to be a discrete quantity, hence the evolution of the graph is observed at integer time instants $t, t+1, t+2, \ldots$. Typically, edge set E_t is assumed to be a random variable. Furthermore, the *probability distribution* governing random variable E_t is typically assumed to be time invariant. In other words, although edge sets at different times are in general different, they are distributed in the graph according to a fixed distribution $f_E()$. Hence, a dynamic graph $G = (V, \{E_t\}_{t\geq 0})$ is a time-stationary stochastic process.

Given a dynamic graph $G = (V, \{E_t\}_{t\geq 0})$, the *broadcasting time* in G is formally defined as follows:

Definition 22.2 *Let S be an arbitrary node in V, which is the broadcast source. Define a node in V to be* colored *if and only if it has already received the message from S. Initially, at time $t = 0$, only S is colored. Given the set $C_t \subseteq V$ of colored nodes at time t, the set of colored nodes at time $t+1$ is defined as*

$$C_{t+1} = C_t \cup \{v \in V | \exists u \in C_t : (u, v) \in E_{t+1}\}.$$

The broadcasting time with source S *is defined as* $B(G, S) = \min\{t : C_t = V\}$. *The* broadcasting time *of dynamic graph* $G = (V, \{E_t\}_{t\geq 0})$ *is defined as* $B(G) = \max_{S\in V} B(G, S)$.

Note that, under the assumption that the edge sets E_t are suitably defined random variables, also $B(G)$ is a random variable. Hence, characterization of broadcasting time in dynamic graphs amounts to estimating bounds for $B(G)$ which, typically, hold with high probability (w.h.p.), where the standard notion of w.h.p. requires that the probabilistic bound must hold with probability at least as large as $(1 - 1/n)$.

The above definition of broadcasting time is appropriate for modeling scenarios in which the rate of information propagation within the network is much higher than the rate at which edges appear/disappear in the network. In fact, the above definition implies that propagation of the message within a connected component of the dynamic graph is *instantaneous*. Note also that, given this assumption of instantaneous message propagation along the edges of the graph, characterizing broadcasting time becomes a trivial problem if the time-stationary graph $G = (V, E_t)$ is connected w.h.p.

Dynamic graphs are a suitable model for opportunistic networks, since:

1. The rate at which information propagates within a connected component of an opportunistic network, even taking into account delays due to multi-hop propagation, is much higher than the rate at which links in the network (corresponding to edges in the dynamic graph) appear/disappear. In fact, in opportunistic networks links appear/disappear mainly due to node mobility, hence they change at the time scale of several seconds or minutes, while information propagation within a network's connected component requires a few seconds.
2. The communication graph representing links in an opportunistic network at an arbitrary time instant is most likely disconnected and very sparsely populated with links – this is indeed a distinguishing feature of opportunistic networks, as described in Chapter 17.

In the following, we will provide asymptotic bounds for the broadcasting time in different classes of dynamic graphs. In order to fairly compare the presented bounds, we define the following metric, which will be kept unchanged across the different classes of dynamic graphs considered:

Definition 22.3 *The* expected degree *of a node u in $G = (V, E_t)$ is defined as the expectation of random variable $deg(u)$, which is defined as follows:*

$$deg(u) = \sum_{v \in V, v \neq u} e^t_{u,v},$$

where $e^t_{u,v}$ is the indicator random variable corresponding to the existence of edge (u, v) in E_t, that is,

$$e^t_{u,v} = \begin{cases} 1 & \textit{if } (u,v) \in E_t \\ 0 & \textit{otherwise} \end{cases}.$$

Note that, if the probability of existence of any edge (u, v) in E_t, for any $t \geq 0$, is the *same* for any pair of nodes u, v, then the expected node degree is the same for all nodes in G. This assumption is valid for all the classes of dynamic graphs considered in the following. Under this assumption, the notion of expected node degree can be extended to the dynamic graph G, and we then use the term *expected node degree in G*.

To be consistent with an opportunistic network scenario, in the following we will present results referring to the case where the expected node degree in G is $\Theta(1)$. In fact, it is known that, in many classes of (static) random graphs, including Erdös–Rény random graphs, geometric random graphs, etc., an asymptotically larger value of the expected degree of $\Theta(\log n)$ is needed to obtain a connected graph w.h.p. Thus, assuming an expected node degree of $\Theta(1)$ is consistent with a scenario in which the dynamic graph $G = (V, E_t)$ is disconnected most of the time.

22.2.2 *Broadcasting with Geometric-Based Mobility*

Dynamic graphs can be used to model the evolution of links in a mobile network, where node mobility occurs in an underlying metric space. In this section, we report a characterization of the broadcasting time in the class of dynamic graphs generated by a set of n nodes moving according to a random walk in a square lattice.

More formally, we assume that n nodes move in a square lattice of $\Theta(n)$ points – that is, the side of the square is formed by $\Theta(\sqrt{n})$ points. Mobility is governed by a time-discrete and space-discrete random walk: at each time step, a node residing on a point p of the lattice with n_p neighboring points (where $n_p = 2, 3, 4$ due to the presence of the border) moves to any of these neighboring points with equal probability $1/5$, while it stays on p with probability $1 - n_p/5$.

Edge sets E_t in the dynamic graph are formed as follows. Each of the n nodes in the network is assumed to have a fixed transmission range $r = r(n)$, and the edge between nodes $u, v \in V$ is assumed to exist at time t if and only if nodes u and v are within distance r at time t, where distance is defined to be the *Manhattan distance* on the lattice.

Pettarin et al. (2011) provide near tight asymptotic bounds for the broadcasting time in dynamic graphs defined as above, under the assumption that the transmission range r is chosen such that the largest connected component of $G = (V, E_t)$ comprises at most a logarithmic number of nodes. To simplify the presentation, in the following we assume $r = \Theta(1)$, which fulfills the requirement of logarithmic size of the largest connected component in G. Note also that, since the density of nodes per unit area in the square lattice is $\Theta(1)$, and the node spatial distribution of random walks is uniform, the assumption $r = \Theta(1)$ implies that the expected node degree in G is $\Theta(1)$.

We start with the following theorem, which establishes an upper bound on the broadcasting time when $r = 0$, that is, when a link between any two nodes is established at time t if and only if the nodes are located on the same lattice point at time t. Clearly, the same bound applies to larger values of r.

Theorem 22.5 *Assume the dynamic graph $G = (V, E_t)$ is defined according to a random walk mobility model as described above, and that $r = 0$. Then, w.h.p.,*

$$B(G) = \tilde{O}\left(\sqrt{n}\right).$$

In this theorem, the notation $\tilde{O}$ is used to hide polylogarithmic factors, namely, $\tilde{O}(f(n)) = O(f(n)\log^c n)$, for some constant $c \geq 0$.

The next theorem states a lower bound on the broadcasting time for the same class of dynamic graphs:

Theorem 22.6 *Assume the dynamic graph $G = (V, E_t)$ is defined according to a random walk mobility model as described above, and that $r = \Theta(1)$. Then, w.h.p.,*

$$B(G) = \Omega\left(\frac{\sqrt{n}}{\log^2 n}\right).$$

By comparing the bounds in Theorems 22.5 and 22.6, we can then conclude that the broadcasting time in the above-defined class of dynamic graphs is $\tilde{\Theta}(\sqrt{n})$, where, as above, the notation $\tilde{\Theta}$ is used to hide polylogarithmic factors.

An upper bound on broadcasting time of $\tilde{O}(\sqrt{n})$ has been proved in Clementi et al. (2011) also for the class of dynamic graphs generated by n nodes moving in a square of side $\sqrt{n}$ according to a space- and time-discretized version of the random waypoint mobility model, under the assumption that the transmission range is set to $r = \Theta(1)$. Note that the expected degree in G as defined above is $\Theta(1)$, as in the case of random walk mobility.

The asymptotic broadcasting performance bounds presented in this section are obtained under the assumption that nodes obey some specific mobility rules, such as random walk and (discrete) random waypoint. It is well known that the inter-contact time distribution generated by these mobility models is exponential (Spyropoulos et al. 2006). In the next section, we will present an asymptotic broadcasting performance bound for a class of dynamic graphs in which the inter-contact time distribution is (experimentally) shown to display the power-law and exponential tail dichotomy observed in real-world data traces (Karagiannis et al. 2007).

22.2.3 *Broadcasting with the Home-MEG Model*

The mobility models used to generate the classes of dynamic graphs considered above are known to generate exponential inter-contact time distributions. Although versions of random walk and random waypoint mobility displaying the power-law and exponential tail dichotomy have been defined in (Karagiannis et al. 2007), the models considered in the theoretical analyses of Clementi et al. (2011) and Pettarin et al. (2011) are based on the original definitions of random waypoint and random walk mobility, respectively, and are hence unable to reproduce the well-known dichotomy in the inter-contact time distributions.

As discussed in Chapter 20, a number of mobility models have been proposed with the aim of faithfully reproducing contact patterns observed in real-world data traces. In this section, we present an upper bound on asymptotic broadcasting time performance for a class of dynamic graphs generated according to one of these mobility models. More specifically, we will consider the Home-MEG contact-based model presented in Section 20.6, and assume that each of the $n(n-1)/2$ possible links in G obeys the Home-MEG link model presented in Section 20.6.

In the following, we consider the class of dynamic graphs obtained when, for any pair of nodes $u, v \in V$, the existence/non-existence of edge (u, v) in E_t is determined (independently of the existence of other edges in the network) by the Markov chain defined in Section 20.6. We recall that the Markov chain has four states – HC, NC, HD, ND – modeling the existence (-C) or non-existence (-D) of the edge when the two nodes are in the "Home" state or "Not-Home" state, respectively. Transitions between the four states are governed by the four parameters of the model, namely (see Section 20.6 for details):

1. the probabilities p_H and p_{NH} of making a transition into the "Home" or "Not-Home" states, respectively; and
2. the probabilities p_h and p_l of making a transition to the edge existence state when in the "Home" and "Not-Home" states, respectively.

The following theorem, proved in Becchetti et al. (2011), establishes an upper bound on the broadcasting time for the class of dynamic graphs described above, under the assumption that the initial set of edges E_0 is chosen according to the stationary distribution of the underlying Markov chains and an additional technical assumption:

Theorem 22.7 *Assume the dynamic graph $G = (V, E_t)$ is defined according to the Home-MEG model of parameters (p_H, p_{NH}, p_h, p_l) as described above, that the initial edge set E_0 is randomly chosen according to the stationary distribution of the underlying Markov chains, and that*

$$\left\lceil \frac{5\Lambda}{n} \right\rceil \leq \min\left\{\frac{1}{p_h}, \frac{1}{4p_{NH}}\right\},$$

where $\Lambda = 4(p_H + p_{NH})/(p_H \cdot p_h)$. Then, w.h.p.,

$$B(G) = O\left(\frac{\log n}{\log(1 + (n/\Lambda))}\right).$$

This theorem requires a technical assumption to hold, namely, that $\lceil 5\Lambda/n \rceil \leq \min\{1/p_h, 1/4p_{NH}\}$, where Λ is defined as above. Is this assumption consistent with the guidelines on Home-MEG parameter setting derived from comparison to real-world data traces, as commented at the end of Section 20.6? We recall the guidelines here for convenience:

1. $p_H \ll p_{NH}$: the link stays more frequently in "Not-Home" states than in "Home" states;
2. $p_h \gg p_l$: establishing a contact between the nodes is much more likely when the link is in one of the "Home" states than when it is in the "Not-Home" states;
3. $p_H + p_{NH} \ll 1$: transitions between "Home" and "Not-Home" states are rare.

By performing some basic algebraic manipulation, it is easy to see that the technical assumption in the theorem holds whenever 1, 2, and 3 are satisfied, plus the additional condition that $p_H \geq p_{NH}/n$. A special case of interest is the one mentioned in the following corollary:

Corollary 22.1 *Under the same assumptions for Theorem 22.7, let p_H, p_{NH}, p_h, p_l be defined as follows:*

$$p_H = \frac{1}{n^{1+\epsilon}}, \quad p_{NH} = \frac{1}{n}, \quad p_h = \frac{1}{n^{1-\epsilon}}, \quad p_l = \frac{1}{n^2}.$$

Then, the broadcasting time in the dynamic graph is $O(\log n)$ w.h.p.

It is important to notice that the settings of p_H, p_{NH}, p_h, p_l as in the statement of the corollary are in accordance with the guidelines for Home-MEG parameter setting derived from comparison to real-world data traces. Also, the expected node degree in G when edge sets in E_t are generated according to the Home-MEG model is

$$deg(G) = (n-1) \cdot \frac{p_H \cdot p_h + p_{NH} \cdot p_l}{p_H + p_{NH}},$$

implying that under the settings of Corollary 22.1 the expected node degree in G is $\Theta(1)$.

Notice also that the setting of Home-MEG parameters assumed in the statement of Corollary 22.1 requires that the parameters of the Markov chain depend on the number of nodes n in the network. This requirement appears reasonable, since a Markov chain in Home-MEG models the existence/non-existence of the link between a *specific* pair of nodes in the network. For instance, this requirement is in accordance with a scenario in which the number of nodes (each with a fixed transmission range) in the network grows, while the density of nodes per unit area remains unchanged (constant density networks).

22.2.4 Discussion

In this section, we have presented the general framework of dynamic graphs, under which asymptotic bounds on broadcasting time can be derived starting from different assumptions concerning the way in which edge sets in the dynamic graph evolve with time. More specifically, we have presented two fundamentally different classes of results: in the first class, edge sets in the dynamic graph are generated as the result of the combination of an underlying geometric mobility model with a notion of transmission range; in the second class, edge sets are generated according to $n(n-1)/2$ independent Markov chains representing the behavior of each single possible link in the network.

It is important to observe that both classes of dynamic graphs display realistic as well as unrealistic features of the induced contact pattern between nodes. Geometric mobility-based dynamic graphs are "realistic" in the sense that the appearance/disappearance of an edge in the dynamic graph is coherent with notions borrowed from the physical world, such as node trajectories, distance between nodes, etc. In other words, if edge (u, v) exists in E_t, this implies that the two nodes are located within distance r of each other in the underlying metric space. Hence, the fact that an edge exists/does not exist in E_t provides some strong information about node positions in the underlying metric space and, in turn, this information strongly impacts the possible existence/non-existence of other edges in E_t, E_{t+1} and so on. Just to give an example, suppose the underlying mobility metric space is a square of side L, and let $c(L, r)$ be the maximum number of non-overlapping circles of radius $r/2$ that can be packed in the square; it is then clear that if the number n of nodes in the network exceeds $c(L, r)$, then edge set E_t will contain at least one edge with probability 1, for any value of t.

The fact that the appearance/disappearance of edges in the dynamic graph is governed by geometric rules is a realistic aspect of the model, since in the real world people do move in a geometric space, and the appearance/disappearance of communication links between them is determined by proximity rules. On the other hand, the specific geometric mobility models used to derive asymptotic bounds on broadcasting time presented in Section 22.2.2 are known to generate inter-contact time distributions different from the one observed in real-world data traces.

Home-MEG-based dynamic graphs, on the other hand, have been shown to be able to faithfully reproduce the shape of the *aggregate* inter-contact time distributions observed in real-world data traces. However, in this class of dynamic graphs there is no geometric notion governing the appearance/disappearance of edges in the dynamic graph. In other words, at any time t, independently of whether edge (u, v) existed at time $t-1$, and independently of whether any other edge (u, x), for any $x \in V$, existed at time $t-1$, there is a non-zero probability of edge (u, v) existing – this is true under the assumption that the setting of parameters in the Home-MEG model is non-degenerate and, in particular, that $p_l > 0$. Clearly this is an unrealistic aspect of the model: in the physical world, if we have nodes u and v located at a distance much greater than r at time $t-1$ (say, a few kilometers), with a transmission range r in the order of tens of meters, the node speed is suitably upper bounded (say, pedestrian velocity), and the time step is reasonably short (say, up to one minute), the probability of having edge (u, v) at time t is 0. Thus, the above property of Home-MEG-based dynamic graphs of always having a non-null probability of edge occurrence between any two nodes in the network is an unrealistic aspect of the model.

It is important to observe, though, that different assumptions on how edges appear/disappear in the dynamic graph, and, consequently, on the observed pairwise inter-contact time distribution, have a profound impact on the asymptotic performance bounds that can be derived in the two classes of dynamic graphs: by comparing the upper bounds derived in Theorem 22.5 and Corollary 22.1, which were obtained under similar assumptions concerning the expected node degree in the dynamic graph, one can conclude that broadcasting is completed in $\tilde{O}(\sqrt{n})$ time with geometric-based mobility, while it requires only $O(\log n)$ time in Home-MEG-based dynamic graphs. As can be seen from Figure 22.2, these two trends of broadcasting time are radically different as the network size increases. Which one of the two is more adherent to the trend that would be observed in an opportunistic network composed of moving individuals is, based on the preceding discussion, still not clear. More work and analysis are needed in order to provide a definite answer to this question.

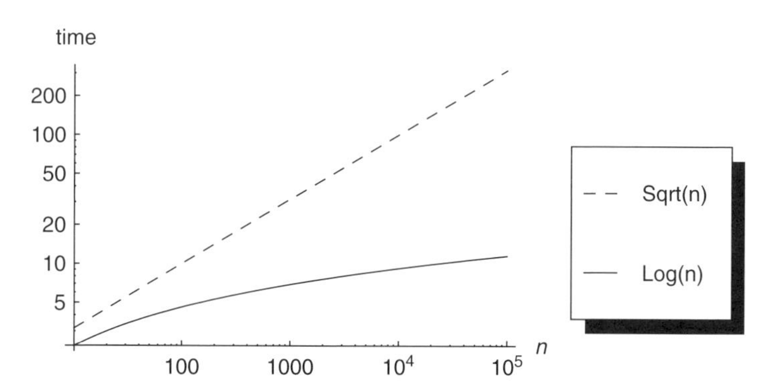

Figure 22.2 Trends of different upper bounds on broadcasting time with geometric-based mobility (Sqrt(n)) and Home-MEG-based dynamic graphs (Log(n)). Both the x- and y-axes are in logarithmic scale.

References

Becchetti L, Clementi A, Pasquale F, Resta G, Santi P and Silvestri R 2011 Information spreading in opportunistic networks is fast. *Internet draft: available at http://arxiv.org/abs/1107.5241*.

Chaintreau A, Hui P, Crowcroft J, Diot C, Gass R and Scott J 2006 Impact of human mobility on the design of opportunistic forwarding algorithms. *Proceedings of IEEE Infocom*, pp. 1–13.

Clementi A, Silvestri R and Trevisan L 2011 Information spreading in dynamic graphs. *Internet draft: available at http://arxiv.org/abs/1111.0583*.

Groenevelt R, Nain P and Koole G 2005 The message delay in mobile ad hoc networks. *Performance Evaluation* **62**, 210–228.

Grossglauser M and Tse D 2002 Mobility increases the capacity of ad hoc wireless networks. *ACM/IEEE Transactions on Networking* **10**(4), 477–486.

Hong S, Rhee I, Kim S, Lee K and Chong S 2008 Routing performance analysis of human-driven delay tolerant networks using the truncated levy walk model. *Proceedings of the ACM Workshop on Mobility Models*, pp. 25–32.

Karagiannis T, Le Boudec JY and Vojnovic M 2007 Power law and exponential decay of inter contact times between mobile devices. *Proceedings of ACM Mobicom*, pp. 183–194.

Kuhn F and Oshman R 2011 Dynamic networks: Models and algorithms. *ACM SIGACT News* **42**, 82–96.

Peres Y, Sinclair A, Sousi P and Stauffer A 2011 Mobile geometric graphs: Detection, coverage and percolation. *Proceedings ACM/SIAM Symposium On Discrete Algorithms (SODA)*, pp. 412–428.

Pettarin A, Pietracaprina A, Pucci G and Upfal E 2011 Tight bounds on information dissemination in sparse mobile networks. *Proceedings of the ACM Conference on Principles of Distributed Computing (PODC)*, pp. 355–362.

Resta G and Santi P 2012 A framework for routing performance analysis in delay tolerant networks with application to noncooperative networks. *IEEE Transactions on Parallel and Distributed Systems* **23**(1), 2–10.

Spyropoulos T, Psounis K and Raghavendra C 2006 Performance analysis of mobility-assisted routing. *Proceedings of ACM MobiHoc*, pp. 49–60.

Spyropoulos T, Psounis K and Raghavendra C 2008 Efficient routing in intermittently connected mobile networks: The multi-copy case. *ACM/IEEE Transactions on Networking* **16**, 77–90.

Vahdat A and Becker D 2000 Epidemic routing for partially connected ad hoc networks. *Technical Report CS-200006, Duke University*.

Zhang X, Neglia G, Kurose J and Towsley D 2007 Performance modeling of epidemic routing. *Computer Networks* **51**, 2867–2891.

Appendix A

Elements of Probability Theory

In this appendix, we report basic notions of probability theory and stochastic process theory. The material of this appendix is based on Feller (1957), on Appendix B of Santi (2005), and on online resources such as Wikipedia and Wolfram MathWorld.

A.1 Basic Notions of Probability Theory

Definition A.1 (Sample space) *A* sample space Ω *is a set representing all possible outcomes of a certain random experiment. A sample space is* discrete *if it is composed of a finite number of elements (e.g., outcomes of a coin toss experiment), or of infinitely many elements that can be arranged into a simple sequence* $e_1, e_2, \ldots$.

Definition A.2 (Random variable) *A* random variable X *is a function defined on a sample space. If the sample space on which* X *is defined is discrete,* X *is said to be a discrete random variable.*

Examples of random variables are the number of heads in a sequence of k coin tosses (discrete random variable), the position of a certain particle in a physical system, the position of a node moving according to a certain mobility model at a certain instant of time, and so on.

Definition A.3 (Probability distribution) *Let* X *be a discrete random variable, and let* $\Omega = \{x_1, x_2, \ldots, x_j, \ldots\}$ *be the set of possible values of* X. *The function*

$$P(X = x_i) = f(x_i) \quad (i = 1, 2, \ldots)$$

Mobility Models for Next Generation Wireless Networks: Ad Hoc, Vehicular and Mesh Networks, First Edition. Paolo Santi.

is called the probability distribution *(also called* probability mass function*) of the random variable X, where, for all i,* $f(x_i) \geq 0$ *and* $\sum_i f(x_i) = 1$.

Definition A.4 (Event and probability) *Let* $\Omega = \{x_1, x_2, \ldots, x_j, \ldots\}$ *be a discrete sample space, and let* $f()$ *be a probability distribution on* Ω*. An* event *is defined as any subset E of the sample space* Ω*. The* probability *of event E is defined as*

$$P(E) = \sum_{x_i \in E} f(x_i).$$

Definition A.5 (Probability density function) *A probability density function (pdf) on* $\mathbb{R}$ *is a function such that*

$$\forall x \in \mathbb{R},\ f(x) \geq 0 \quad \text{and} \quad \int_{-\infty}^{+\infty} f(x)\,dx = 1.$$

Similarly, with a fixed integer $d > 1$*, a pdf on* $\mathbb{R}^d$ *is a function such that*

$$\forall (x_1, \ldots, x_d) \in \mathbb{R}^d,\ f(x_1, \ldots, x_d) \geq 0 \quad \text{and}$$

$$\int_{-\infty}^{+\infty} f(x_1, \ldots, x_d)\,dx_1 \ldots dx_d = 1.$$

Definition A.6 (Continuous random variable) *A random variable X taking values in* $\mathbb{R}$ *is* continuous *if and only if there exists a pdf f on* $\mathbb{R}$ *such that*

$$P(a < X \leq b) = \int_a^b f(x)\,dx,$$

for any $a < b$*. Function f is called the* density *of the random variable X. A similar definition applies to random variables taking values in* $\mathbb{R}^d$*, for some integer* $d > 1$.

Definition A.7 (Distribution function) *Let* $X = (X_1, \ldots, X_d)$ *be a continuous random variable taking values in* $\mathbb{R}^d$*, for some integer* $d \geq 1$*. The function*

$$\begin{aligned} F(x_1, \ldots, x_d) &= P(X_1 \leq x_1, \ldots, X_d \leq x_d) \\ &= \int_{-\infty}^{x_1} \cdots \int_{-\infty}^{x_d} f(y_1, \ldots, y_d)\,dy_1 \ldots dy_d, \end{aligned}$$

where $f(y_1, \ldots, y_d)$ *is the density of X, is called the* (cumulative) distribution function *of the random variable X. Function f is called the density of the random variable X.*

Definition A.8 (Complementary cumulative distribution function) *Let $X = (X_1, \ldots, X_d)$ be a continuous random variable taking values in $\mathbb{R}^d$, for some integer $d \geq 1$. The function*

$$F_c(x_1, \ldots, x_d) = P(X_1 \geq x_1, \ldots, X_d \geq x_d) = \int_{x_1}^{+\infty} \cdots \int_{x_d}^{+\infty} f(y_1, \ldots, y_d)\, dy_1 \ldots dy_d,$$

where $f(y_1, \ldots, y_d)$ is the density of X, is called the complementary cumulative distribution function *of the random variable X.*

Definition A.9 (Support of a pdf) *The support of a pdf f on $\mathbb{R}^d$, for some integer $d \geq 1$, is the set of points in $\mathbb{R}^d$ on which f has positive value. Formally,*

$$supp(f) = \{(x_1, \ldots, x_d) \in \mathbb{R}^d : f(x_1, \ldots, x_d) > 0\}.$$

Clearly,

$$\int_{supp(f)} f(x_1, \ldots, x_d)\, dx_1 \ldots dx_d = 1.$$

Definition A.10 (Asymptotic distribution) *A sequence $X_1, X_2, \ldots, X_n, \ldots$ of continuous random variables, with distribution functions $F_1, F_2, \ldots, F_n, \ldots$, is said to* converge in distribution *to a certain random variable X with distribution F if and only if*

$$\lim_{n\to\infty} F_n(x) = F(x)$$

at every continuity point x of $F(x)$. If sequence $\{X_n\}$ converges in distribution to a certain random variable X with distribution F, we say that F is the asymptotic distribution *of $\{X_n\}$.*

Definition A.11 (The a.a.s. and w.h.p. event) *Let E_n be a random variable representing a random event which depends on a certain parameter n. We say that the event represented by E_n holds* asymptotically almost surely (a.a.s.) *if*

$$\lim_{n\to\infty} P(E_n) = 1.$$

We say that the event holds with high probability (w.h.p.) *if*

$$P(E_n) \geq 1 - \frac{1}{n}.$$

Note that the fact that an event holds w.h.p. implies that the same event holds a.a.s.

Definition A.12 (Stochastic process) *Given a sample space Ω, a* stochastic process *is a collection of random variables taking value in Ω indexed by a set T of time instants. Formally, a stochastic process S is a collection $\{S_t : t \in T\}$ where random variables S_t take values in Ω.*

Examples of stochastic processes are the sequence of trip lengths of a mobile node, the sequence of statuses of a communication link (active or inactive) between two nodes in a network, and so on. A special class of stochastic processes are *Markov chains*, which will be formally defined in Section A.3.

Definition A.13 (Stationary process) *Let $S = \{S_t : t \in T\}$ be a stochastic process, and let $F(s_{t_1+\tau}, \ldots, s_{t_k+\tau})$ be the cumulative distribution function of the joint distribution of variables $\{S_t\}$ at time $t_1 + \tau, \ldots, t_k + \tau$. Process S is said to be* stationary *if and only if, for all $k \geq 1$, for all $\tau > 0$, and for all $t_1, \ldots, t_k$,*

$$F(s_{t_1+\tau}, \ldots, s_{t_k+\tau}) = F(s_{t_1}, \ldots, s_{t_k}).$$

Intuitively speaking, a stochastic process is stationary if the joint probability distribution of the observed random variables, and hence statistical properties such as mean and variance, do not change with time. A related notion for stochastic processes is that of *ergodicity*, which refers to the fact that the statistical properties of a (stationary) stochastic process can be deduced from sampling a large group of identical, independent instances of the observed process at a *single* time instant. Below we present the formal definition of *mean ergodicity*, a property used, for example, in the characterization of the node spatial distribution of mobility models. The more general notion of ergodicity is cumbersome to define, and it is therefore not reported here.

Definition A.14 (Mean ergodic process) *Let $S = \{S_t : t \in T\}$ be a stochastic process, and let $E[S_t]$ denote the expected value of S_t computed in two possible ways:*

$$E[S_t] = \lim_{h\to\infty} \frac{\sum_{t=1}^{h} S_t}{h}$$

and

$$E_e[S_t] = \lim_{k\to\infty} \frac{\sum_{i=1}^{k} S_t^i}{k},$$

where, for any fixed *value of t, S_t^i denotes the ith independent instant of process S at time t. If process S satisfies the* mean ergodic property, *then $E[S_t] = E_e[S_t]$.*

A.2 Probability Distributions

Definition A.15 (Discrete uniform distribution) *Let X be a random variable taking values in $\Omega = \{x_1, \ldots, x_k\}$, for some integer $k > 0$. Random variable X is said to have* uniform distribution *if and only if its probability mass function satisfies $P(X = x_i) = 1/k$, for each $i = 1, \ldots, k$. A discrete random variable with uniform distribution is called a* (discrete) uniform random variable.

Definition A.16 (Continuous uniform distribution) *Given an interval $[a, b]$ on $\mathbb{R}$, with $a < b$, the* uniform distribution *on $[a, b]$ is defined by the following probability density function:*

$$f(x) = \begin{cases} 0 & \text{for } x < a \\ 1/(b-a) & \text{for } a \leq x \leq b \\ 0 & \text{for } x > b \end{cases} .$$

The uniform distribution on arbitrary d-dimensional rectangles is defined similarly. A continuous random variable with uniform distribution on a certain (d-dimensional) interval is called a (continuous) uniform random variable.

Definition A.17 (Poisson process and distribution) *Let us consider a discrete random variable $X(t)$, counting the number of events (e.g., arrival of telephone calls) occurring in the time interval $[0, t]$. If the following properties hold:*

(a) the probability of occurrence of the observed events does not change with time, and
(b) the probability of occurrence of the observed events does not depend on the number of events occurred so far,

then the corresponding random process is called a Poisson process. In a Poisson process, the number of events counted after time t follows the probability function

$$P(X(t) = x) = e^{-\lambda t} \frac{(\lambda t)^x}{x!} \quad \text{for } x = 0, 1, 2, \ldots,$$

for some constant $\lambda > 0$. Parameter λ is called the intensity *of the Poisson process. The above probability function is called a Poisson distribution of parameter λ. A random variable with a Poisson distribution is called a Poisson random variable.*

Definition A.18 (Normal distribution) *The* normal (or Gaussian) distribution *on $\mathbb{R}$ of parameters μ (mean) and σ (standard deviation) is defined by the following probability density function $\mathcal{N}(\mu, \sigma)$:*

$$\mathcal{N}(\mu, \sigma)(x) = \frac{1}{\sigma\sqrt{2\pi}} e^{-(x-\mu)^2/(2\sigma^2)}.$$

The normal distribution on $\mathbb{R}^d$, for some integer $d > 1$, is defined similarly. A random variable with a normal distribution is called a normal random variable.

Definition A.19 (Log-normal distribution) *The* log-normal distribution *on $\mathbb{R}$ of parameters μ (mean) and σ (standard deviation) is defined by the following probability density function:*

$$f(x) = \frac{1}{\sigma\sqrt{2\pi x}} e^{-(\ln x-\mu)^2/(2\sigma^2)}.$$

The log-normal distribution on $\mathbb{R}^d$, for some integer $d > 1$, is defined similarly. A random variable with a log-normal distribution is called a log-normal random variable.

The shape of the log-normal distribution with $\mu = 1$ and different values of the standard deviation σ is shown in Figure A.1.

Definition A.20 (Exponential distribution) *The* exponential distribution *on $\mathbb{R}$ of parameter λ (called* rate parameter*) is defined by the following probability density function:*

$$f(x) = \begin{cases} \lambda e^{-\lambda x} & \textit{if } x \geq 0 \\ 0 & \textit{otherwise} \end{cases}.$$

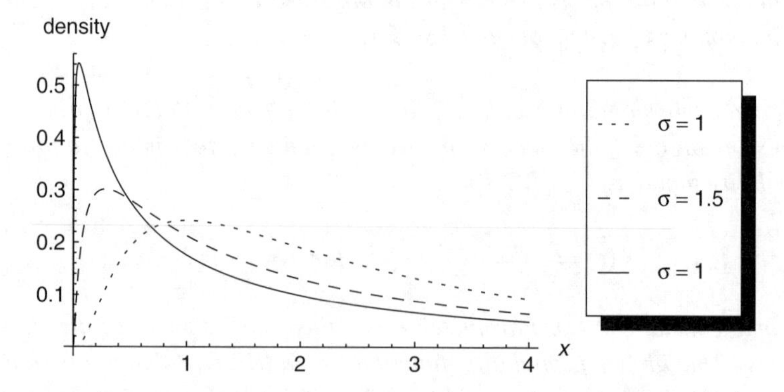

Figure A.1 Shape of the log-normal distribution with $\mu = 1$ and different values of the standard deviation σ.

A random variable with an exponential distribution is called an exponential random variable.

The exponential distribution and Poisson process are related as follows: the time between two consecutive events in a Poisson process of intensity λ is an exponentially distributed random variable of rate λ.

Definition A.21 (Power-law distribution) *In the most general sense, a* power-law distribution *on* $\mathbb{R}$ *is characterized by a probability density function (or probability mass function in the discrete case) of the following form:*

$$f(x) \propto L(x)x^{-\alpha},$$

where $\alpha > 1$ *is the slope parameter and* $L(x)$ *is a slowly varying function, that is, a function such that* $\lim_{x\to\infty} L(tx)/L(x) = 1$*, for any constant* t*.*

The most interesting property of a power-law distribution is *scale invariance*, which states that scaling the argument of the function results in a proportional scaling of the density function itself. In formulas, if f is a power law of slope parameter α, then

$$f(cx) = c^{-\alpha} f(x).$$

Examples of probability distributions belonging to the class of power laws are presented in the following.

Definition A.22 (Power-law distribution with exponential cutoff) *A* power-law distribution with exponential cutoff *on* $\mathbb{R}$ *is characterized by a probability density function (or probability mass function in the discrete case) of the following form:*

$$f(x) \propto L(x)x^{-\alpha}e^{-\lambda x},$$

where $\alpha > 1$ *is the slope parameter,* $L(x)$ *is as defined above, and* $\lambda > 0$ *is the decay parameter of the exponential cutoff.*

Note that in the power-law distribution with exponential cutoff the exponential decay term $e^{-\lambda x}$ overwhelms the power-law behavior for large values of x, implying that the scale invariance property no longer holds.

Definition A.23 (Zipf's law) *Let* X *be a discrete random variable defined on sample space* $\Omega = \{1, \ldots, N\}$*, and assume without loss of generality that elements in* Ω *are ordered from the most probable to the less probable outcomes of* X*; that is,* $Prob(X = 1) \geq Prob(X = 2) \geq \ldots$*. We say that variable*

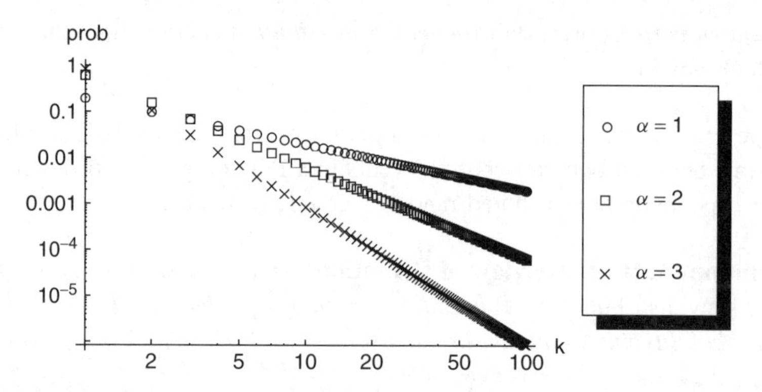

Figure A.2 Shape of the Zipf's law distribution with $N = 100$ and different values of the slope parameter α.

X obeys Zipf's law *of slope parameter* α *if and only if*

$$P(X = k) = f(k) = \frac{1}{k^{\alpha} H_{N,\alpha}},$$

where $H_{N,\alpha} = \sum_{i=1}^{N} 1/i^{\alpha}$ *is the* N*th generalized harmonic number.*

Zipf's law is typically used to model the uneven popularity of interests in a population, such as the popularity of multimedia files in a peer-to-peer file sharing application. Examples of Zipf's law with different slope parameters are given in Figure A.2. Notice that both axes are in logarithmic scale. In fact, log–log plots are commonly used to display power laws, since a power-law function corresponds to a linear function in log–log scale.

Definition A.24 (Pareto distribution) *Let X be a random variable taking values in* $[x_m, +\infty)$, *for some* $x_m > 0$. *The random variable X is said to obey a* Pareto distribution *if and only if its probability density function is defined as follows:*

$$f(x) = \begin{cases} \alpha x_m^{\alpha} / x^{\alpha+1} & \text{if } x \geq x_m \\ 0 & \text{otherwise} \end{cases},$$

where $\alpha > 0$ *is the slope parameter. A random variable with a Pareto distribution is called a* Pareto random variable.

The Pareto distribution can be considered as the continuous counterpart of the Zipf's law distribution. Examples of a Pareto distribution with $x_m = 1$ and different slope parameters are given in Figure A.3.

Definition A.25 (Truncated Pareto distribution) *Let X be a random variable taking values in* $[x_m, x_M]$, *for some* $0 < x_m < x_M$. *The random variable X is said to obey a* truncated Pareto distribution *if and only if its probability*

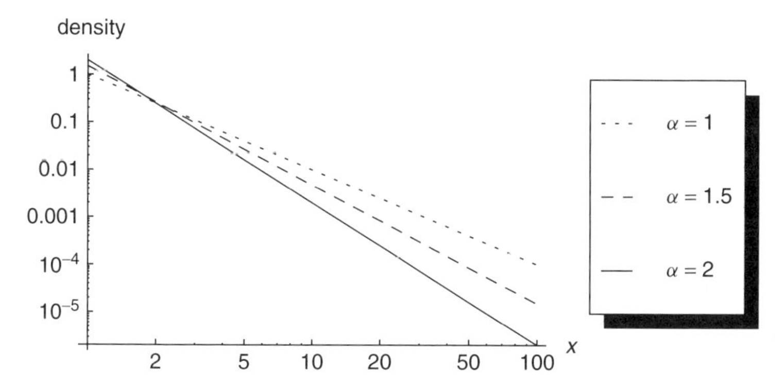

Figure A.3 Shape of the Pareto distribution with $x_m = 1$ and different values of the slope parameter α.

density function is defined as follows:

$$f(x) = \begin{cases} \frac{1}{1-(x_m/x_M)^\alpha} \cdot \frac{\alpha x_m^\alpha}{x^{\alpha+1}} & \text{if } x_m \le x \le x_M \\ 0 & \text{otherwise} \end{cases},$$

where $\alpha > 0$ is the slope parameter.

A.3 Markov Chains

Definition A.26 (Markov chain) *A* Markov chain *is a stochastic process where the random variables $X_1, X_2, \ldots$ in the sequence represent a discrete set $\mathcal{S} = \{S_1, \ldots, S_n\}$ of possible* statuses, *and where the probabilities governing transitions between states satisfy the* Markov property *(also known as* memoryless property*). The set $\mathcal{S}$ of possible states is called the* state space, *and it is the sample space for each of the X_i. The Markov property states that the probability of making a transition into any state S_i in the state space at time t depends only on the status of the chain at time $(t-1)$. Formally,*

$$P(X_t = x | X_1 = x_1, \ldots, X_{t-1} = x_{t-1}) = P(X_t = x | X_{t-1} = x_{t-1}),$$

for any time $t > 0$, where $x, x_i \in \mathcal{S}$.

Note that the state space in the Markov chain can be formed by a finite number or an infinite but countable number of elements.

Definition A.27 (Time-homogeneous Markov chain) *A* time-homogeneous Markov chain *is a Markov chain where transition probabilities do not change with time. Formally,*

$$P(X_{t+1} = x | X_t = y) = P(X_t = x | X_{t-1} = y),$$

for any time $t > 0$, where $x, y \in \mathcal{S}$.

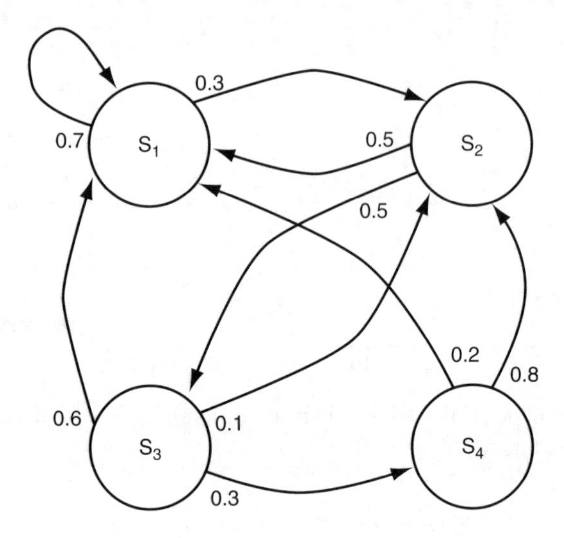

Figure A.4 Pictorial representation of a four-state Markov chain.

Time-homogeneous Markov chains with finite state spaces can be pictorially described by a directed graph $G = (\mathcal{S}, E)$, where directed edge $(x_i, x_j) \in E$ is labeled with transition probability $p_{ij} = P(X_1 = x_j | X_0 = x_i) > 0$ (edges are omitted for transitions occurring with zero probability). The pictorial representation of a four-state Markov chain is shown in Figure A.4. Note that the sum of the weights on the outgoing edges of each node is 1. In formulas,

$$\sum_{x_j \in \mathcal{S}} p_{ij} = 1,$$

for any $x_i \in \mathcal{S}$. The transition probabilities can be summarized in the *transition matrix* $\Pi = (p_{ij})$. Since the transition probabilities are time-invariant, the k-step transition probabilities can be computed as Π^k, that is, the kth power of the transition matrix Π.

Definition A.28 (Accessible state and communicating class) *A state $S_i \in \mathcal{S}$ is said to be* accessible *from state $S_j \in \mathcal{S}$ if a chain started in state S_j has non-zero probability of making a transition into state S_i at some point in time. Formally, a state S_i is accessible from S_j if and only if*

$$P(X_t = i | X_0 = j) > 0,$$

for some $t > 0$. A state S_i is said to communicate *with state S_j if and only if S_i is accessible from S_j and vice versa. A set of states $\mathcal{C} \subseteq \mathcal{S}$ is a* communicating class *if, for every $S_i, S_j \in \mathcal{C}$, S_i and S_j communicate, and no state in $\mathcal{C}$ communicates with any state in $\mathcal{S} - \mathcal{C}$.*

Definition A.29 (Irreducible Markov chain) *A Markov chain is* irreducible *if its state space is a single communicating class.*

Informally speaking, a Markov chain is irreducible if it is possible to make a transition into any state, starting from any state. The Markov chain represented in Figure A.4 is irreducible. In fact, starting from S_1 it is possible to make a transition into S_1 (in one step), into S_2 (in one step), into S_3 (in two steps, through state S_2), and into S_4 (in three steps, through states S_2 and S_3). The same transition applies starting from any other state.

Definition A.30 (Periodicity) *A state $S_i \in \mathcal{S}$ has period k if, starting from S_i, any return to state S_i must occur in multiples of k time steps. If $k = 1$, the state is said to be* aperiodic. *Otherwise, it is said to be* periodic of period k. *A Markov chain is said to be* aperiodic *if all its states are aperiodic.*

For instance, state S_1 in the Markov chain of Figure A.4 is aperiodic, since there is a non-null probability of making a transition into state S_1 starting from S_1. State S_2 is periodic of period 2, since once in state S_2 the shortest sequence of transitions leading back to state S_2 has length 2 (going through either state S_1 or S_3). Similarly for state S_3. State S_4 is periodic of period 3, since once in state S_4 the shortest sequence of transitions leading back to state S_4 has length 3 (going through states S_2 and S_3).

Definition A.31 (Absorbing Markov chain) *A state $S_i \in \mathcal{S}$ is said to be* absorbing *if it is impossible to leave this state. Formally, state S_i is absorbing if $p_{ii} = 1$ (which implies that $p_{ij} = 0$ for any $S_j \neq S_i$). If an absorbing state can be reached with non-zero probability starting from any state in $\mathcal{S}$, then the Markov chain is said to be an* absorbing Markov chain.

Definition A.32 (Recurrent Markov chain) *A state $S_i \in \mathcal{S}$ is said to be* transient *if, starting from S_i, there is a non-zero probability of never reaching S_i again. Formally, denoting by $p_{ii}^t = P(R_i = t)$ the probability that, starting from S_i at time 0, the first return to state S_i is at time t, we have that a state is transient if and only if*

$$\sum_{t=1}^{+\infty} p_{ii}^t < 1.$$

A state is said to be recurrent *if it is not transient. A Markov chain is said to be* recurrent *if all its states are recurrent.*

Definition A.33 (Positive recurrent Markov chain) *The* mean recurrence time *is defined as the expected value of the random variable R_i, that is,*

$$M_i = E[R_i] = \sum_{t=1}^{+\infty} t \cdot p_{ii}^t.$$

A state S_i is said to be positive recurrent *if M_i is finite. A Markov chain is said to be* positive recurrent *if all its states are positive recurrent.*

Definition A.34 (Stationary distribution) *Given a time-homogeneous Markov chain, the* stationary distribution *of the Markov chain is a vector $\pi = (\pi_i)$ such that:*

1. $\forall i, \pi_i \geq 0$.
2. $\sum_i \pi_i = 1$.
3. $\pi_i = \sum_{S_j \in \mathcal{S}} \pi_i p_{ij}$.

An irreducible Markov chain has a stationary distribution if and only if it is positive recurrent. In that case, the stationary distribution π is uniquely defined, and is related to the mean recurrence time as follows:

$$\pi_i = \frac{1}{M_i}.$$

If the state space is finite, the stationary distribution π satisfies the equation

$$\pi = \pi \Pi,$$

that is, it is the (normalized) left eigenvector of the transition matrix Π associated with the eigenvalue 1.

Definition A.35 (Continuous-time Markov chain) *A* continuous-time Markov chain *is a stochastic process (X_t) defined for any continuous time value $t \geq 0$. The random variables $\{X_t\}$ in the sequence represent a discrete set $\mathcal{S} = \{S_1, \ldots, S_n\}$ of possible* statuses. *Random variable X_t, taking values in* state space *$\mathcal{S}$, represents that status of the chain at time t. The probabilities governing transitions between states satisfy the (continuous)* Markov property *(also known as* memoryless property*). The (continuous) Markov property states that the probability of finding the chain in any state S_j in the state space at time t depends only on the status of the chain at the most recent time prior to t. Formally,*

$$P(X_t = x_j | X_s = x_i, X_{t_{n-1}} = x_{i_{n-1}}, \ldots, X_{t_1} = x_{i_1}) = P(X_t = x_j | X_s = x_i),$$

where $0 \leq t_1 \leq t_2 \leq t_{n-1} \leq s \leq t$ is any non-decreasing sequence of $n+1$ terms and $x_j, x_i, x_{i_h} \in \mathcal{S}$, for any integer $n \geq 1$.

Continuous-time Markov chains are the time-continuous version of Markov chains, where transitions between states, instead of occurring at regular time steps, are themselves a random process, with *exponentially distributed transition times*.

In a continuous-time Markov chain, transitions between states are governed by *transition rates*, which measure, given the state of the chain at a certain time t, how quickly transition to a different state is likely to occur. Formally, given that the chain was in state S_j at time t, we have

$$P(X_{t+h} = S_j | X_t = S_i) = q_{ij}h + o(h),$$

where q_{ij} is the *transition rate* between states S_i and S_j, and h is a small enough time interval (for the use of order notation in the above formula, see Appendix B). In a *time-homogeneous* continuous-time Markov chain, the transition rates q_{ij} do not change with time.

From the above formulas it follows that, over a sufficiently small time interval, the probability of observing any particular transition in the chain is proportional to the length of the time interval (up to lower order terms). Similar to the discrete case, if the state space is finite the transition rates can be summarized in a square matrix called the *transition rate matrix* Q, containing in position (i, j) the transition rate between state S_i and state S_j.

Definition A.36 (Semi-Markov process) *A continuous-time stochastic process is called a* semi-Markov process *if the process reporting which values the process takes–the X_t random variables as defined above–is a Markov chain, and the holding times $DT_i = T_i - T_{i-1}$ denoting the times between transitions are distributed according to an arbitrary probability distribution, which may depend on the two states between which the move is made.*

The difference between a semi-Markov process and a continuous-time Markov chain is that, while in the latter holding times between transitions are exponentially distributed, in the former holding times obey a general probability distribution.

References

Feller W 1957 *An Introduction to Probability Theory and its Applications*. John Wiley & Sons, Inc., New York.

Santi P 2005 *Topology Control in Wireless Ad Hoc and Sensor Networks*. John Wiley & Sons, Ltd, Chichester.

Appendix B

Elements of Graph Theory, Asymptotic Notation, and Miscellaneous Notions

In this appendix, we start by introducing the asymptotic notation, and then report basic definitions and concepts from graph theory that have been used in this book. We will also formally define miscellaneous notions used in the book, such as linearity of operators, convexity and concavity, and so on. Most of the material presented in this appendix are based on Bollobàs (1985) and Bollobàs (1988), on Appendix B of Santi (2005), and on online resources such as Wikipedia and Wolfram MathWorld.

B.1 Asymptotic Notation

Definition B.1 (Big-O notation) *Let $f(n)$ and $g(n)$ be two functions defined on $\mathbb{R}$. One can write $f(n) = O(g(n))$ if and only if there exist a positive real number c and a real number n_0 such that*

$$f(n) \leq c \cdot g(n) \quad \textit{for all } n \geq n_0.$$

The big-O notation is used to express the fact that, as the argument n of the function grows larger, function $f(n)$ is *upper bounded* by function $g(n)$, up to a constant factor.

Definition B.2 (Big-Omega notation) *Let $f(n)$ and $g(n)$ be two functions defined on $\mathbb{R}$. One can write $f(n) = \Omega(g(n))$ if and only if there exist a positive real number c and a real number n_0 such that*

$$f(n) \geq c \cdot g(n) \quad \textit{for all } n \geq n_0.$$

Mobility Models for Next Generation Wireless Networks: Ad Hoc, Vehicular and Mesh Networks, First Edition. Paolo Santi.

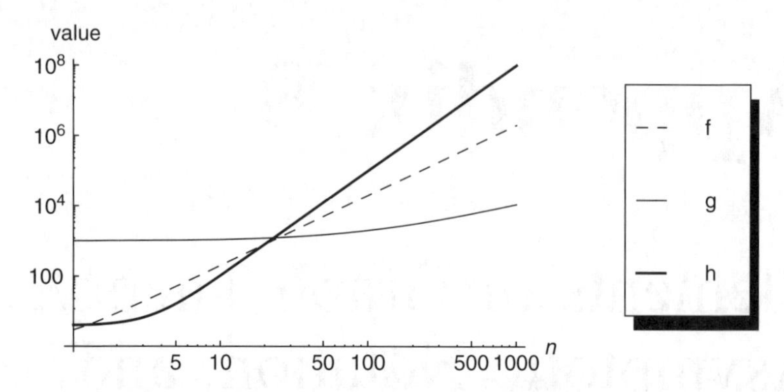

Figure B.1 Pictorial representation of the intuition behind the big-O and big-Omega notation. Notice the logarithmic scale on both axes.

The big-Omega notation is used to express the fact that, as the argument n of the function grows larger, function $f(n)$ is *lower bounded* by function $g(n)$, up to a constant factor.

The intuition behind the big-O and big-Omega notation is pictorially represented in Figure B.1. The figure shows the behavior for increasing values of n of the three functions

$$f(n) = 2n^2 + 1, \quad g(n) = 10n + 1000, \quad h(n) = 0.1n^3 + 4.$$

For small values of n, function $g(n)$ is larger than both $f(n)$ and $h(n)$, due to the large constants in the polynomial. However, as n grows larger, the dominating terms in the three functions become n^2 for $f(n)$, n for $g(n)$, and n^3 for $h(n)$. This explains why, for large enough values of n (say, $n > 30$), we have $h(n) > f(n) > g(n)$. Using asymptotic notation, we have

1. $f(n) = \Omega(g(n))$ and $f(n) = O(h(n))$;
2. $g(n) = O(f(n))$ and $g(n) = O(h(n))$;
3. $h(n) = \Omega(f(n))$ and $h(n) = \Omega(g(n))$.

Definition B.3 (Big-Theta notation) *Let $f(n)$ and $g(n)$ be two functions defined on $\mathbb{R}$. One can write $f(n) = \Theta(g(n))$ if and only if there exist positive real numbers c_1, c_2 and real number n_0 such that*

$$c_1 \cdot g(n) \leq f(n) \leq c_2 \cdot g(n) \quad \text{for all } n \geq n_0.$$

Alternatively, one can define that $f(n) = \Theta(g(n))$ if and only if $f(n) = O(g(n))$ and $f(n) = \Omega(g(n))$.

The big-Theta notation is used to express the fact that, as the argument n of the function grows larger, function $f(n)$ is both *upper* and lower bounded by function $g(n)$, up to a constant factor.

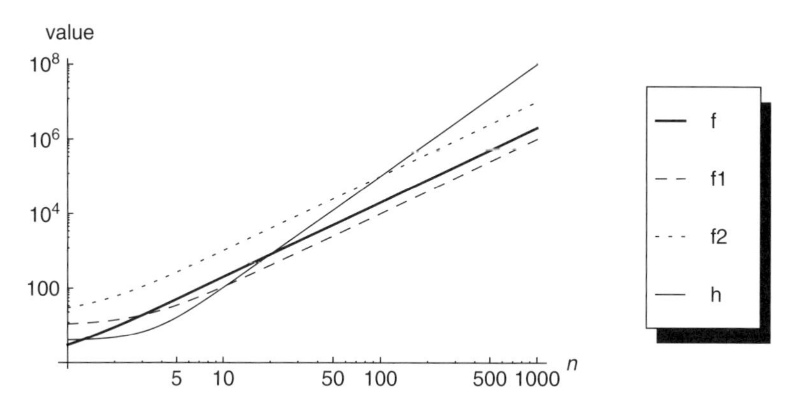

Figure B.2 Pictorial representation of the intuition behind the big-Theta notation. Notice the logarithmic scale on both axes.

The intuition behind the notion of big-Theta notation is pictorially represented in Figure B.2. The figure shows the behavior for increasing values of n of function $f(n)$ as defined above, and of the two functions

$$f1(n) = n^2 + 10, \quad f2(n) = 10n^2 + 20.$$

Although functions f, $f1$, and $f2$ differ in their constant terms, their asymptotic behavior is dominated by the n^2 term, hence it is equivalent (up to a constant factor). This fact can clearly be seen from Figure B.2 which shows, for comparison, also the plot of a function with different asymptotic behavior (namely, function $h(n)$ as defined above).

Definition B.4 (Tilde notation) *Let $f(n)$ and $g(n)$ be two functions defined on $\mathbb{R}$. One can write $f(n) = \tilde{O}(g(n))$ if and only if there exists a positive real number c such that $f(n) = O(g(n) \log^c n)$. Notation $\tilde{\Omega}$ and $\tilde{\Theta}$ can be defined similarly.*

Definition B.5 (Small-o notation) *Let $f(n)$ and $g(n)$ be two functions defined on $\mathbb{R}$. One can write $f(n) = o(g(n))$ if and only if for every positive real number ϵ there exists a real number n_ϵ such that*

$$|f(n)| \leq \epsilon |g(n)| \quad \text{for all } n \geq n_\epsilon.$$

If $g(n)$ is non-zero (at least for sufficiently large n), we can equivalently define $f(n) = o(g(n))$ to hold if and only if

$$\lim_{n \to \infty} \frac{f(n)}{g(n)} = 0.$$

The small-o notation is used to express the fact that function $f(n)$ is *asymptotically dominated* by function $g(n)$.

Definition B.6 (Small-omega notation) *Let $f(n)$ and $g(n)$ be two functions defined on $\mathbb{R}$. One can write $f(n) = \omega(g(n))$ if and only if for every positive real number k there exists a real number n_k such that*

$$|f(n)| \geq k|g(n)| \quad \text{for all } n \geq n_k.$$

If $g(n)$ is non-zero (at least for sufficiently large n), we can equivalently define $f(n) = \omega(g(n))$ to hold if and only if

$$\lim_{n\to\infty} \frac{f(n)}{g(n)} = \infty.$$

The small-omega notation is used to express the fact that function $f(n)$ *asymptotically dominates* function $g(n)$.

B.2 Elements of Graph Theory

Definition B.7 (Graph) *A* graph *G is an ordered pair of disjoint sets (V, E), where $E \subseteq V \times V$. Set V is called the vertex, or node, set, while set E is the edge set of graph G. Typically, it is assumed that self-loops (i.e., edges of the form (u, u), for some $u \in V$) are not contained in a graph.*

Definition B.8 (Order of a graph) *The* order *of graph $G = (V, E)$ is the number of nodes in G, that is, the cardinality of set V.*

Definition B.9 (Directed and undirected graph) *A graph $G = (V, E)$ is* directed *if the edge set is composed of ordered node pairs. A graph is* undirected *if the edge set is composed of unordered node pairs.*

Examples of directed and undirected graphs are reported in Figure B.3 Unless otherwise stated, in the following by *graph* we mean *undirected graph*.

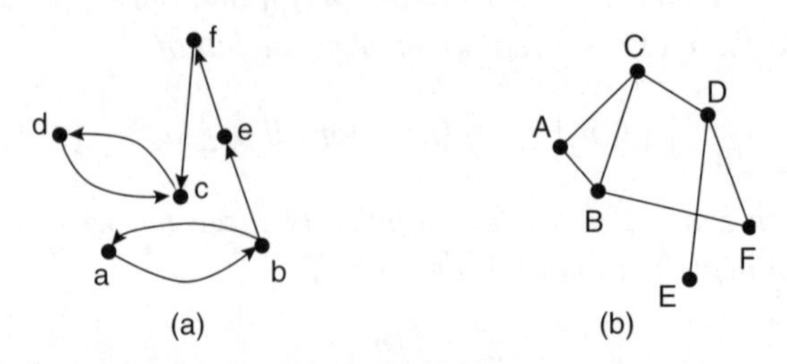

Figure B.3 Examples of directed graph (a) and undirected graph (b).

Definition B.10 (Weighted graph) *A* weighted graph *is a graph in which edges, or nodes, or both, are labeled with a weight.*

Definition B.11 (Neighbor nodes) *Given a graph $G = (V, E)$, two nodes $u, v \in V$ are said to be* neighbors, *or* adjacent nodes, *if $(u, v) \in E$. If G is directed, we distinguish between* incoming neighbors *of u (those nodes $v \in V$ such that $(v, u) \in E$) and* outgoing neighbors *of u (those nodes $v \in V$ such that $(u, v) \in E$).*

Definition B.12 (Node degree) *Given a graph $G = (V, E)$, the* degree *of a node $u \in V$ is the number of its neighbors in the graph. Formally,*

$$deg(u) = |\{v \in V : (u, v) \in E\}|.$$

If G is directed, we distinguish between in-degree *(number of incoming neighbors) and* out-degree *(number of outgoing neighbors) of a node.*

For instance, node A in the undirected graph in Figure B.3 has degree 2, and its neighbors are nodes B and C. Node b in the directed graph in Figure B.3 has in-degree 1 and out-degree 2; its incoming neighbor is node a, while its outgoing neighbors are nodes a and e.

Definition B.13 (Regular graph) *A graph $G = (V, E)$ is* regular *if all its nodes have the same degree. If the graph is directed, it is required that all the nodes in the graph have the same in-degree and out-degree. A regular graph where all nodes have degree k is called a k-*regular *graph.*

Definition B.14 (Complete graph) *The* complete graph *$K_n = (V, E)$ of order n is such that $|V| = n$, and $E = V \times V$, that is, $(u, v) \in E$ for any two distinct nodes $u, v \in V$. Otherwise stated, the complete graph of order n is the unique $(n - 1)$-regular graph of order n.*

Definition B.15 (Square grid) *A* toroidal square grid of side *m is a 4-regular graph $G = (V, E)$ in which nodes are arranged in the Euclidean plane at integer coordinates $(0, 0), (0, 1), \ldots, (m, m)$. The node at coordinates (i, j) in the grid has an edge to the four nodes at coordinates $(i + 1, j)$, $(i - 1, j)$, $(i, j + 1)$, and $(i, j - 1)$, where all the operations are modulo m. In a* simple square grid *of side m, wrap-around edges are omitted.*

The toroidal and simple square grids of side 4 are shown in Figure B.4. Notice that the simple square grid is *not* a regular graph, since the corner nodes have degree 2, boundary nodes have degree 3, and internal nodes have degree 4.

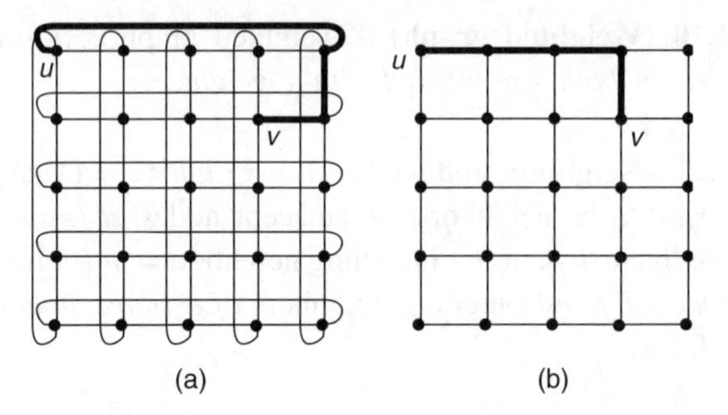

Figure B.4 The toroidal square grid (a) and simple square grid (b) of side 4. Some nodes and paths in the graph are highlighted to clarify the notion of Manhattan distance.

Definition B.16 (Path) *Given a graph $G = (V, E)$, and given any two nodes $u, v \in V$, a* path *connecting u and v in G is a sequence of nodes $\{u = u_0, u_1, \ldots, u_{k-1}, u_k = v\}$ such that for any $i = 0, \ldots, k-1$, $(u_i, u_{i+1}) \in E$. The* length *of the path is the number of edges in the path.*

Definition B.17 ((Hop) distance) *Given a graph $G = (V, E)$ and any two nodes $u, v \in V$, their* distance *$dist(u, v)$ (also called* hop distance*, especially within the networking community) is the minimal length of a path connecting them. If there is no path connecting u and v in G, then $dist(u, v) = \infty$.*

Definition B.18 (Manhattan distance) *Let $u = (u_1, \ldots, u_d)$, $v = (v_1, \ldots, v_d)$ be any two vectors in $\mathbb{Z}^d$, for some $d \geq 1$. The* Manhattan distance *between u and v is defined as follows:*

$$d_M(u, v) = \sum_{i=1}^{d} |u_i - v_i|.$$

The notion of Manhattan distance can be equivalently defined on simple square grids $G = (V, E)$ as follows: given any two nodes $u, v \in V$, the Manhattan distance *between nodes u and v is the hop distance between the two nodes in the graph.*

The Manhattan distance between nodes u, v in the simple square grid of Figure B.4 is 4; a minimum length path connecting u and v is shown in bold in the figure. Note that the notion of Manhattan distance is no longer equivalent to the notion of hop distance in toroidal square grids. In fact, the hop distance between nodes u and v in the toroidal square grid of Figure B.4 is 3 (see path in bold), while we have

$$d_M(u, v) = |0 - 3| + |4 - 3| = 4.$$

Definition B.19 (Graph diameter) *The* diameter *of graph $G = (V, E)$ is the maximum possible distance between any two nodes in G. Formally,*

$$diam(G) = \max_{u,v \in V} dist(u, v).$$

The distance between nodes a and e in the directed graph in Figure B.3 is 2, and the diameter of the graph is 5. The diameter of the undirected graph in Figure B.3 is 3.

Definition B.20 (Sub-graph) *Given a graph $G = (V, E)$, a* sub-graph *of G is any graph $G' = (V', E')$ such that $V' \subseteq V$ and $E' \subseteq E$. Given any subset V' of the nodes in G, the sub-graph of G* induced *by V' is defined as $G_{V'} = (V', E(V'))$, where $E(V') = \{(u, v) \in E : u, v \in V'\}$, that is, $G_{V'}$ contains all the edges of G such that both endpoints of the edge are in V'.*

Definition B.21 (Connected and strongly connected graph) *A graph $G = (V, E)$ is* connected *if for any two nodes $u, v \in E$ there exists a path from u to v in G. If G is directed, we say that G is* strongly connected *if for any two nodes $u, v \in E$ there exist a path from u to v and a path from v to u in G.*

The undirected graph in Figure B.3 is connected, while the directed graph in the same figure is not strongly connected. In fact, there exist a pair of nodes for which connecting paths in both directions do not exist. For instance, there is no path (neither direct, nor inverse) connecting node d with node f.

Definition B.22 (Tree) *A* tree *$T = (V, E)$ is a connected graph with n nodes and $n - 1$ edges. That is, a tree is a minimally connected graph.*

Definition B.23 (Rooted tree) *A* rooted tree *$T = (V, E)$ is a tree in which one of the nodes is selected as the tree root. Once the root node r is chosen, the other nodes in the tree can be classified as either* internal node *or* leaf node. *An internal node u is such that there exists $v \in V$ such that $(u, v) \in E$ and $dist(u, r) < dist(v, r)$. A leaf node l is such that, for any $v \in V$ such that $(l, v) \in E$, then $dist(l, r) > dist(v, r)$.*

Figure B.5 displays examples of a tree and a rooted tree.

Definition B.24 (Random graph) *Two methods are commonly used to define random graphs. In the first method, due to Gilbert (1959), a $G(n, p)$* random graph *is built on n nodes by including each of the $n(n - 1)/2$ possible edges in the graph independently with probability $p = p(n)$. In the second method, due to Erdos and Rény (1959) and ■, a $G(n, M)$* random graph *of n nodes is selected uniformly at random from the collection of all graphs with n nodes and $M = M(n)$ edges.*

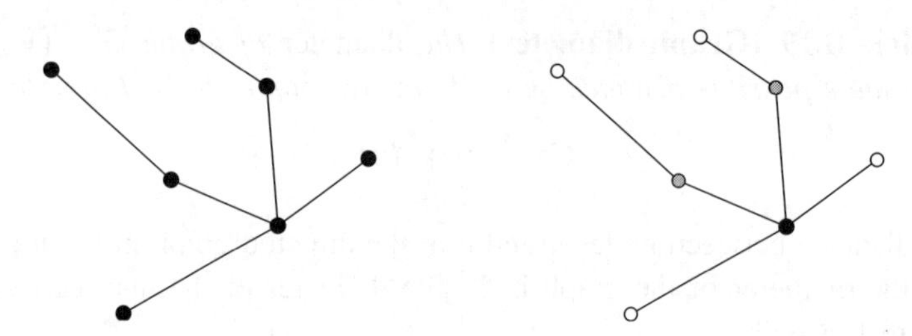

Figure B.5 Examples of tree (left) and rooted tree (right). In the rooted tree, the root node is black, internal nodes are gray, and leaf nodes are white.

Note that, due to the law of large numbers, the number of edges in a $G(n, p)$ graph is approximately $p \cdot n(n-1)/2$, under the condition that $pn^2 \to \infty$. Then, under this condition, graph $G(n, p)$ behaves similarly to graph $G(n, p \cdot n(n-1)/2)$. Several asymptotic properties of random graphs such as connectivity, diameter, node degree, etc., have been studied in the literature. The interested reader is referred to Bollobás (1985).

B.3 Miscellaneous Notions

Definition B.25 (Voronoi diagram) *Let $\mathcal{P} = \{p_1, \ldots, p_k\}$ be a set of points in a metric space S. For definiteness, suppose S is the Euclidean plane with the associated Euclidean distance. The* Voronoi diagram *associated with set $\mathcal{P}$ is a tessellation of S induced by a collection of* Voronoi cells $\{C_i\}_{i:p_i \in \mathcal{P}}$, *where Voronoi cells are defined as follows. The Voronoi cell C_i associated with point $p_i \in \mathcal{P}$ is the set of points in S whose distance to p_i is not larger than the distance to any other point $p_j \in \mathcal{P}$. Formally,*

$$C_i = \{x \in S : \forall p_j \in \mathcal{P}, j \neq i, d(x, p_i) \leq d(x, p_j)\},$$

where $d()$ denotes distance in S.

An example of a Voronoi diagram computed starting from a set of 16 points in the Euclidean plane is shown in Figure B.6.

Definition B.26 (Convex function) *A function $f : X \to \mathbb{R}$ defined on an interval X is called* convex *if, for any two points $x_1, x_2 \in X$ and any $t \in [0, 1]$,*

$$f(tx_1 + (1-t)x_2) \leq tf(x_1) + (1-t)f(x_2).$$

The function is said to be strictly convex *if*

$$f(tx_1 + (1-t)x_2) < tf(x_1) + (1-t)f(x_2),$$

for any t, with $0 < t < 1$.

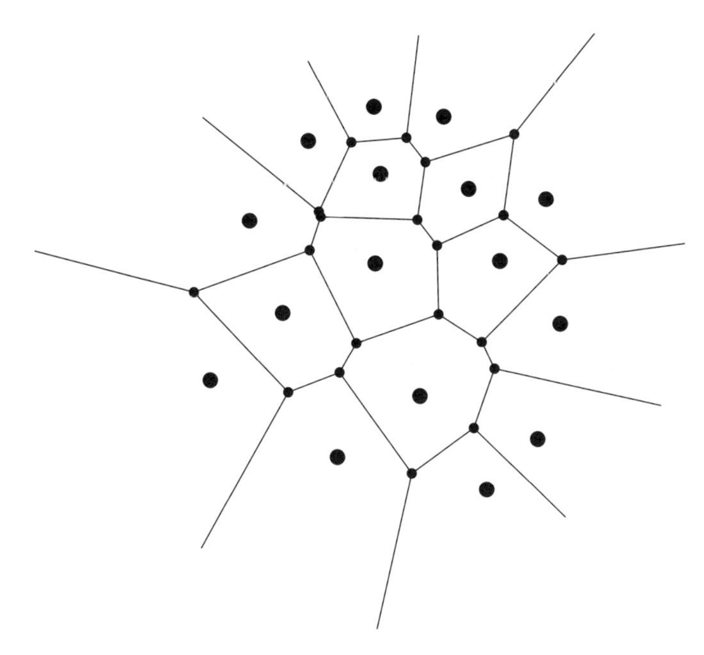

Figure B.6 Example of Voronoi diagram associated with a set of 16 points in the Euclidean plane (in bold).

Definition B.27 (Concave function) *A function $f : X \to \mathbb{R}$ defined on an interval X is called* concave *if function $g(x) = -f(x)$ is convex in X.* Strict concavity *is defined similarly.*

Informally speaking, a function is convex if the graph of the function lies *below* the line segment joining any two points in the graph. A concave function, instead, lies *above* the line segment joining any two points in the graph. See Figure B.7 for an example of convex and concave functions.

Definition B.28 (Convex set) *A set $C \subseteq \mathbb{R}^2$ is said to be* convex *if, for any $x, y \in C$ and any $t \in [0, 1]$, the point $(1-t)x + ty$ is in C. A set $C \subseteq \mathbb{R}^2$ that does not satisfy this condition is called a* non-convex *set.*

Informally speaking, a set C is convex if, for any two points x, y chosen in C, the line segment connecting points x and y is also within C. Examples of convex and non-convex sets are shown in Figure B.8. Notice that Voronoi cells are convex polygons – see Figure B.6. Notice also that the above definition of set convexity can be immediately extended to other vector spaces such as $\mathbb{R}^d$ for $d > 2$, etc.

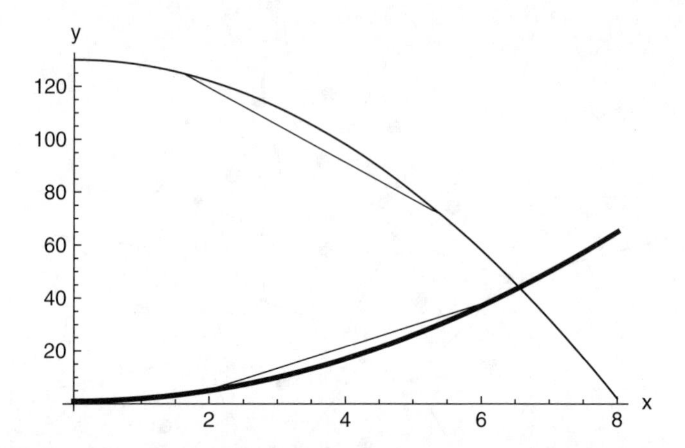

Figure B.7 Example of a convex (thick line) and concave (thin line) function.

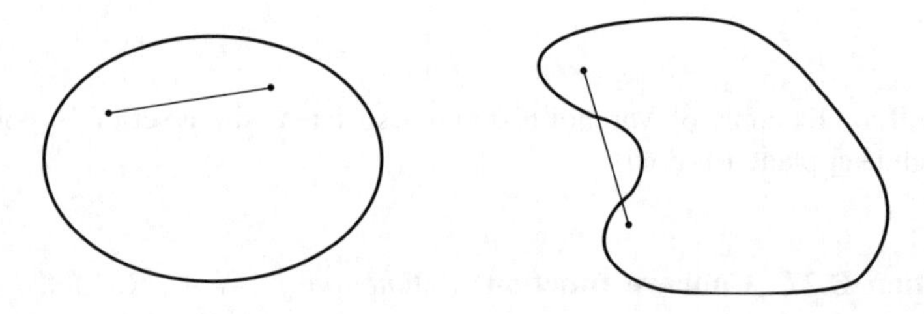

Figure B.8 Example of a convex (left) and non-convex (right) set.

Definition B.29 (Linear function) *A function* $f : \mathbb{R}^d \rightarrow \mathbb{R}$ *is* linear *if, for any* $\mathbf{x}_1, \ldots, \mathbf{x}_k \in \mathbb{R}^d$, *and any* $\alpha_1, \ldots, \alpha_k \in \mathbb{R}$,

$$f(\alpha_1 \mathbf{x}_1 + \cdots + \alpha_k \mathbf{x}_k) = \alpha_1 f(\mathbf{x}_1) + \cdots + \alpha_k f(\mathbf{x}_k).$$

The above definition of linearity can be easily extended to functions between two arbitrary vector spaces.

Definition B.30 (Pearson's correlation coefficient) *Let* $(X_1, Y_1), \ldots, (X_n, Y_n)$ *be a collection of samples of random variables X and Y. The* Pearson's correlation coefficient *between X and Y is defined as follows:*

$$r = \frac{\sum_{i=1}^{n}(X_i - \bar{X})(Y_i - \bar{Y})}{\sqrt{\sum_{i=1}^{n}(X_i - \bar{X})^2}\sqrt{\sum_{i=1}^{n}(Y_i - \bar{Y})^2}},$$

where $\bar{X}$ *and* $\bar{Y}$ *denote the mean value of X and Y, respectively.*

The Pearson's correlation coefficient takes values in $[-1, 1]$, with -1 and 1 representing maximum correlation between random variables X and Y (*inverse* and *direct* correlation, respectively), and 0 representing *statistical independence*.

References

Bollobás B 1985 *Random Graphs*. Academic Press, London.
Bollobás B 1998 *Modern Graph Theory*. Springer, New York.
Erdos P and Rény A 1959 On random graphs. *Publicationes Mathematicae* **6**, 290–297.
Gilbert E 1959 Random graphs. *Annals of Mathematical Statistics* **30**, 1141–1144.
Santi P 2005 *Topology Control in Wireless Ad Hoc and Sensor Networks*. John Wiley & Sons, Ltd, Chichester.

Index

Mobility Models for Next Generation Wireless Networks: Ad Hoc, Vehicular and Mesh Networks, First Edition. Paolo Santi.